Environmental Engineering

Second Edition

PROF. M.P. POONIA
VICE CHAIRMAN,
ALL INDIA COUNCIL FOR
TECHNICAL EDUCATION (AICTE)

S.C. SHARMA
FORMER GENERAL MANAGER
THDC INDIA LIMITED

DR. SANTOSH KUMAR
FORMER UNIVERSITY PROFESSOR
BIHAR COLLEGE OF ENGINEERING
PATNA UNIVERSITY (NOW NIT PATNA)

KHANNA BOOK PUBLISHING CO. (P) LTD.
Publisher of Engineering and Computer Books
4C/4344, Ansari Road, Darya Ganj, New Delhi-110002
Phone : 011-23244447-48 **Mobile:** +91-9910909320
E-mail : contact@khannabooks.com
Website : www.khannabooks.com

Dear Readers,

To prevent the piracy, this book is secured with HIGH SECURITY HOLOGRAM on the front title cover. In case you don't find the hologram on the front cover title, please write us to at contact@khannabooks.com or whatsapp us at +91-99109 09320 and avail special gift voucher for yourself.

Specimen of Hologram on front Cover title:

Moreover, there is a SPECIAL DISCOUNT COUPON for you with EVERY HOLOGRAM.

How to avail this SPECIAL DISCOUNT:

Step 1: Scratch the hologram
Step 2: Under the scratch area, your "coupon code" is available
Step 3: Logon to www.khannabooks.com
Step 4: Use your "coupon code" in the shopping cart and get your copy at a special discount
Step 5: Enjoy your reading!

Thanks,
Khanna Publishing House

Price: ₹ 499.00

ISBN: 978-93-90779-07-9

Environmental Engineering
by Prof. M.P. Poonia, S.C. Sharma,
Santosh Kumar

Second Edition: 2023

To view complete list of books, please scan the QR Code:

Published by:
Khanna Book Publishing Co. (P) Ltd.
CIN: U22110DL1998PTC095547

Visit us at: www.khannabooks.com
Write us at: contact@khannabooks.com

Printed in India by: SPS Printers

Preface

The need for a suitable concise and comprehensive book was being felt on the subject covering the syllabi of "Environmental Engineering" and "Public Health Engineering". The book covers the engineering aspects of the environmental issues, which include Water Supply Engineering, Sewage Engineering, Air Pollution Engineering, Rural Water Supply and Sanitation, Plumbing Systems in Buildings, Municipal Solid Wastes, Industrial Wastes, Air Pollution Control Devices, Noise Pollution Environment Impact Assessment etc.

The book though designed for the students of Diploma and Degree students to cover the syllabi prescribed by most of Indian Universities, it is likely to be useful for practising engineers who are engaged in finding the control methodologies of air and water pollution. The book also covers the model syllabus on "Environmental Engineering" prescribed by AICTE for undergraduate students in Civil Engineering.

Constructive suggestions from all corner will be thankfully acknowledge and appreciated.

—Authors

Contents

Introduction to Environmental Engineering

1.1. DEFINITIONS

Environmental Engineering is that part of the engineering which deals with the technique of finding out :

(*i*) The causes and extent of various types of pollution.

(*ii*) The adverse effects of pollution, and

(*iii*) The control and remedial measures of environmental pollution.

Environmental Engineering is the branch of engineering concerned with the application of science and engineering principles for protection of human population from the effects of adverse environmental factors, protection of environment, both local and global from potential deleterious effects of natural and human activities and improvement of environmental quality.

Environment Engineering can also be described as a branch of applied science and technology that addresses the issues of energy preservation, production asset and control of waste from human and animal activities. Furthermore, it is concerned with finding plausible solutions in the field of healthop such as waterborne diseases, implementing laws which promote adequate solution in urban, rural, and recreational areas. It involves waste water management, air pollution control, recycling waste disposal, radiation protection, industrial hygiene, environment sustainability, public health and environmental law. It also includes studies on the environmental impact of proposed construction projects.

Environmental engineering can also be defined as a profession that applies mathematics and science to utilize the properties of matter and sources of energy in the solution of problems of environmental sanitation. These include the provision of safe, palatable, and ample public water supplies; the proper disposal of or recycle of waste water and solid wastes; the adequate drainage of urban and rural areas for proper sanitation; and the control of water, soil, and atmospheric pollution, and the social and environmental impact of their solutions, Furthermore, it is concerned with engineering problems in the field of public health areas such as control of waterborne diseases, the elimination of industrial health hazards, and the provision of adequate sanitation in urban, rural, and recreational areas, and the effect of technological advances on the environment.

Environmental engineering is not concerned primarily with heating, ventilating or air conditioning (HVAC), nor it is concerned primarily with landscape architecture. It should also not be confused with the architectural and structural engineering functions associated with built environments such as homes, offices and other work places.

1.2. SCOPE (FUNCTIONS) OF ENVIRONMENTAL ENGINEERING

1. Solid waste management
2. Environmental impact, assessment and mitigation.
3. Water supply treatment
4. Waste water treatment.
5. Air pollution management

1.3. SCOPE OF THE BOOK

The book has been designed to discuss the major topics of environmental engineering, such as :

(*i*) Water supply engineering
(*ii*) Sewage disposal
(*iii*) Rural water supply and sanitation
(*iv*) Industrial water disposal
(*v*) Solid waste management
(*vi*) Air pollution
(*vii*) Odour pollution, and
(*viii*) Noise pollution

1.4. HISTORICAL BACKGROUND

Even in olden days, water purification was in practice in India. In "Sushurta Samhita", a Sanskrit medical book, written in 2000 year B.C., there is a mention about "Keeping drinking water in copper vessels, and exposing it to sun-light and filter it through charcoal'.

In another book compiled in the same period of time, there is a mention about "Polluted water to be treated first by boiling and exposing to sunlight, then by dipping several times into it a piece of hot copper, and lastly filter it through clean sand."

In the Indus Valley Civilization of Mohan-jo-daro there are sufficient indications to prove that public, water supply and drainage system existed even about 5000 years ago.

In recent history in India it was in Calcutta (now Kolkata) where first municipal water treatment plant with slow sand filter was installed in 1880.

1.5. DUTIES OF ENVIRONMENTAL ENGINEERS

Environmental engineers should be conversant with the subject matters as mentioned under the heading 'Scope of Industrial Engineering". His most important task is to deal with the **water supply** system. As a water works engineer, he is required to take up the works related from planning till commissioning and operation. As proper planning will lead to feasibility and economy of the scheme. A better design will increase the efficiency of the plant. Efficient supervision and execution during construction will bring up the plant with perfection at comparatively lower construction cost. Proper operation of water works will lead to safe palatable water supply; control over quantity and quality of water supply, distribution system and the water source from any pollution.

As we know, the drinking water may convey waterbrone diseases, such as cholera, typhoid, dysentery etc. Chemical and biological means are available for controlling them. This means help in ensuring, safe, pleasant, taste and smell, and hygienic water.

Currently, more attention being paid in the field of **waste water engineering**. A **sanitary engineer** must have knowledge and experience about proper planning, economic design and efficient execution for successful sewerage and solid waste management projects. It is expected from a sanitary engineer to help in keeping a clean and healthy environment by introducing modern methods of collection, treatment and disposal of waste matters.

Adverse effects of pollution on environment and the remedial and control methods are also within the scope of environment engineering. The duty of an environmental engineer, as a **sanitary engineer** is to take care of the solid waste management aspect. He is required to assess the quantity and quality of the wastes for the purpose of its proper disposal. This requires adequate knowledge of different treatment and disposal methods such as sanitary landfill, composting, incineration, waste to energy etc.

Since nature can absorb the pollution load to a certain extent only. If this load exceeds beyond permissible limits, the environment is going to be affected and ecology will be disturbed. Therefore, environmental engineer must also have some knowledge of ecology. Hence, the air pollution control is also within the scope of environmental engineering.

Measures to be taken for controlling air pollution are also within the scope of environmental engineer. These measures may be minimisation of quantity of pollutants, to arrest the pollutional load at the source prior to discharging to the atmosphere, and purifying the pollutants.

1.6. IMPORTANCE OF ENVIRONMENTAL ENGINEERING

Major challenges for the development being faced are :

1. Providing clean water, sanitation and electricity to all so that poverty gets reduced.
2. More and more energy needs to be generated with degradation of environment. Degradation may be of following types :
   (*a*) *Locally* : particulate matter, and smog.
   (*b*) *Regionally* : acid deposition/acid rain.

(c) *Globally* : climate change, global warning.

3. Urban population is increasing very fast and is subjected to problem of inadequate potable drinking water, poor basic sanitation, inadequate solid waste and polluted air containing particulates, sulphur and lead.

Major environmental problems being faced are: unhygienic slums, lack of sanitation, polluted water supply, air and water pollution due to construction and industrial projects, hazardous waste disposal, waterlogging in irrigation projects, pesticide pollution, run-off into streams, automobile pollution, etc. Many problems are created by the current pattern of development, lifestyle and energy use mix etc.

Improper handling and disposal of solid wastes can lead to infections of nose, throat, and lungs, cause allergy or asthma. Poorly maintained landfill sites may lead to ground water contamination. Open dumping of garbage can facilitate breeding of disease vectors such as flies, mosquitoes, cockroaches, rats and other pests.

Pollution of the atmosphere affects the lives of millions of people, especially those living in large industrialised cites with heavy motor vehicles traffic. The ill-effects are unpleasant fumes and odours, reduced visibility, injury to human health and crops and other forms of vegetation etc. Thus, there is increasing public demand for the control and abetment of air pollution.

A large number of people die annually due to water related diseases. About two-thirds of the people on the earth have no choice but to drinks it, cook with it and bathe in it. In this regard, World Health Organisation (WHO) states: "One hospital bed out of four in the world is occupied by a patient who is ill because of polluted water... Provision of a safe and convenient water supply is the single most important activity that could be undertaken to improve the health of people living in rural areas of the developing world.

All the above mentioned problems can be minimised or mitigated by using the knowledge of environmental engineering.

1.7. ENVIRONMENTAL MANAGEMENT SYSTEMS OVERVIEW

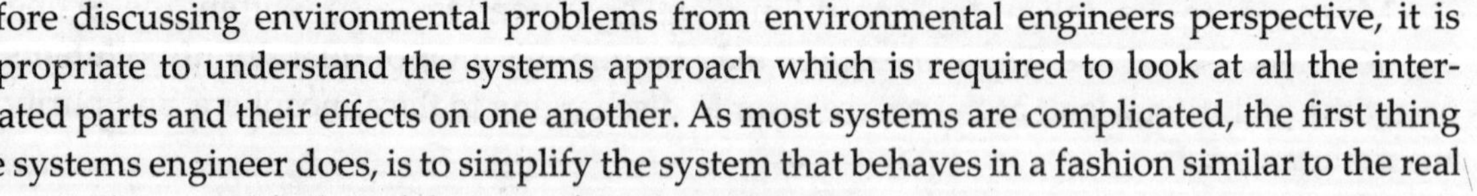

Before discussing environmental problems from environmental engineers perspective, it is appropriate to understand the systems approach which is required to look at all the inter-related parts and their effects on one another. As most systems are complicated, the first thing the systems engineer does, is to simplify the system that behaves in a fashion similar to the real system. This simplified model gives a fair approximation.

Systems

A 'systems' may be defined as a group of inter-related elements or parts designed to achieve a particular goal. A 'system' always exists within an environment which provides resources as inputs into the system, and that accepts outputs of the system.

A 'sub-system' is usually a completely system functioning within a large system.

A 'system' can also be defined as an orderly arrangement of components (or elements) that are inter-related and act and interact with one another to problem task or function in a particular environment.

A system of management is a system that integrates the functions of management, resources flow (men, materials, money and machinery), organisation and the various techniques of management. A component of the system – management information system – is also added to provide information the planning and furnishes the essential feedback information necessary to achieve stability through control.

Thus, a 'systems approach' a systematic and organised approach to get the task accomplished more efficiently and economically.

Environmental Management

Environmental management is "the reduction or minimisation of the impact of human activities on the environment, thus on endeavour to avoid overuse, misuse and abuse of environmental resources."

Environmental management can be defined as an inter-disciplinary approach to resource conservation and recycling, and it acts a regulatory force on human greed in resource exploitation and resources wasting.

Environmental management involves:

(*i*) Judicious and gainful utilisation of natural resources for socio-economic development, and

(*ii*) Maintenance of environmental quality by pursuing ecological principles.

The main tasks of environmental management are:

1. Environmental planning considering laws/prescribed norms, and environmental status evaluation.
2. Environmental impact assessment (EIA).
3. Minimizing adverse environmental impacts of activities, products and services.
4. Management of disposal of wastes.

Environment management must regulate the demands and activities of man in such a way that the ability of environment to sustain future development remains unimpaired. Environmental management, is therefore related to the rational adjustment of man with nature involving judicious exploitation and utilization of natural resources without disturbing the ecological balance and ecosystem equilibrium.

Here we are discussing following environmental management systems :

1. Water resource management system.
2. Air resource management system.
3. Solid waste management system.
4. Multimedia pollution problems. The environmental problems which are not confined to one of the above systems but cross the boundaries from one system to the other are known as multimedia pollution problems.

1.7.1. Water Resources Management Systems

Water is our most abundant resource and covers about 71% of the earth's surface. About 97% of the earth's total water is ocean water and is too salty for drinking, irrigation and most industrial purposes. Out of the remaining, more than three–fourths is unable for use because it lies too far under the earth's surface or is locked up in glacier, polar ice caps etc. Ultimately, only about 0.003% of the world's water supply is available as fresh water for human use. This supply is also unevenly distributed.

Human beings have corrected these imbalances with varying degrees of success by capturing fresh water in reservoirs behind dams, transferring fresh water in rivers and streams from one area to another, tapping underground supplies, and attempting to reduce water use, waste and contamination.

With the growth of human population, there is an increasing need for large amounts of water to fulfil everyone's needs. One of the greatest challenges facing the world is the management of water resources. Water sources are overused or wasted to such an extent that they locally run dry. Water sources are also becoming heavily polluted by sewage and toxic substances that it becomes impossible to use this water. The mismanagement of water resources has resulted in inequitable distribution of water. Some use more water than they need while others do not have access to clean water to all.

In short, water resource management is the process of managing both the quantity and quality of the water usedd for human benefit without destroying its availability and purity.

Water Supply Subsystem

The two major sources of water used to supply community needs are surface water and ground water. Surface water sources consists of streams, lakes, and rivers. Ground water sources are those pumped from wells.

The nature of the water source decides the planning, design, and operation of the collection, purification, (treatment, transmission and distribution works/structure).

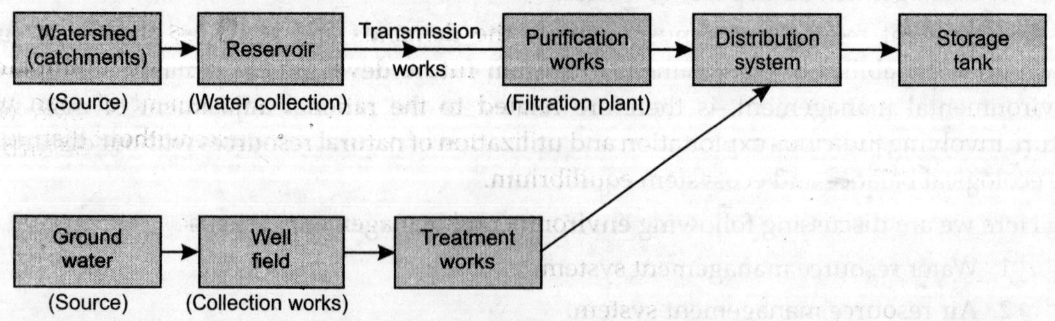

Fig. 1.1. Water supply resource system.

Fig. 1.1 shows a water supply resources system showing watershed surface water source and ground water source for water supply. The source of water may be either anyone or both of the systems depending upon the quantity of water available at the source and demand.

The source of water supply in each case determines the type of treatment works. Distribution system in the city is pipe network. The pipes are sometimes known as **water mains,** wherein the water is kept at under pressure. The steady tank may be either elevated or at ground level. The stored water is used to meet high demand during the supply period. It also serves an emergency backup in case of a fire.

Water consumption generally varies depending upon watering in the lawns and gardens, and other activities at home and in industry. The **main factors which influence water consumption** are:

1. Climate
2. Industrial activity
3. Consumption charges and metering system
4. Standard of living
5. System management
6. Availability of private wells

Waste Water Disposal Subsystem

In order to protect the health of individual, the family, and the community and also to prevent the occurrence of certain nuisances, it is necessary of all human water safety. Following conditions are necessary with regard to disposal of human wastes so as to achieve satisfactory results:

(*i*) It should not contaminate any drinking water supply.

(*ii*) It should not create public hazard and should not be accessible to vectors (insects, rodents or other possible carriers) that may come into contact with food or drinking water.

(*iii*) It should not be accessible to children so as not to give rise to health hazard.

(*iv*) It should follow the laws and regulations governing waste pollution, and sewage disposal.

(*v*) It should not pollute the water to be used for any other purpose for human beings, fisheries, and recreation purposes.

(*vi*) It should not create a nuisance such as bad odour or unsightly appearance.

The source of waste water may be either industrial waste water or domestic sewage or both. (Domestic sewage is sometimes called as sanitary sewage). The sewers are classified into three categories, namely: sanitary, storm, and combined.

Sanitary sewers (municipal waste water) are designed to carry numerical waste water from homes and commercial establishments. Industrial wastes may also be discharged in these sewers after proper pretreatment.

Storm sewers are defined to handle excess rainwater to prevent flooding of low areas. The storm sewers generally discharge into rivers and streams, while sanitary sewers convey waste water to treatment plants.

Combined sewers accommodate both municipal waste water and storm water. Many communities are in the process of replacing the combined sewers with separate systems for sanitary and storm flow.

Sewage treatment is performed at the waste water treatment plant (WWTP) also known as sewage treatment plant (STP). The effluent from the plant may be discharged into an ocean, lake, or river (called the receiving body). Alternatively, it may be discharged onto the ground or be processed for reuse for irrigation or gardening purposes. The by-product sludge from the WWTP may be disposed off in an environmentally acceptable manner.

A simplified form of waste water management subsystem is shown in Fig. 1.2.

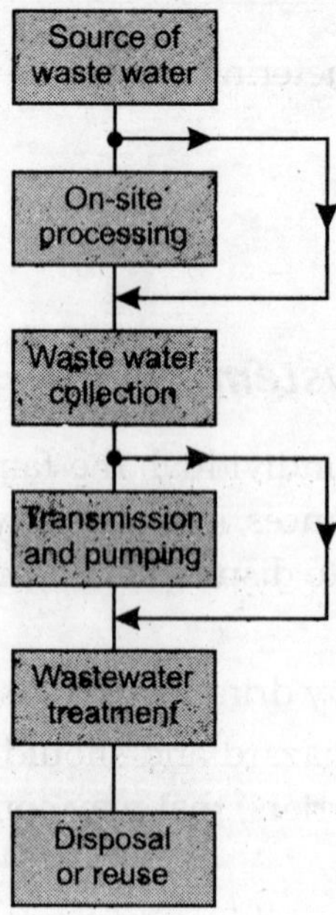

Fig. 1.2. Waste water management subsystem.

1.7.2. Air Resource Management System

Air is delivered by our atmosphere, free of charge in whatever quantity we desire. But unlike water, which can be treated before its use, it is impracticable to use gas mask all the time to treat impure air. Similarly we cannot use air plugs all the times to keep out the noise.

Air resource management programmes are necessary because air quality is deteriorating very fast and there is immediate need for correction.

Clean air is very essential for living system, particularly human life for their health and survival. Industries, power plants, smelters, mines, and refineries besides motor vehicles and construction activities are the **prime sources of air pollution.**

The common air contaminants and their main sources are :

(*i*) Oxides of carbon, sulphur, and nitrogen.

(*ii*) Dust.

(*iii*) Gases, vapours, odours – emissions from furnitures, carpets, and paints.

(*iv*) Dust mites – from carpets, fabric, foam, chair/sofa cushions.

(*v*) Microbial contaminants, fungi, bacteria from dump areas, stagnant water.

Air pollution can be controlled by following approaches :

1. Preventive techniques.
2. Effluent control.

Common devices for removal of pollutants from flue gases through scrubbers, closed-collection recovery systems, the use of the dry and wet collectors, filters, and electorstatic precipitators.

Higher smoke stacks facilitate the discharge of pollutants a far away from the ground as possible. Industries should be carefully located in order to minimise the effects of pollution after considering the topography and wind direction.

Vehicular emissions are the main causes of pollution in the cities. The use of cleaner fuels (such as CNG) and other measures such as banning old vehicles, improving traffic management, implementing the higher level emission standards, introducing improved vehicular technology, and use of battery/electric operated vehicles can reduce the levels of NO_2, SO_2, CO_2, and CO.

Other Measures for Preventing Air Pollution

1. Adequate dust supervision measures such as regular water sprinkling on construction sites, haul, and unpaved roads particularly near settlement may be undertaken to control fugitive dust.
2. Trucks carrying sand, soil, stone etc., may be duly covered to avoid spilling.
3. Air quality monitoring may be conducted at vulnerable locations,

1.7.3. Solid Waste Management System

Waste is a substance arising from human or animal activities, that has to be discarded as useless or unwanted. Waste can be any trash, rubbish, garbage, refuse, filth, bottles, glass, crockery, cans cartons, scrap metal, junked vehicles, appliances or hardware, waste soil, animal careasses, discarded matter or any industrial, commercial, mixing, agricultural or other waste.

Waste management is the collection, transportation, processing and disposal of waste materials in an effort to reduce their effect on human health or local aesthetics. A sub-focus in recent decades has been to reduce waste materials' effect on the natural world and the environment.

A simplified block diagram of a solid waste management system is shown in Fig. 1.3.

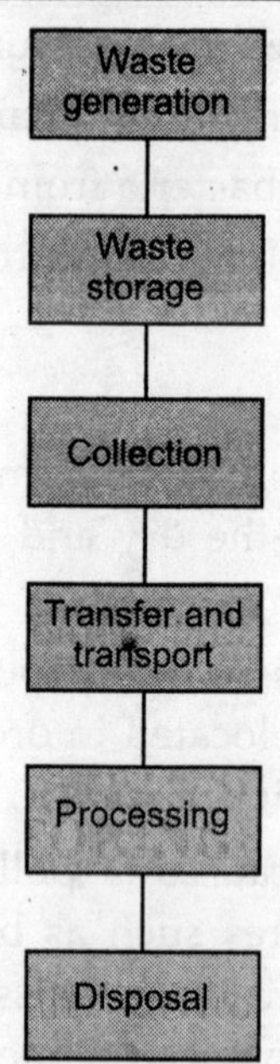

Fig. 1.3. A simplified block diagram of a solid waste management system.

Different waste management and disposal methods and strategies are followed depending on the nature and volume of waste. Waste matter is classified as solid, liquid or sludge based on its consistency and hazardous or non-hazardous based on whether or not they have properties such as ignitability reactivity and reachable toxic components or are bio-hazards.

Types of solid wastes		Source of generation
1. Municipal solid waste	:	(*a*) Residential
(*b*) Commercial		
(*a*) Open areas		
2. Industrial waste	–	Chemicals, textile by-products etc.
3. Agricultural waste	–	Agricultural by-products and residues etc.
4. Hazardous waste	–	Chemicals and oil from refineries, waste from ordinance and fire work factories etc.
– *Hospitals*: bio-medical waste		

Waste Management Techniques

Managing domestic, industrial and commercial waste has, traditionally, consisted of collection, followed by disposal. Some of the common methods for effective waste management are given hereunder:

1. **Sanitary Landfills.** It involves disposing municipal solid wastes by compacting it and then covering with soil. These should be well-located, designed, operated, monitored, closed, cared for after closure and cleaned up when necessary.

2. **Composting.** It is the process of controlled decomposition of biodegradable organic wastes such as food any yard wastes, into humus. It is nature's way of recycling organic waste into new soil composting keep organic wastes out of landfills and instead, puts it to use in gardens, where it provides nutrients and increases beneficial soil organisms.
3. **Incineration.** It is controlled burning process to reduce waste volume, combustors when properly equipped, can also convert waste into steam for heating system. Burning waste at extremely high temperature also destroys disease-causing bacteria and breaks down many harmful chemical compounds.
4. **Reduction in wastage.**
5. **Reusing.** It refers to using items by repairing/donating after repairing as charity.
6. **Recycling.**

The lack of space for dumping solid waste has become a serious problem in several cities and towns. Dumping and burning waste in also not an acceptable practice from an environment or a health perspective.

1.7.4. Multi-media Systems

Many environment problems cross air-water-soil boundary and effect adversely at different levels. Acid-rain is an example of such problems. That results from the emission of sulphur oxide and nitrogen oxide into the atmosphere.

Acid rain is commonly defined as "a condition in which natural precipitation become acidic after reacting with pollutions in the air". Acid rain is formed when industrial emissions of gases including sulphur dioxide, nitrogen oxides and hydrogen chloride combine with water droplets in the atmosphere. This chemical reaction forms weak sulphuric acid and nitric acid that cause acid rain.

The ill-effects of acid rain can be found on vegetation, soil, marine resources, monuments as well as on man. The ill-effects of acid rain have no national limits as nature known no geographical boundaries.

In this case of acid rains pollutants are washed out of the atmosphere, thus cleansing it, but in turn polluting water and changing the soil chemistry, which ultimately results an adverse effect on fisheries and vegetation.

The lesson learnt from this is that, it is dangerous to develop models that are two simplistic, therefore, environmental engineers must work with a multidisciplinary team to solve environmental problems. The best solution to environmental problems is to waste minimization. If waste is not produced there will be no need for its treatment and disposal.

1.8. ENVIRONMENTAL LEGISLATIONS

Environmental legislations are major considerations in the selection of technology and the design of treatment processes to remove contaminants.

Major legislations directly dealing with the protection of environment are:

1. The Forest (Conservation) Act, 1980.
2. The Water (Prevention and Control of Pollution) Act, 1974.
3. The Water (Cess) Act, 1977.
4. The National Environment Appellate Authority Act, 1977.
5. The Air (Prevention and Control of Pollution) Act, 1981.
6. The Environment (Protection) Act, 1986.
7. The National Environment Tribunal Act, 1995.
8. The Biological Diversity Act, 2002.
9. The Hazardous Wastes (Management and Handling) Amendment Rules, 2003.
10. The Factories Act, 1948.

1.9. ENVIRONMENTAL MONITORING

Environmental monitoring is one of the most important component for sustainability of any project and is an integral part of any environmental assessment process. Monitoring will determine whether the mitigation measures undertaken are achieving their objectives. A monitoring plan is framed to monitor the implementation of activities provided in Environmental Management Plan (EMP). However, a third party monitoring for the implementation of the EMP may also be conducted by an agency of or a committee of experts who will visit the project site, twice a year to oversee and evaluate the EMP implementation and suggest improvements.

Environmental monitoring is essential for:

1. Ensuring that impact do not exceed the established legal standards.
2. Checking the implementation of mitigation measures in the manner described in the EIA (Environment Impact Assessment) report.
3. Providing early warning of potential environmental damage.

An Environmental Monitoring Programme should have the following objectives:

(*a*) Assess the changes in environmental conditions, if any, during construction and operation of the project.

(*b*) Monitor the effective implementation of mitigation measures.

(*c*) Warning of any significant deterioration in environmental quality so that additional mitigation measures may be planned in advance.

From the monitoring point of view, the important parameters are: water quality, ecology, air quality, noise, soil erosion and siltation, incidence of water-related diseases.

S.N	Monitoring Component	Parameters
1.	Air quality	SPM, SO_2, NO_X, at different locations.
2.	Water quality	Physical, chemical and bacteriological parameters and heavy metals at different locations (surface water and ground water) including pH, temperature etc.
3.	Noise and vibration levels	At different locations in d_B (A).
4.	Meteorology	Wind speed and direction, rainfall, temperature, relative humidity, solar radiation.
5.	Effluent from septic tanks	pH, BOD, COD, TSS, TDS (Total dissolved solids).
6.	Water-related diseases	Identification of water-related diseases, adequacy of local vector control and curative measure etc at labour camps and colonies, three times a year.
7.	Soil	Parameters related to agricultural and forestation potential. Soil erosion rates, stability bank embankment etc. Status of afforestation programme, green belt development programme.

QUESTIONS

1. What is the importance of public water supply schemes in the present day civil life?
2. Why is it necessary to control the water quality in nature and also in community?
3. Write briefly the historical background of water treatment and water supply engineering.
4. Explain the connection between "Public Health" and "Engineering", and also between "Water Supply" and 'Diseases".
5. Define 'environmental engineering'. Discuss issues dealt in 'environmental engineering'.
6. Enumerate and discuss the scope (functions) of environmental engineering.
7. Write a detailed note on duties and functions of "Environmental Engineers".
8. Discuss the importance of 'environmental engineering.
9. Write a detailed note on 'environmental management system.
10. Enumerate and discuss the major 'environmental management systems.
11. Write short notes on the following:
    (*a*) Water resource management system.
    (*b*) Air resource management system.
    (*c*) Solid waste management system.
12. Write a detailed note on 'Environmental Monitoring'.
13. Discuss 'water resource management system' and its sub-systems.

PART – 1

Water Supply Engineering

Water Demands

2.1. INTRODUCTION TO WATER SUPPLY

Water is a chemical compound and may occur in a liquid form or in a solid form or in a large in a gaseous form. All these three forms of water are extremely useful to man, providing him the luxuries and comforts, in addition to fulfilling his basic necessities of life.

Water is absolutely essential for survival of human beings, and also for animals, plants and all other living beings. It is necessary that the water required for their needs must be good, and should not contain unwanted impurities or harmful chemical compounds or bacteria.

In order to ensure the availability of sufficient quantity of good quality water, it is imperative, to plan and build suitable water supply schemes, so as to provide potable water to the various sections of community in accordance with their demands.

Such a scheme shall help in supplying safe ***wholesome**** water to the people for drinking, cooking, bathing, washing, etc. so as to keep the diseases away and thereby promoting health; and would also supply water for fountains, gardens, etc., This helps in maintaining better sanitation and beautification of surroundings, and *reduce environment pollution*.

***Note :** Meaning of Wholesome Water : The water which is safe and potent for drinking to human health is referred to as **wholesome water**. Such a water will *neither* contain any chemicals or impurities in excess of which may they may harm a human body *nor* any harmful bacteria or micro-organisms.

Water is also required for safety against fire by supplying sufficient quantity of water to extinguish it. The existence of such a water supply scheme shall help in attracting industries (since industries require large amounts of water), consequently reducing unemployment and ensuring better living standards.

The municipalities or the municipal corporations are responsible for providing public water supply system which includes collection of water from source of supply, giving necessary treatment to make it hygienically safe and potable and finally distribution of water through a network of piping work. Water from the street main is supplied to the individual building through a service connection. Within the building the water is distributed to different fixtures through pipes.

2.2. WATER DEMANDS

In order to design a water supply scheme for a particular section of the community, it becomes imperative, to first of all, evaluate the amount of water demanded by the public and the amount of water available. *In fact, the first study is to consider the demand, and then the second requirement is to find sources to fulfil that demand.* Many a times, a compromise between the two becomes necessary.

For planning a water supply scheme, it is necessary to find out total yearly water demand and also to assess the required average rate of flow and the variations in these rates. The following quantities are, therefore, generally considered.

(*i*) Total annual volume (V) in litres or million litres.

(*ii*) Annual average rate of flow in litres per day, *i.e.* $V/365$.

(*iii*) Annual average rate of flow in litres per day per person (*i.e.* litres per capita per day), called ***per capita demand (q).***

(*iv*) Average rate of draft in litres per day per service, *i.e.*

$$\frac{V}{365} \times \frac{1}{\text{No. of services}}$$

(*v*) Fluctuations in flows expressed in terms of percentage ratios of maximum or minimum yearly, monthly, daily or hourly rates to their corresponding average values.

There are many variable factors affecting water consumption. Certain thumb rules and empirical formulas are generally used to assess this quantity, which may give fairly accurate results.

2.3 SECTORWISE WATER CONSUMPTION IN INDIA

On account of rapid economic growth and demographic changes, the water demand in all sectors is showing increasing trend. The following table indicates the estimated water demand in India for various sectors :

Table 2.1. Projected Water Demand in India (By Different Use)

Sector	Water Demand in Billon Cubic Meter (BCM)								
	Standing Sub-Committee (MOWR)			National Commission on Integrated Water Resources Development (NCIWRD)					
	2010	2025	2050	2010		2025		2050	
				Low	High	Low	High	Low	High
Irrigation	688 (84.6)	910 (83.26)	1072 (74.1)	543 (78.24)	557 (78.45)	561 (71.6)	611 (72.50)	628 (64.55)	807 (68.4)

Drinking Water	56 (6.9)	73 (6.68)	102 (7.0)	42 (6.05)	43 (6.06)	55 (7.0)	62 (7.35)	90 (9.25)	111 (9.4)
Industry	12 (1.5)	23 (2.10)	63 (4.4)	37 (5.30)	37 (5.21)	67 (8.5)	67 (7.95)	81 (8.30)	81 (6.9)
Energy	5 (0.6)	15 (1.37)	130 (9.0)	18 (2.60)	19 (2.67)	31 (4.0)	33 (3.90)	63 (6.50)	70 (5.9)
Others	52 (6.4)	72 (6.59)	80 (5.5)	54 (7.81)	54 (7.61)	70 (8.9)	70 (8.30)	111 (11.40)	111 (9.4)
Total	**813 (100)**	**1093 (100)**	**1447 (100)**	**694 (100)**	**710 (100)**	**784 (100)**	**843 (100)**	**973 (100)**	**1180 (100)**

Source : *Basin Planning Directorate, CWC, XI Plan document.*

1 Billion = 100 Crore;

MOWR : Ministry of Water Resources;

% in brackets.

Table 2.2. Water Use by Different Industries in India

Industrial Sector	Annual Water Consumption (Million Cubic meters)	Proportion of total Water Consumed in Industry (%)
Thermal Power Plants	35,157.4	87.87
Engineering	2019.9	5.05
Pulp and Paper	905.8	2.265
Textiles	829.8	2.075
Steel	516.6	1.29
Sugar	194.9	0.49
Fertilizer	73.5	0.18
Others	314.2	0.78
Total	**40,012.1**	**100**

Source : *Centre for Science and Environment (CSE) 2004.*

Table 2.3. Water Requirements of Institutions and Commercial Establishments

S.N.	Type of Unit/Building		Domestic Use (Litres per head per day)	Flushing (Litres per head per day)	Total Consumption (Litres per head per day)
1.	Factories including Canteen and bathrooms		30	15	45
2.	Factories with Canteen but no bathrooms		20	10	30
3.	Hospital (excluding laundry and kitchen)				
	(a)	Number of beds ≤ 100	230	110	340
	(b)	Number of beds > 100	300	150	450
	(c)	Out Patient Department (OPD)	10	5	15
4.	Nurses Homes and Medical Quarters		90	45	135
5.	Hostels		90	45	135
6.	Hotels (upto 3 Star) excluding laundry, kitchen, staff, water bodies		120	60	180
7.	Hotels (4 Star and above) excluding laundry, kitchen, staff, water bodies		260	60	320
8.	Offices (including canteen)		25	20	45
9.	Restaurants and Food Courts including Kitchen				
	(a)	Restaurants	55 per seat	15 per seat	70 per seat
	(b)	Food Courts	25 per seat	10 per seat	35 per seat
10.	Club House		25	20	45
11.	Stadiums		4	6	10
12.	Cinemas, Concert Halls, Theatres and Multiplex		5 per seat	10 per seat	15
13.	Schools/ Educational Institutions				
	(a)	Without boarding facilities	25	20	45
	(b)	With boarding facilities	90	45	135

14.	Shopping and Retail (Mall)				
	(a)	Staff	25	20	45
	(b)	Visitors	5	10	15
15.	Traffic Terminal Stations				
	(a)	Airports	40	30	70
	(b)	Railway Stations (Junction) with bathing facility	40	30	70
	(c)	Railway Stations (Junction) without bathing facility	25	20	45
	(d)	Railway Stations (Intermediate) with bathing facility	25	20	45
	(e)	Railway Stations (Intermediate) without bathing facility	15	10	25
	(f)	Interstate Bus Terminals	25	20	45
	(g)	Interstate Bus Terminals/Metro Stations	10	5	15

Source : National Building Code 2016, BIS.

From the tables above it can be inferred that overall water requirement will go up 40 to 50% in next three decades.

2.4 AVAILABILITY OF WATER IN INDIA

India is blessed with good rainfall. The average annual rainfall in the country is 1170 mm. The total available sweet water in the country is 4000 billion cubic meters per year. Out of this usable water is only 1123 billion cubic meters. Out of the total usable water 728 billion cubic meters is contributed from surface water and 395 billion cubic meters from the groundwater.

Due to the rapidly increasing population, urbanization and industrialization the water demand is rapidly increasing. Consequently, the per capita water availability is rapidly decreasing.

Water-stress Condition

Based on the average requirement of water for various purposes, the situation is considered of **Water Stress** condition when per capita water availability ranges from 1000 to 1700 cubic meter per capita per year, and it is considered **Water Scarcity** condition when availability reduces to 1000 cubic meters per year. If present trend continues most of India will reach water-stress condition very soon and water scarcity condition by mid 21st century.

Figure 2.1 shows the observed and projected decline in per capita average annual fresh-water availability and growth of population from 1951 to 2050. This clearly indicates the *two-sided* effect on water resources. As a result, availability of water is bound to reach critical levels sooner or later.

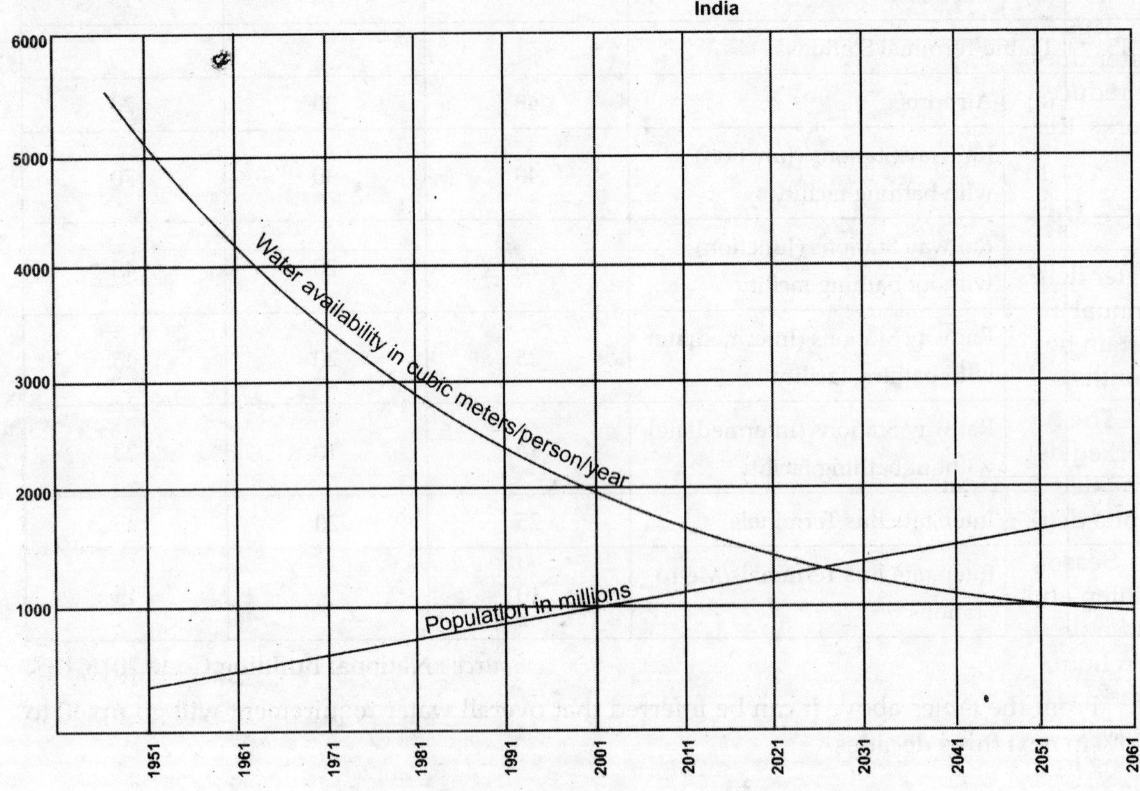

Observed and Projected decline in per capita average annual Freshwater availability and growth of population censuswise

Fig. 2.1

Fire Demand

Fires when break out may lead to serious damages, if not controlled effectively. Big cities maintain full fire-flighting squads, which require sufficient quantity of water, so as to throw it over the fire at high speeds.

The quantity of water required for extinguishing fires should be easily available and kept always stored in storage reservoirs. Fire hydrants are usually fitted in the water mains at about 100 to 150 metres apart. Fire-fighting pumps are immediately connected into them by the fire brigade personnel, as soon as a fire breaks out. These pumps then throw water on the fire at very high pressures, so as to quickly bring it under control. The minimum water pressure available at fire hydrants should be about 100 to 150 kN/m^2 (10 to 15 m of water head) and should be maintained even after 4 to 5 hours of constant use of fire hydrant.

Although the rate at which water is required for fire-fighting is very large, the total amount of water consumption hardly amounts to 1 litre/head/day, *which is generally ignored while computing the total water requirement for a city*. However, for cities having populations exceeding 50,000, the water required in kilo litres may be computed by using the relation

$$\text{Kilo litre of water required} = 100\sqrt{P} \quad ...(2.1)$$

where P is population in thousands.

Water Required to Compensate Losses

This includes the water lost in leakage due to bad plumbing or damaged metres, stolen water due to unauthorised water connections, and other losses and wastes. These losses can be reduced by careful maintenance and universal metering. Even in the best managed water works, this amount may be upto 15% of the total consumption.

2.5. VARIATION IN THE PER CAPITA DEMAND

Water demand is not constant throughout. It will have annual, seasonal and hourly variation. Annual average daily consumption, though useful, does not give a complete picture. During certain hours of the day the water demand is found to be maximum. Similarly during certain months of year it may be found water consumption is high.

The total requirement of water for a town (or city) in litres/day required by a town can be worked but by multiplying the probable number of persons who are going to use the facilities provided by the scheme, and the annual average daily consumption of each person (called per capita demand).

Seasonal variations occur due to large use of water in summer season, lesser use in winter, and much less in rainy season. The water consumption is generally more on Sundays and holidays, on days of dust storms, etc. The consumption in the early hours of morning (0 to 6 hours—say) is generally small, increases sharply as the day advances, reaching a peak value between about 8 to 11 AM, then decreases sharply upto about 1 PM, remains constant upto about 4 PM, again increases in the evening reaching a peak between 7 to 9 PM, finally falling to a law value in the late hours of night, as shown in Fig. 2.2. The night flow (excluding industrial consumption) generally represent the magnitude of losses and wastes, since there is no appreciable domestic consumption during this period.

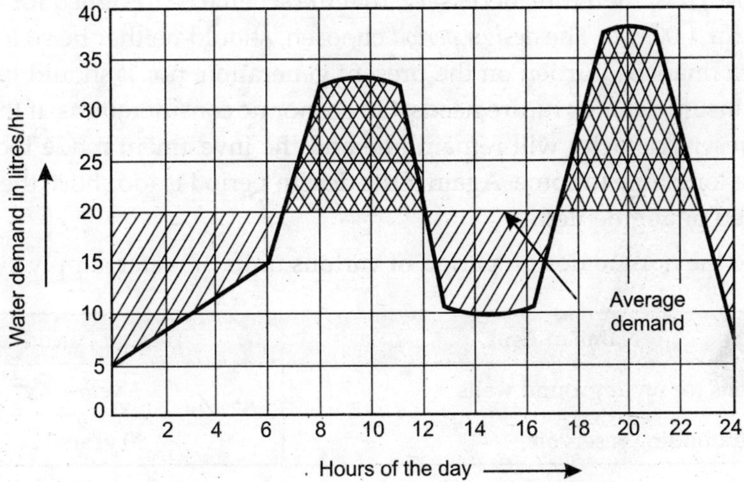

Fig. 2.2. Hourly rate of water consumption.

These normal variations in the demand should be assessed to design supply pipes, service reservoirs, distributary pipes, etc. A suitable allowance must also be made for sudden and heavy demand required for fire-fighting.

Assessment of Normal Variations

The smaller the town, the more variable is the demand. The shorter the period of demand, the greater is the departure from the mean. The maximum demands (monthly, daily or hourly) are generally expressed as ratios of their means. These ratios may vary considerably for different communities.

Following figures are generally adopted :

1. *Maximum daily consumption* is generally taken as 180 per cent of the average daily consumption
   Therefore, Maximum daily demand
   = 1.8 (*i.e.* 180 per cent) × Annual average daily demand
   = 1.8 q
2. *Maximum hourly consumption* is generally taken as 150 per cent of its average hourly consumption.

2.6. DESIGN PERIOD

The plant which is being planned, designed and executed today will serve the future population also. Therefore, it should have adequate capacity to cope up with increase in population.

The future period for which a provision is made in the water supply scheme is known *as the* **design period.** Say for example, let the population of a certain area be, 1,00,000 in 2016 ; but in the year 2046, it may rise to 2,00,000. In order to ensure a reliable water supply even after 30 years (*i.e.* upto 2046) it is, therefore, necessary, that the scheme be designed for a population of 2,00,000 and not for 1,00,000. The *design period* choosen, should neither be so long as to throw unnecessary extra financial burden on the present generation, nor it should be so short as to make the design insufficient for future needs. On economic considerations, if this period is too long, then the present capacity will remain idle and the investment made today will not be utilised fully till a long time to come. Again if the design period is too short, the system cannot be easily expanded or augmented.

Following are the normal design period of various units of water supply system :

	Name of Unit	Design Period
1.	Wells for underground wells	5 years
2.	Impounding reservoir	30 years
3.	Water pumping station	10 to 15 years
4.	Distribution lines from source to city	20 years

5.	Water treatment plant	10 to 15 years
6.	Service reservoirs	30 years

2.7. FACTORS AFFECTING PER CAPITA DEMAND

The annual average demand for water (*i.e.* per capita demand) considerably varies for different towns or cities. This figure generally ranges between 100 to 360 litres/capita/day for Indian conditions. These variations in total water consumption of different cities or towns depend upon various factors as discussed hereunder:

1. **Size of the City:** The per capita demand for big cities is generally large as compared to that for smaller towns. This is because of the fact that big cities are generally sewered, and as such require large quantities of water. Similarly, in a big city, commercial and industrial activities are generally more, thus requiring more water.
   Sometimes even a smaller town may have high water consumption, if it is fully industrialised or is having some industry requiring tremendous quantities of water, or if rich affluent people are living in it. On an average, the per capita demand for Indian towns may vary with the population, as shown in Table 2.4.
2. **Climatic Conditions:** At hotter and dry places, the consumption of water is generally more. Similarly, in extremely cold countries, more water may be consumed, because the people may keep their taps open to avoid freezing of pipes, and there may be more leakage from pipe joints, because metals contract with cold.
3. **Types of Gentry and their Habits:** Rich and upper class communities generally consume more water due to their affluent living standards. Middle class communities consume average amounts, while the poor slum dwellers consume very low quantity. The amount of water consumption is thus directly dependent upon the economic status and living styles of the consumers.

Table 2.4. Variations in per capita demand with population in India

S. No.	Population	Per capita demand in litres / day
1.	Less than 20,000	110
2.	20,000–50,000	110–150
3.	50,000–2 lakhs	150–240
4.	2 lakhs–5 lakhs	240–275
5.	5 lakhs–10 lakhs	275–335
6.	Over 10 lakhs	335–360

4. **Industrial and Commercial Activities:** Many industries require huge amounts of water and as such, increase the water demand considerably. The industrial water demand is having no direct connection with the population or the size of the city,

but more industries are generally situated in big cities, thereby increasing the per capita demand for big cities. However, for a properly planned zoned city, the water requirement can be more accurately predicated for the industrial and commercial demands separately for each zone.

5. **Quality of Water Supplies:** If the quality and taste of the supplied water is good, it will be consumed more, because in that case, people will not use other sources such as private wells, hand pumps, etc. Similarly, certain industries such as boiler feeds, etc., will use public supplies, provided the supplied water is upto their required standards.
6. **Development of Sewerage Facilities:** The water consumption will be more, if the city is provided with 'flush system' and shall be less if the old 'conservation system' of latrines is adopted.
7. **Cost of Water:** If the water rates are high, lesser quantity may be consumed by the people. However, affluent and rich people are little affected by such policies.
8. **Policy of Metering and Method of Charging:** When the supplies are unmetered and the charges are fixed, people generally do not practice economy in the use of water, because they think that they have to pay only a fixed amount irrespective of the quantity of water used by them. This leads to high wastage and high consumption of water. However, when the supplies are metered, people use only that much of water as much is required by them.

2.8. FACTORS AFFECTING LOSSES AND WASTES

Enormous amount of good water is lost in leakages, wastes, thefts, etc., in every water supply scheme. In order to control these losses and to reduce them to minimum, it is desirable to thoroughly study the factors on which these losses depend. The various factors on which these losses depend and the measures to control them are as under:

1. **Water Tight Joints:** The joints in the water mains and pipes generally leak due to bad plumbing, leading to high wastage of water. The leakage of water can be reduced by careful and better plumbing with constant maintenance, and keeping the joints water tight.
2. **Pressure in the Distribution System:** High pressure in the distribution pipes leads to higher leakage losses. An increase in pressure from 20 m head of water to 30 m head of water may increase the leakage by about 20 to 30 per cent. Hence, in order to keep the leakage losses less, it is desirable to keep the pressure in the pipes to a minimum possible value. Multistoreyed buildings may be encouraged to develop their own storage and pumping.
3. **System of Supply:** In intermittent system of supplies, the leakage losses are reduced they do not occur for all the 24 hours but occur only for a few hours during the period the supply is restored. However, in intermittent supply, people generally waste more water by leaving their taps open.

4. **Metering:** When the supplies are metered, wastage is considerably reduced, because people become more careful in using water, as they have to pay for the volume of water consumed by them.
5. **Unauthorised Connections:** Water is sometimes stolen through unauthorised connections, which must be detected and controlled by heavily punishing the defaulters.

2.9. POPULATION FORECASTING

Design of any water supply scheme is done on the basis of projected population at the end of design period. Otherwise the scheme will become inadequate in near future :

Following are the common methods for population forecasting :

1. **Arithmetical Increase Method:** In this method the average increase in population per decade is calculated from the past census reports. This rate of increase is also considered for finding out the population in next decade. This method of forecasting gives satisfactory results for old and large cities whose scope for future expansion has reached to saturation limit.
2. **Geometrical Increase Method:** In this method the percentage increase is assumed to be the rate of growth of the town under consideration. Average percentage increase is used to find out the future increment in population. This method is applied to new industrial town at the beginning to development for new decades.

   The basic difference between arithmetic and geometric progression methods of forecasting future populations is that : Whereas, in arithmetic method no compounding is done; in geometric method, compounding is done every decade. The computations in two methods are, thus, comparable to simple and compound interest computations, respectively.

   For example, if population increases from 1 lakh to 1.1 lakh in a past decade *i.e.* @ 10%, then the population in the next decade will increase from 1.1 lakh to 1.21 lakh (*i.e.* 10% increase over the present decade population of 1.1 lakh). In simple arithmetic method, this value would have been 1.20 lakh only (*i.e.* 10% simple increase over 1 lakh for 2 decades).
3. **Incremental Increase Method:** For an average size town under normal condition the growth rate is found to be in increasing order. Therefore, there will be an increment in increase of population in each decade. The increase in increment is determined from post two decades for calculating the incremental increase value in future decade.
4. **Simple Graphical Method:** In this method, a graph is plotted from the available data, between time and population. The curve is then smoothly extended upto the desired year. The method gives very approximate results, as the extension of the curve is done by the intelligence of the designer.

   All the methods described above are based on the assumption that factors and conditions which were responsible for population increase in the past will continue even in future also, with the same intensity.

QUESTIONS

1. How will you estimate the quantity of water required by a town while arranging water supply scheme for the same?
2. What are the various purposes for which provision should be made in the average daily per capita demand of water in a water supply scheme?
3. Describe how you would arrive at the total quantity of water to be supplied to a metropolitan area.
4. What is meant by term 'per capita demand'? How is it estimated? Mention and discuss the factors that influence per capita demand.
5. Write short notes on the following :
   (*a*) Fire demand in water supply.
   (*b*) Wastage of water in public supplies.
   (*c*) Detection and prevention of water wastage.
   (*d*) Peak-hour demand and its effect on design of water supply scheme.
   (*e*) Intermittent versus continuous water supply.
   (*f*) Metered versus non-metered water supply.
6. Discuss briefly as to how the requirements of water vary from season-to-season, day-to-day, and hour-to-hour, and in what way does it affect the design capacities of various components of a water supply scheme.
7. Why is the population forecast necessary in the design of public water supply schemes? Discuss the different methods employed for the purpose and their comparative merits and demerits.
8. Explain how per capita rate is arrived in a water supply scheme of an industrial town.
9. What are the basis for estimation of water demand and would you adopt suitable measure for fire fighting.
10. Discuss how the future population of a town is estimated. What are the assumptions made for forecasting the population of a city.
11. Discuss the various factors which influence the water consumption in a commonly in a community.
12. Write briefly about the fluctuations in the rate of water consumption.

Sources of Water

3.1. INTRODUCTION

Water is a most important resource for the mankind, since no life is possible without water. Water may occur in nature in liquid, solid, as well as gaseous form. It can be put to various uses due to its unique properties. Since water is a limited resource, its efficient use is basic to the survival of the ever increasing population of the world. Conservation and development of water resource include the availability of water, its quality, location, distribution, climate condition, socio-economic conditions etc.

About 97% of the total available water on earth is contained in oceans, and is saline in nature. Out of the balance 3%, which is available as fresh water, about 2% is contained as ice on poles and 0.75% as ground water. Out of the remaining about 0.25%, only about 0.01% is available in lakes and rivers, and rest occurs as glaciers and snow. Therefore, the surface water which can really be utilised by the society is very small. Even the surface run-off that flows in rivers, mostly goes waste, as it felt does to the oceans in the absence of proper storage.

Water is required by the community for fulfilling its several needs, and for surviving not only human life but also animals and plant life. The water is generally required for:

(*i*) Municipal and industrial water supply.

(*ii*) Irrigation.

(*iii*) Hydropower generation, navigation, recreation etc.

Hydrology

Hydrology is a science which deals with the occurrence, distribution and movement of water over and under the surface of the earth. It deals with the solid, liquid, and vapour forms of water. The knowledge of this science enable to design with greater reliability, the irrigation and flood control works, municipal and industrial water supply schemes, power projects etc.

Hydrology helps in the process of planning and execution of water resources projects, as it answers the following:

1. The quantum of peak flood flows expected at the proposed reservoir site.

2. Minimum long term yield of a given catchment to decide the required reservoir capacity.
3. Effects produced on river water levels, by the construction of proposed dams, embankments and other controlling hydraulic structures.

After estimating the water requirement for the proposed water supply scheme, the planners of the scheme must search nearby water sources, so as to supply the required amount of water. If the available supplies of water in the vicinity of the area are not sufficient to meet the demand, then it becomes imperative to search some other distant source or sources and to make arrangements for bringing the water from there.

Various sources of water available on the earth can be classified into the following two categories :

(A) **Surface Sources**

(*i*) *Ponds and lakes* ;

(*ii*) *Streams and rivers* ;

(*iii*) *Storage reservoirs* ; and

(*iv*) *Oceans*, generally not used for water supplies, at present.

(B) **Sub-surface Sources** *or* **Underground Sources**

These are available generally in the form of springs, wells and tube-wells.

Most of the earth's water sources get their water supplies from precipitation, which may fall in various forms, such as, rain, snow, etc. Rains no doubt, form the principal and the major part of the resultant supplies.

3.2. HYDROLOGIC CYCLE

Water in the oceans evaporates due to the heat energy provided by solar radiation. The water vapour moves upwards and forms cloud. Much of the cloud condenses and falls back to the ocean as rain. A part of the cloud is driven to the land area by winds. The condenses and precipitates on the land mass as rain, hail, storm etc. It may infiltrate the surface of the ground or may flow over the land surface as runoff. A portion of the infiltrated water is used by vegetation, while the balance becomes a part of the deep ground water supply or moves slowly to streams and to the ocean.

The precipitation and evaporation continue forever, and thereby a balance is maintained between the two. This process is known as hydraulic cycle and is shown in Fig. 3.1. Thus, we can also say that hydrologic cycle is involved in the transfer of moisture from the sea to the land and back to the sea again.

Knowledge of the hydraulic cycle helps in the design and operation of projects dealing with water supply, irrigation and drainage etc.

Various earth's water-sources get their supplies from precipitation, while the precipitation in itself is the evaporation from these sources. Water is lost to the atmosphere as vapour from the earth, which is then precipitated back in the form of rain, snow, hail, dew, sleet or frost, etc.

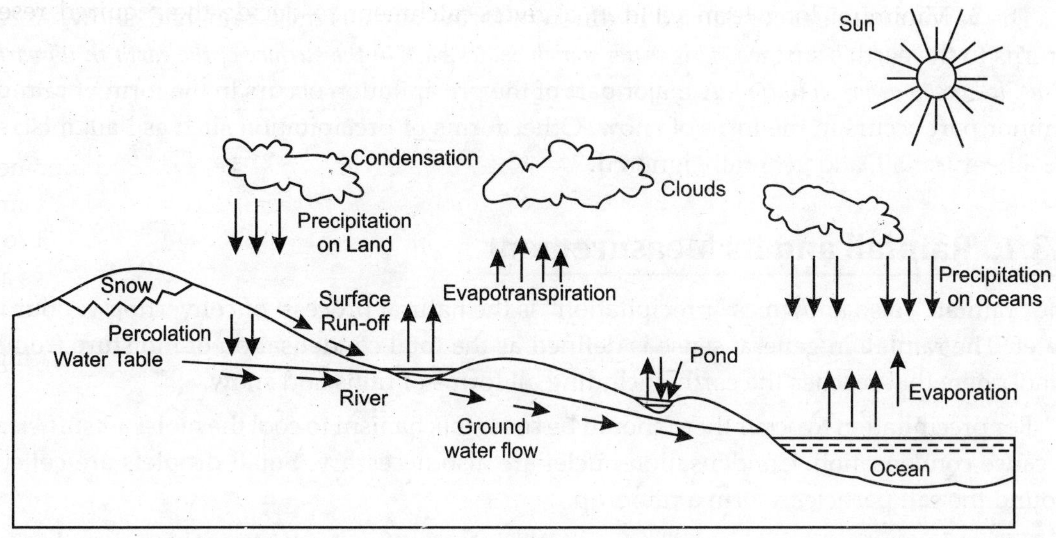

Fig. 3.1. The hydrologic cycle.

Hydrologic equation is expressed as follows :

Precipitation (P) = Run-off (R) + Evaporation (E).

1. **Precipitation** includes all the water that falls from atmosphere to the earth surface. Mainly precipitation is of two types :
   (*a*) Liquid precipitation (Rainfall).
   (*b*) Solid precipitation (Hail storm).
2. **Run-off.** Run-off is that portion of precipitation which makes its way towards streams, lakes or oceans. Run-off occurs only if the rate of precipitation exceeds the rate at which water infiltrates into the soil and after depressions, small and large, on the soil surface get filled with water. Rainfall duration , its intensity, and a real distribution influence the rate and volume of run-off.
3. **Evaporation:** Evaporation of water takes place from the surface of ocean, rivers, lakes, and the moist soil due to the heat of sun.
4. **Evapotranspiration:** Process by which water is transferred from land to atmosphere by evaporation from the soil and other land surfaces and by transpiration from plants.

   Yield of Catchment: It is the net quantity of water available for storage, after all losses, for the purpose of water resources and utilisation.

3.3. PRECIPITATION

As we know that, water evaporates from water surfaces like streams, rivers, oceans, ponds, etc., and also from the land and plants in the form of water vapour. The water vapour get collected in the atmosphere. As the evaporation continues, the amount of atmospheric vapour goes on increasing. Since a space can hold only a certain fixed amount of water vapour, a stage is reached when any further addition of vapour will get condensed on the surfaces.

The vapour may get condensed in different forms, such as mist, rain, hail, snow, etc, and returns to the earth's surface. *This water which comes back to the surface of the earth in its various forms, is known as precipitation.* A major part of the precipitation occurs in the form of rain, and a minor part occurs in the form of snow. Other forms of precipitation such as hail, mist, etc., are all very small, and generally ignored.

3.3.1. Rainfall and its Measurement

The "rainfall", also known as "precipitation" is the natural process' of converting vapour into water. The rainfall in general sense is defined as the total condensation of moisture from the atmosphere that reaches the earth, including all forms of rains and snow.

For precipitation to occur there should be some mechanism to cool the moist air sufficiently to cause condensation. Condensation nuclei are also necessary. Small droplets are collected around the salt particle to form a raindrop.

The precipitation is measured as the vertical depth of water that would accumulate on a level surface, if the entire precipitation remained where it fell, assuming there were no losses by evaporation or runoff.

Since the amount of precipitation varies from place to place, it is necessary to install measuring devices at various key points. The simplest method of measuring precipitation is by setting up gauges with a horizontal circular aperture of known area and collecting and measuring at regular intervals the precipitation collected in them.

The two types of precipitation (*i.e.* rain and snow are measured separately by measuring devices called gauges and snow gauges respectively).

Main rainfall variables are :

1. **The average rainfall over the area.** The data from a single nearby rain gauge are often sufficient representation to allow their use in the design of small projects.

   Since the rainfall over a large area is not uniform, average rainfall is estimated from rainfall records of raingauge network. The three methods used to evaluate average rainfall are : Arithmetic Mean Method. Thissen Polygon Method and Ishyetal Method. (contours of equal rainfall are called Isohyets).
2. **Intensity.** Intensity is a measurer of how hard it rains. The unit of measures is mm/h.
3. **Duration.** How long it rains at any given intensity.
4. **Frequency.** How often it rains at any given intensity and duration.

Relative Humidity (RH)

The amount of water which a given quantity of air can hold varies with temperature. Warm air can hold a larger amount of moisture, but if the temperature of the air mass is lowered, a certain temperature is reached when the moisture tends to separate out. This temperature is called **Dew Point** and at this point the air is said to be **saturated.**

The wetness of air mass under any condition is given by a factor called **Relative Humidity** (RH). It is defined as,

$$RH = \frac{\text{Actual Quantity of Water Present in given volume of air}}{\text{Amount of water required to saturate it at the same temperature}}$$

It is expressed in percent. At saturation RH = 100%. Any further fall in temperature below Dew Point results in release of moisture and consequent precipitation.

3.3.2. Snowfall and its Measurement

Snow is the second important part of precipitation. Although it does not yield immediate run-off, but is generally responsible for delayed supplies of water. Most of our Himalayan, rivers are snow fed and are, therefore, provide continuous perennial supply of water.

The snowfall is generally measured by a non-recording type of rain gauge. The snow is melted and measured. Non-recording rain gauges do not record the rain, but only collect the rain. The collected rain is then measured by means of graduated cylinders.

To measure the snow, snow surveys are also conducted. These snow surveys are generally conducted at the start of summer season, when snow starts melting. These surveys are useful in estimating the amount of available water expected due to the melting of snow in summer.

3.4. RUN-OFF

It is that part of the precipitation as well as any other flow contribution, which appears in surface streams of either perennial or intermittent form. Runoff may consist of following components depending upon the source it is derived,

(i) **Surface Runoff :** It is that part of the total runoff which travels over the ground surface and through channels to reach the basin outlet.

(ii) **Subsurface Runoff (Interflow)** : It is the water which infiltrates the soil surface and then moves laterally through the upper soil strata above the main groundwater table (also known as Quick Return Flow).

(iii) **Groundwater Flow (Base Flow) :** It is that part of the runoff which is due to deep percolation of the infiltrated water which has joined the groundwater Table. With this rise in groundwater table it flows back into the stream (also known as Delayed Runoff).

The Surface and Sub-surface runoff together are known as **Direct Runoff**.

Run-off Measurement

The run-off or stream flow is generally expressed as the instantaneous rate or average rate of flow for short periods (such as cubic meters per sec), but when it is expressed over long periods, such as total volume of water flowing per year, as million cubic metres per year.

The discharge of the stream at the intended site of the water supply intake point can be measured by various methods, as mentioned hereunder :

1. Velocity-area method.

2. By the use of a weir or a spillway of any existing dam.
3. By using venturi flumes.

3.5. SURFACE SOURCES OF WATER

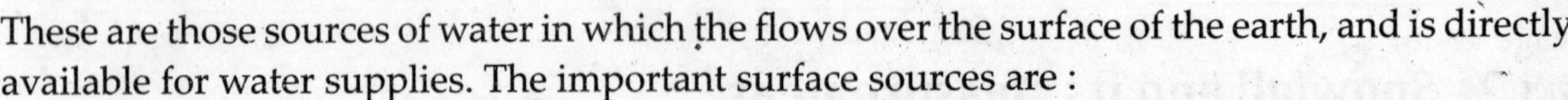

These are those sources of water in which the flows over the surface of the earth, and is directly available for water supplies. The important surface sources are :

(*i*) Natural ponds and lakes;
(*ii*) Streams and rivers; and
(*iii*) Storage reservoirs.

These sources are discussed here under :

3.5.1. Ponds and Lakes

When natural large sized depression formed within the surface of the earth gets filled up with water, is known as a pond or a lake. The difference between a pond and a lake is only that of size. If the size of the depression is comparatively small, it may be termed as a *pond*, and when the size is larger it may be termed as a *lake*.

Generally, the surface run-off from the catchment area enters the lake through small drains or streams. Sometimes, the underground water through some spring, also enters natural depressions and gets collected into the water ponds and lakes.

The water in a lake generally have good quality and does not need much purification. Larger and older lakes, however, provide comparatively purer sedimentation of suspended matter, makes the lake's water purer. On the other hand, in still waters of lakes and ponds, the algae, weed and vegetable growth take place freely, imparting bad smells, tastes and colours to their waters.

The amount of water available from lakes depends upon the catchment area of the basin, annual rainfall, and geological formations. Due to the smaller quantity of water available, lakes are usually not considered as principal sources of water supplies. They are, therefore, useful for only small towns and hilly areas. However, when no other available, larger lakes may become the principal sources of supplies.

3.5.2. Streams and Rivers

Streams feed their waters to the lakes or rivers. Quantity of water available in them is generally very small, therefore, are useful as source of water only for small villages, especially in hilly regions. Larger and perennial streams may, however be used as source of water, by providing storage reservoirs, barrages, etc.

Rivers are the most important sources of water for water supply schemes. In perennial rivers water is available throughout the year. Such rivers are generally fed by rains during rainy season and by snow during summer season. These rivers are used as sources of water supplies directly. The non-perennial rivers can be used as sources of public supplies by providing storage on the upstream of the intake works. The head works, such as a barrage or a weir, may also be constructed on those perennial rivers, where supplies are considerably reduced during dry weather periods.

The *quality* of water obtained from rivers is generally not reliable, as it contains large amounts of silt, sand and a lot of suspended matter. The disposal of the untreated or treated sewage into the rivers further contaminate their waters. *The river waters must, therefore, be properly analysed and well treated before supplying to the public.*

3.5.3. Storage Reservoirs

A water supply scheme drawing water directly from a river or a stream may not satisfy the consumer demands during extremely low flows ; while during high flows, it may again become difficult to carry out its operations due to devastating floods. A barrier in the form of a dam may, therefore, sometimes be constructed across the river. The pool or artificial lake formed on the upstream side of the dam is known as the storage reservoir. The water stored in the reservoir can be used easily for water supplies.

The *surface supplies are generally contaminated* and cannot be used with minor treatment or without any treatment. They need proper water treatment plants (WTPs) and other connected works before being used for public supplies. They are useful for big cities and large industrial town where huge quantities of water are required by the public.

3.5.4. Wetlands

A wetland is a distinct ecosystem that is flooded with water, either permanently or seasonally. The primary factor that distinguishes wetlands from other water bodies is the characteristic vegetation of aquatic plants. Wetlands play a number of functions, including water purification, **Water Storage**, processing of carbon and nutrients etc. The main wetland types are swamp, marsh, bog and fen.

3.6. UNDERGROUND (SUB-SURFACE) SOURCES

The water stored in the ground water reservoir through infiltration, is known as the **underground water.** This water is generally pure, because it percolates through the soil pores. These waters are less likely to be contaminated by bacteria. These are generally rich in dissolved salts, minerals, gases, etc. The extent of the salts and minerals present in the ground water depends upon the type and extent of geological formations through which the water is passing before joining the watertable. Theses waters are trapped by artificial means by constructing wells, tube wells, etc.

The rainfall that percolates below the ground surface, passes through the voids of the rocks, and joints the watertable. These voids are generally interconnected, permitting the movement of the ground water. But in some rocks, they may be isolated, and thus, preventing the movement of water.

The volume of water contained in the ground water reservoir in any localised area, *i.e.* **the water storage capacity of the ground water** is dependent upon;

(*i*) the porosity and permeability of the rocks;

(*ii*) the rate at which water is added to it by infiltration; and

(*iii*) the rate at which water is lost from it by evaporations, transpiration, seepage to surface courses, and withdrawal by man.

3.6.1. Groundwater Recharge

Groundwater is an important source of water supply throughout the world. Groundwater usage has increased at an accelerating rate and the indications are that the trend will continue. Main reasons for excessive use of groundwater are its good quality and round the year availability.

With increasing use of groundwater for various needs, the annual extractions of groundwater are far in excess of net average recharge from natural sources. Consequently, groundwater storage is depleting fast and groundwater table is going down.

Principal sources of natural recharge of groundwater include - precipitation, streamflow, lakes, reservoirs and wetlands. In many instances the overdraft of groundwater is due to the diminishing opportunity for natural recharge of groundwater storage. This may be due to impervious drains, lining of canals, urbanisation and diversion of water.

In order to increase the natural supply of groundwater artificial recharge (or replenishment) has to be practiced. Artificial recharge is defined as, "Augmenting the natural infiltration of precipitation or surface water into underground formations by some method of construction, spreading of water or by artificially changing natural conditions".

Methods of Artificial Recharge

Various methods of **artificial recharge** of groundwater are in use. These are classified as : Spreading Methods, Injection Method and Induced Recharge.

(i) **Spreading Method :** The method comprise increase in the surface area of infiltration. This method is suitable where large area of basin is available and aquifer is **unconfined** with greater permeability.

(ii) **Injection Method :** Injection method is especially useful to feed deep confined aquifers. The water is injected into the aquifer through shafts, wells and horizontal infiltration galleries. This method is suitable where land is expensive and surface permeability is low.

(iii) **Induced Recharge :** It implies withdrawal of water near any surface water source, such as stream or lake so that seepage rate increases due to lowering of water level. Induced recharge has the advantage of simplicity and instantaneous recharge.

3.7. SELECTION OF A SOURCE OF WATER

Following factors are generally considered in selecting a source for supplying water to a city or a town.

1. **Quantity of Available Water.** The quantity of water available at the source must be sufficient to meet the various demands during the entire design period of the scheme. If sufficient quantity of water is not available in the vicinity of the area, we may have to think of bringing water from distant sources.
2. **Quality of Available Water.** The water available at the source must not be toxic, poisonous or in any other way injurious to health. The impurities present in the water should be as less as possible, and should be such as to be removed easily and economically by normal treatment methods.
3. **Distance of the Source of Supply.** The source of water must be situated as *near the city* as possible. Because when the distance between the source and the city is less, lesser will be the cost of the scheme.
4. **General Topography of the Area.** The area between the source and the city should not contain deep valleys or high mountains and ridges. In such uneven topographies, the cost for carrying water pipes in valleys and that of constructing tunnels in mountains, shall be very high.
5. **Elevation of the Source of Supply.** The source of water must be at a level higher than the city of town to be supplied with water, so as to make the gravity flow possible. When the water is available at lower levels than the average city level, pumping has to be resorted to, which involves huge operation and maintenance cost.

3.8. USE OF WATER RESOURCES

Water is important in almost every sphere of human activity including drinking, irrigation, thermal power generation, industries, livestock management, hydroelectric generation, fisheries, nevigation, recreation, etc.

Water is required by the commonly for fulfilling its several needs, and for surviving not only human life but also animal and plant life. Water should not only be in sufficient quantity, but should also be of good quality, so as not to cause any harmful effects on such lives. Water is generally required for the following important uses:

1. **Domestic Use:** This include the water required for drinking, cooking, bathing, washing, lawn sprinkling, gardening, sanitary toilatary uses etc. The amount of domestic water consumption per person vary according to the living conditions (lifestyle) of the consumers. On an average, the domestic consumption under normal conditions in an Indian city is expected to be around 135 litres per day per person against those developed countries like U.S.S. to be around 340 litres per day per person.
2. **Industrial and Commercial Use:** This include the water used in factories, hospitals, hostels, hotels, restaurants, offices, schools etc. In thermal power plants, water is required for boiler make-up water, for cooling in cooling towers etc.

3. **Public Uses:** This includes the water required for public purposes, such as watering of public parks, gardening, washing and sprinkling on roads, use in public fountains, fire fighting purposes.
4. **Irrigation Needs:** Irrigation is generally defined as an application of water to soil for the purpose of supplying moisture essential for plant growth. The main purpose of irrigation are:
   (*a*) supply essential moisture for plant growth,
   (*b*) save crops against short duration droughts, and
   (*c*) wash or dilute salts in the soil.

   Every crop require a certain quantity of water after a certain fixed interval, throughout the period of growth. If the natural rain is sufficient and timely so as to satisfy both these requirements, no irrigation water is required for raising that crop.
5. **For Hydropower Generation:** Hydroelectric power is generated by running the turbine blades by the energy of the flowing water, and thus producing electricity from the generators coupled to the turbines. In order to ensure the continuous production of hydropower, it is essential to store water to a certain height and then let a discharge flow through the turbines. Production of hydropower is an excellent use of water, as no water is consumed in it.
6. **For Navigation and Recreation:** Navigation is another important use of water, which provides as a cheap means of transport without any consumption of water. Although no water is consumed in this use of water, however, sufficient quantity of water must remain available throughout the year, so as to enable the ships, motorboats, steamers etc. to safely float over it with minimum drag.

QUESTIONS

1. What are the common sources of water for a water supply scheme? State the factors that govern the final selection?
2. Give a list of different sources of water, which can be considered for large supplies, bringing out the quality that can be expected in each case, and the treatment that may have to be undertaken in each case.
3. Write short notes on following :
   (*i*) Hydrological cycle,
   (*ii*) Mass curve,
   (*iii*) Precipitation and its types,
   (*iv*) Rainfall and its distribution.
4. Enumerate the various surface sources of water discuss about the quality and quantity of water supplies that may be available from these sources.
5. Discuss various factors that may govern the selection of a particular source of water informulating a township water supply schemes.

6. Write short notes on:
   (*a*) Dams and selection of site.
   (*b*) Storage capacity of reservoir.
   (*c*) Selecting a suitable size for a storage reservoir.
   (*d*) Run-off.
   (*e*) Intakes for collecting surface water.
7. Enumerate various forms in which ground water occurs in nature. Discuss briefly the exploration of its most important source.
8. What are springs, and how are they formed?
9. What is a 'river intake' ? What are the factors which govern the location of an intake structure on a river?
10. What is an 'intake structure'? Enumerate different types of intakes.
11. Discuss the advantages and disadvantages of underground and surface sources of public water supply.
12. What factors are considered in choosing between ground water and river water as a source of water supply to a town?
13. Write a note on springs as a source of water. Draw a sketch of a spring with necessary protection in collection of water.
14. Discuss various uses of water resources.
15. Differentiate between: Evaporation, Transpiration and Evapotranspiration.
16. Write a note on Groundwater recharge.

Water Quality and Characteristics

4.1. INTRODUCTION

Pure water is a chemical compound with each of its molecule containing two hydrogen atoms and one oxygen atom, and nothing else (H_2O being the chemical formula for water). However, pure water can never be available in nature.

The precipitation, at the instant of its formation, contains no impurities, but during the process of formation and fall through the earth's atmosphere, it may dissolve certain gases, traces of minerals, and other substances. When precipitation reaches the earth's surface, many more chances are available for the introduction of various physical, chemical, or bacterial impurities in it.

When rain water flows over the surface of the earth, it picks up or dissolves particles of soil, garbage, sewage, pesticides and other human, animal or chemical wastes. It may also pick up and dissolve certain decayed organic materials, such as plants or dead animals. When this water seeps into the "Ground-Water Reservoir," most of the suspended particles are filtered out, but on the other hand, the water dissolves the minerals and salts present in the earth's layers, through which it travels.

Certain minerals such as iron, calcium, magnesium, fluorine, etc. in small quantities may be good for the health of the people, because human beings need a certain amount of these elements in their bodies. But when these materials are dissolved in large amounts or in certain combinations, the water may become unfit or less useful for municipal, industrial and other uses.

Sometimes the water may contain toxic or poisonous substances such as arsenic, barium, cadmium, chromium, cyanides, lead, copper, etc., which may be very harmful to the public health, even if present in low quantities. The water may contain too much of common salt, thereby rendering it *brackish* and making it undrinkable and less useful for cloth washing or for irrigation and farming. Sometimes the water may contain harmful bacteria, the presence of which may cause diseases such as cholera, typhoid, dysentery, gastroenteritis, infectious hepatitis etc.

For the purpose of safety to public health, economy and utility in industries and other uses, it becomes imperative to thoroughly *check, analyse,* and treat the raw water to safe and permissible limits, before supplying to the public. This must be strictly adhered to, when water is supplied for drinking, bathing, washing, etc. When water is required solely for irrigation, the quality standards may vary depending upon the requirement of each industry of farming and irrigation.

Water Quality Parameters (Criteria)

The municipal water produced should meet the following criteria :

1. Hygienic with a minimum of pathogens and toxins.
2. Aesthetically acceptable (colourless and sparkling clear) and potable.
3. Non-corrosive to the distribution system.
4. Free from excessive amounts of mineral and organic matters.
5. Water can be produced for the least cost.
6. No vegetable stains should be present.
7. Free from taste and odour.
8. Water should be soft and should not be hard.
9. Free from radioactive substances.

To meet these conditions, several sets of standards have been proposed and are continually being reversed. The general types of criteria for water quality are discussed in subsequent paragraphs.

4.2. CHARACTERISTICS OF WATER

To be potable, water should be significant free from colour, turbidity, bad taste and odour, and should be moderate temperature during summer and winter. Water becomes impure when it contains one or more substance in excess concentrations, or toxic chemicals which might have adverse effects on the user. Impurities, generally found in water are classified according to their characteristics.

Following four types of characteristics (criteria) are used to describe drinking water quality:

1. **Physical characteristics.** These relate to the quality of water for domestic use and are usually associated with the appearance of water, its colour or turbidity, temperature, and taste and odour.
2. **Chemical characteristics.** These are sometimes evidenced reach turns, such as the comparative performance of hard and soft waters in laundering. But, generally, these are not visible.
3. **Microbiological characteristics.** Bacteria available in water may be harmful or not harmful. Waterborne diseases are caused by pathogenic micro-organisms, like viruses, bacteria, protozoa and helminthic worms.
4. **Radiological characteristics.** These factors are considered in areas where there is a possibility that the water may have come in contact with radioactive substances.

For the purpose of the development of a water supply system, it is necessary to examine carefully all these factors (criteria) that might adversely affect the desired use of a water supply source. All these factors are being discussed hereunder:

4.3. PHYSICAL CHARACTERISTICS

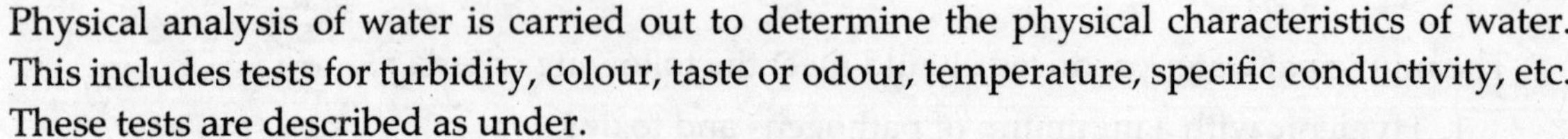

Physical analysis of water is carried out to determine the physical characteristics of water. This includes tests for turbidity, colour, taste or odour, temperature, specific conductivity, etc. These tests are described as under.

1. **Turbidity:** When suspended matter such as clay, silt or some other finely divided organic materials are present in water, it will appear to be *muddy* or *cloudy* (*i.e. turbid*) in appearance. The turbidity depends upon the finesse and concentration of particles present in water. Although, these may not be harmful to health, yet they are to be removed or reduced for aesthetic and psychological reasons. The turbidity is measured by a *turbidity rod* or by a *turbidimeter* with optical observations, and is expressed as the amount of suspended matter in mg/L or ppm. A turbidity in excess of 5 units is easily detectable in a glass of water, and is thus, usually objectionable. Turbidity in a clear lake water is about 25 units.

   The insoluble particles of soil, organics, micro-organisms and other materials, impeding the passage of light through water by scattering and absorbing rays, thus causes, turbidity, which can be measured as the interference to the passage of light through the given water. Turbidity can be measured in the field by using turbidity rods and turbidimeters.

2. **Colour:** Dissolved organic matter from decaying vegetation or some inorganic materials, such as coloured soils, etc., may impart colour to the water. The excessive growth of algae and aquatic micro-organisms may also sometimes impart colour of the water. The presence of colour in water is not objectionable from health point of view, but may spoil the colour of the clothes being washed in such waters, and is also objectionable from aesthetic and psychological point of view.

   For public supplies, the colour number measured by tintometer on cobalt scale should not exceed 20, and should preferably be less than 10.

3. **Tastes and Odours:** The dissolved organic materials or the inorganic salts, or the dissolved gases may impart tastes and odours to the water. Taste and odour may be caused by the presence of : *dissolved gases* such as H_2S, CH_4, CO_2, O_2 etc., combined with organic matter; *mineral substances* like NaCl, iron compounds, carbonates and sulphates of other elements; and *phenols* and other tarry or oily matter, especially after chlorination.

   Some tastes, such as those imparted by dissolved oxygen and dissolved carbon dioxide are generally desirable. For drinking purposes, the water must not contain any undesirable or objectionable taste and odour. In order to remove the taste and odour from the raw water, the chemical composition of water must first be ascertained, so as to determine as to what treatment, if any, is required to make its taste and odour acceptable.

The odour is generally determined when water is cold; as it may change with the change in temperature. In thorough analysis of water, however, odour should be determined at different temperatures, and results produced in tabulated form.

4. **Specific Conductivity of Water:** Electrical conductivity of water is a measure of water's ability to conduct electricity. *Pure water is a poor conductor of electricity*; but it shows significant conductivity when ions of dissolved salts are present in it. The electrical conductivity, called the **specific conductivity** or the **specific conductance** of water. The elements whose ionic forms contribute to these measures include to these measures include Calcium (Ca^{++}), Magnesium (Mg^{++}),. Potassium (L^{+}), Bicarbonate (HCO_3^{-}), Sulphate (SO_4^{--}) and Chloride (Cl^{-}), etc. The higher is the concentration of such ionic (dissolved) constituents, the higher will be the conductivity.

4.4. CHEMICAL CHARACTERISTICS

Chemical analysis of water is carried out to determine the chemical characteristics of water. This involves tests for determining the total solids and suspended solids, pH value, hardness, chloride content, nitrogen content, iron, manganese and other metal contents dissolved gases etc. These tests are described hereunder :

1. **Total Solids and Suspended Solids:** The total amount of solids (suspended as well as dissolved solids) present in water can be determined by evaporating a sample of water and weighing the dry residue left. The suspended solids can be found be filtering paper. The difference between the total solids and the suspended solids will then represent the dissolved solids. The total permissible amount of solids in water is generally limited to 500 ppm to 1000 ppm.
2. **Acidity, Alkalinity and pH Value:** Acidity is caused by CO_2, mineral acids, and salts from strong acids. Alkalinity is due to carbonates and hydroxides. It is the capacity to neutralise acid. pH value denotes the nature and intensity of water towards acidity or alkalinity. pH value of 7 indicates neutrality, 0 – 7 acidity and 7 – 14 alkalinity. pH value of 6 – 8 is desirable for drinking water.
3. **Hardness of Water:** Hardness in water is that characteristic which prevents the formation of sufficient leather or foam, when such hard waters are mixed with soap. It is usually caused by the presence of Calcium and Magnesium salts present in water. Hard waters are undesirable because they may lead to greater soap consumption, scaling of boilers, causing corrosion and incrustation of pipes, etc.

   If bicarbonates and carbonates of calcium and magnesium are present in water, the water is rendered hard *temporarily*, as this hardness can be removed to some extent by simple boiling or to full extent by adding lime to the water. Such a hardness is known as ***temporary hardness*** or ***carbonate hardness***.

   If sulphates, chlorides and nitrates of calcium or magnesium are present in water, they cannot be removed at all by simple boiling, and, therefore, such waters require special treatment for *softening*. Such a hardness is known as ***permanent hardness*** or ***non-carbonate hardness.***

4. **Chloride Contents:** Chlorides of Ca, Mg and Na are present commonly in natural waters as they are highly soluble. Chlorides are generally present in water in the form of sodium chloride (common salt) and may be due to : leaching of marine sedimentary deposits, pollution from sea water, brine or industrial and domestic wastes, etc. Their concentrations above 250 mg/L produce a noticeable salty taste in drinking water and are thus objectionable. The presence of high quantity of chloride (or sodium chloride) in river or stream water may indicate pollution of water due to sewage and other human or industrial wastes.
5. **Nitrogen Contents:** Decomposed organic matter (from sewage) or agricultural (from fertiliser) drainage may join water sources and result in excessive nitrates. The presence of nitrogen in water is an indication of the presence of the organic matter, and may occur in one or more of the following forms :
   (*a*) Free ammonia or Ammonia nitrogen;
   (*b*) Albuminoid or Organic nitrogen;
   (*c*) Nitrites; and
   (*d*) Nitrates.
6. **Fluorides:** Ground water is likely to have excessive concentrations if fluoride bearing rock exist. Industrial effluents also contribute fluoride compounds to water sources. Dental caries (decay of tooth enamel) in children due to drinking fluoride free water, and fluorosis (mottling of teeth and bone/skeletal softness) due to excessive fluorides (more than 1.5 mg/L) occur when used such water over a long period.
7. **Metals and Other Chemical Substances:** Chemicals containing zinc, copper, lead barium, cadmium, arsenic, mercury, nickel, selenium etc., cause damage to internal organs of the human body (kidney, brain, lever etc.)when present in water. The amounts of these minerals or metals in potable waters should be limited to their permissible standard values, so as to avoid their harmful effects on human life.

   Various organic compounds, pesticides, vinyl chloride, cyanides, phenol etc, are highly toxic. When iron is present in water, it leaves strains on clothes, imparts bad tastes and is also responsible for bacterial growth.
8. **Dissolved Gases:** The various gases which may get dissolved in water due to its contact with the atmosphere or the ground surface may be nitrogen, methane, hydrogen sulphide, carbon dioxide, and oxygen. The contents of these dissolved gases in water may be worked out, so as to get the idea of the extent of contamination of the water and the treatment required for purifying the same.

The nitrogen gas is not very important. The concentration of methane gas is to be studied for its explosive tendency. The hydrogen sulphide gas (H_2S), if present even in small amounts, may give bad taste and odour to the water and should, therefore, be removed. Presence of carbon dioxide gas indicates biological activity and may impart bad taste and odour to the water and may also cause corrosion.

Oxygen gas is generally absorbed by water from the atmosphere, but is being consumed by unstable organic matter for their oxidation. Hence, if the oxygen present in water found to be less than its saturation level, it indicates presence of organic matter.

Natural water has a **dissolved oxygen (D.O.)** content of 8–10 mg/L depending on its temperature. Dissolved oxygen makes the water taste good. A minimum of 4–5 mg/L is essential for fish and aquatic life. Surface water has more D.O. due to greater surface area in contact with atmosphere.

4.5. BIOLOGICAL CHARACTERISTICS

1. Bacteria and Other Micro-organisms

Water used for drinking and cooking purposes must be free from disease-producing organisms (known as *pathogens*). These organisms include viruses, bacteria, protozoa, and worms. ***Waterborne diseases*** are caused by these pathogenic micro-organisms. These microorganisms remain in the intestines of sick human beings and when discharged through fecal/oral route, contaminate water sources.

Common waterborne diseases are : gastroenteritis, typhoid, cholera, dysentry, polio-myelitis, infective hepatitis (**jaundice**) etc. These **bacteria** multiply at a very rapid rate. The bacteria are the minute single cell organisms possessing no defined nucleous and having no green material to help them manufacture their own food. They are reproduced by binary fission and may be of various shapes. They are generally present in raw or contaminated waters. Most bacteria range in size from 0.2 to 5 μm in width or diameter, and upto 1–10 μm in length. 1 micron (μ) = 10^{-6}, hence 1 μm (micro meter = 10^{-6} meter).

Most bacteria are harmless, and under certain conditions beneficial to human beings, animals and crops. Such bacteria or micro-organisms are called **non-pathogenic bacteria** or ***non-pathogens.*** However, certain other bacteria are the deadly foes of man and animals and may enter tissues, causing serious water borne diseases, such as cholera, typhoid, infectious hepatitis, etc. Such harmful bacteria or organisms are known as **pathogenic bacteria** or ***pathogens.*** The rain water, as it falls down through the atmosphere, collects bacteria and other types of parasitic organisms, such as viruses, etc., from the dry dust or smog present in the air or the atmosphere. Since the initial rains wash away most of these dusts, they are likely to be highly contaminated with bacteria, virus and other organisms.

Although many bacteria are useful and helpful in cleaning raw waters, certain other bacteria, often found in polluted waters, produce total diseases.

2. Microscopic Plants

In addition to the bacteria and other micro-organisms described above, certain microscope plants. such as algae, plankton, fungi, etc., are also sometimes present in water.

(*a*) **Algae** is a plant that grows in water and flourishes in the presence of sunlight. *Seven types of algae* have been distinguished by colour and shape. All of them contain the green substance ***chlorophyll,*** which permits them to utilise sunlight for their growth by conversion of carbon dioxide, water and other simple substances into organic carbon compounds.

(*b*) **Plankton.** These are the microscopic plants and animal life that either swim weakly or float in water, and serve as a food for small sea creatures and fish. In most of the cases, they are not injurious to health but they affect the palatability of waters by secreting certain oils, which give bed tastes and odours to the waters. Large quantities of plankton may impart green, brown or yellow hue colour of the water.

Certain types of plankton, if present in water, may cause stains on white porcelain fixtures in kitchens or bathrooms. They may also affect the industries like laundries, dye works, photographic paper etc., due to colouring tendencies.

(*c*) **Fungi.** These plants grow without sunlight and live on other plants or animals, dead or alive. Some fungi may be fairly large, while other fungi are microscopic. They may have to be removed with chlorine treatment.

4.6. RADIOLOGICAL CHARACTERISTICS

The effect human exposure to radiation or radioactive materials are harmful, therefore, unnecessary exposure should be avoided. The amount of radiation to which the individual is normally exposed varies with the amount of background radioactivity.

Use of atomic energy as a power source and the mining of radioactive materials, as well as naturally occurring radioactive materials necessitated to establish its limiting concentrations for the intake into the body including drinking water.

Dangerous radioactive substance, such as uranium, thorium, strontium, caesium, etc. from the wastes of the atomic reactors or from the atomic explosions may sometimes pollute the water sources. The radioactive pollutions of water sources must, therefore, be controlled by properly controlling their waste discharge. A lot of research is being undertaken for safety discharging these wastes, and thus, to prevent any major health hazard.

4.7. WATERBORNE DISEASES AND THEIR CONTROL

Waterborne disease are those diseases which spread primarily through contaminated waters. The important waterborne diseases are :

1. Diseases caused by **bacterial infections :**
   (*i*) Typhoid fever.
   (*ii*) Cholera.
   (*iii*) Dysentery.
2. Diseases caused by **viral infections :**
   (*i*) Infectious hepatitis or infectious jaundice (caused by hepatitis virus)
   (*ii*) Poliomyelitis (caused by polio virus).
3. Diseases caused by **protozoal infections :**
   (*i*) Amoebic dysentry (caused by *Entamoeba hystolytica* germ).

Organisms like bacteria, viruses, and protozoa are known as **pathogenic organisms.**

The germs or micro-organisms responsible for spreading these diseases, enter the human body with food or water which has been contaminated by the excreta of patients with the diseases, or of carriers of the organisms. Shallow well waters in villages are more likely to get such contaminations due to the leakage of sewage. Sometimes in cities too, the sewer pipe lines may leak into a water pipe through some faulty joint, thereby containing the same.

All such situations may contaminate the water and may spread any such water borne diseases among its consumers. The persons attending such patients should always be careful to thoroughly wash his/her hands with soap and water every time after attending the patient and before laking his food or drink. In order to prevent the spread of water borne diseases and especially in the form of epidemics, the various preventive or precautionary measures as recommended should be taken. Some of the measures are :

1. The water supplies of the town or the city must be checked thoroughly and disinfected.
2. The water pipe lines should be frequently tested and checked for detecting any leakage. The leaking joints must be properly sealed.
3. While laying or designed the water distribution system, keep the sewer lines and water lines as far away as possible.
4. General habit of cleanliness must be inculcated among the people.
5. As soon as some cases of such diseases are reported, the people should be educated to follow the following precautions. While attending their patients :
   (*a*) Least or no direct touch with the patient,
   (*b*) Cleaning and washing of hand as stated above.
   (*c*) Making available proper medical care and nutrients to the patient.
   (*d*) Maintaining proper sanitation around the patient.

4.8. WATER POLLUTION

Water is essential to all forms of life. 20% of the surface of the earth is land and the rest is all covered by water. Nothing could survive on earth without water. Conservation and keeping up of the good quality of water are of prime importance, since water as a natural resource is 'limited' in a sense, because 97% of the water is contained in oceans as salt water add additional about 0.3 % of salt water is buried under ground. Of the remaining 2.5 % about 2.2% is frozen in the polarice caps and in glaciers. The remaining small quantity of about 0.3% is available for our use.

Water pollution can be defined as the adverse change in composition or condition of water so that it becomes less suitable for the purpose for which it would be suitable in its natural state. These changes may be chemical, physical, biological or radioactive. Thus, water pollution means addition of any thing to water which changes the natural quality of water so that people residing in downstream does not receive the natural water of the steam.

Man has polluted much of this limited quality water with sewage, industrial wastes and synthetic chemicals. We are harming our planet to the point where organisms are dying at a very alarming rate, in addition, our drinking water has become greatly affected. Water pollution occurs when a body of water is adversely affected due to the addition of large amounts of undesired materials of the water. When it is unfit for its intended use, water is considered as polluted.

Types of Water Pollutants

1. Oxygen-demanding wastes.
2. Disease-causing agents.
3. Synthetic organic compounds.
4. Plant nutrients.
5. Inorganic chemicals and mineral substances.
6. Suspended solids.
7. Radioactive wastes.
8. Thermal discharges.

4.8.1. Sources of Water Pollution

1. *Point Source Pollution.* These occur when harmful substances are emitted directly into a body of water like industrial discharge and discharge of untreated or poorly-treated sewage/ waste water.
2. *Non-point or diffused pollution.* Non-point or diffused pollution an arise from many sources. These are mainly related to the way we use and manage land and soil. Ground waters are vulnerable from, and affect by, leaching of pollutants from the land surface and from areas of contaminated land, while surface waters are affected by rainfall that washes over and off the land (run-off). Run-off has increased as agricultural activities have intensified and as we have built more roads and houses, particularly where we have degraded the natural permeability of the landscape and reduced its capacity to retain water.

Water Contaminants (Impurities)

Contaminants may include organic and inorganic substances.

1. ***Organic water pollutants:***
   - Insecticides and herbicides, and other chemicals;
   - Bacteria from sewage or live-stocks operations;
   - Food processing waste;
   - Tree and brush debris from logging operations.
2. ***Inorganic water pollutants:***
   - Heavy metals including acid mine drainage;
   - Acidity caused by industrial discharges;
   - Chemical waste as industrial by-products;
   - Fertilizers in runoff from agriculture.

- Silt in surface runoff from construction sites, logging, burn practices or land clearing,

3. ***Additional Pollutants:***
   - Petroleum;
   - Radioactive pollutants;
   - Heat, It increases temperature resulting in the death of many organism.

4.8.2. Effects of Water Pollution

1. ***Ecological Impacts:***
   - Rise in Biochemical Oxygen Demand (BOD).
   - Lowering of dissolved (D.O.)
   - *Algal boom.* Chemical run-off from farmers' fields bring extra nutrients into a lake; allowing increasing growth of algae in the water.
   - *Bio-accumulation and Bio-magnification.* Inside a biological system, non-essential and harmful substances may either be stored or are reacted upon by the enzyme system and excreted.
2. ***Chemicals Affecting the Human Health***

   A large number of chemicals that either exists naturally in the land or are added due to human activity dissolve in the water, thereby contaminating it and leading to various diseases.
   - *Fluoride.* Though it protects the dental caries, but higher levels can have an adverse effect on health. Excess fluorides can cause yellowing of the teeth and damage to the spinal cord.
   - *Arsenic.* High concentration of arsenic in water can have an adverse effect on health
   - *Lead.* Higher levels of lead in the water may affect the central nervous system. Children and pregnant women are most at risk.
   - *Petrochemicals.* These may cause cancer.
   - *Other heavy metals.* These contaminants come from mining/quarying waste, landfills, or hazardous waste dumps.
   - *Chlorinated solvents.* Metals and plastic effluents, fabric cleaning, etc., are often discharged and contaminate ground water.
   - *Pesticides.*
   - *Nitrates.* It restricts the amount of oxygen that reaches the brain.
3. ***Waterborne Diseases***

   These are the infectious diseases which spread through primarily through contaminated water. These may be:

   (a) *Bacterial infections.* Typhoid, Cholera, Paratyphoid. Bacillary dysentery.

   (b) *Viral infections.* Infectious Hepatitis (Jaundice), Poliomyelitis.

   (c) *Protozoan infections.* Amoebic dysentery.

4.8.3. Control of Water Pollution

1. Water Pollution Treatment

(*a*) Treatment of municipal waste water.

(*b*) Treatment of industrial discharge.

These aspects have been discussed in detail in the chapter on "Purification of Water Supply".

2. Preventive Measure

In order to prevent the spread of water-borne infectious diseases, people should take adequate precautions. Water pipes of city water supply should be regularly checked for leaks and cracks. The water at home should be boiled or filtered to ensure that it is free from infection.

We should learn ways for disposing harmful house hold wastes so that they don't end up in sewage treatment plants that cannot handle them or landfills not designed to receive hazardous materials.

4.9. WATER QUALITY STANDARDS

Municipal water required for domestic uses, particularly the water required for drinking, must be colourless, odourless and tasteless. It should be free from turbidity, and excessive or toxic chemical compounds. Harmful micro-organisms and *radio activity*" must be absent. The quality of water for municipal supplies is, therefore, generally controlled throughout the world, and even World Health Organisation (W.H.O.) has laid down its standards for potable waters.

When these water standards are not fulfilled, the water may not be 100 per cent fit for drinking, and may be termed as "contaminated". It may not be healthy and tasty, or may cause the spread of some diseases.

Following tables show Water Quality Standards as prescribed by ISI

Water Quality Characteristic, Acceptable Limit, Permissible Limit, Risks or effects, Sources, Treatment and Health Effects of Chemical Parameters are listed below

S. No.	Characteristic	Acceptable Limit	Permissible Limit	Risks or Effects	Sources	Treatment
1.	Colour (Hz)	5	15	Visible tint, acceptance decreases	Tannins, Iron, Copper, Manganese, Natural deposits, organic waste	Filtration, Distillation, Reverse osmosis, Ozonisation
2.	Odour	Agreeable	Agreeable	Rotten egg, musty, chemical smell	Chlorine, Hydrogen sulfide, Organic matter, Septic contamination, Methane gas	Activated carbon, Air stripping, oxidation, Filtration

3.	pH	6.5-8.5	6.5-8.5	Low pH -corrosion, metallic taste High pH –bitter/ soda taste, deposits	Natural	Increase pH by soda ash Decrease pH with white vinegar / citric acid
4.	TDS mg/l, Max	500 mg/l	2000 mg/l	Hardness, scaly deposits, sediment cloudy coloured water, staining, salty or bitter taste, corrosion of pipes and fittings	Livestock waste, septic system Landfills, nature of soil Hazardous waste landfills Dissolved minerals, iron and manganese	Reverse Osmosis, Distillation, deionization by ion exchange
5.	Turbidity NTU Max	1 NTU	5 NTU	Cloudiness or haziness	High sediment deposition due to pollution and industrial activities like construction, mining, quarrying, agriculture, due to presence of organic matter such as phytoplankton	Settling or filtration processes using sand filtration, settling tanks, and clarifiers.
6.	Aluminium (as Al), mg/l, Max	0.03 mg/l	0.2 mg/l	High risk associated with dialysis patients	Rock and soil leaching	Portable Cation Exchange*** Distillation, Reverse Osmosis
7.	Ammonia (as total ammonia-N), mg/l, Max	0.5 mg/l	No relaxation	None proposed for humans but toxic for aquatic life	Disinfection with chloramines, wastes, fertilizers & natural processes	Ion Exchange with zeolite Sodium alumino silicate zeolites Distillation
8.	Barium (as Ba), mg/l, Max	0.7 mg/l	No relaxation	Difficulties in breathing, increased blood pressure, changes in heart	Mineral deposits Disposal of drilling wastes Smelting of	Cation Exchange Reverse Osmosis Distillation

				rhythm, stomach irritation, brain swelling, muscle weakness, and damage to the liver, kidney, heart, and spleen.	copper Motor vehicle parts manufacturing	
9.	Iron (as Fe), mg/l, Max	0.3mg/l	No relaxation	Brackish color, rusty sediment, bitter or metallic taste, brown-green stains, iron bacteria, discolored beverages	Leaching of cast iron pipes in water distribution systems Natural	Oxidizing Filter, Green-sand Mechanical Filter
10.	Manganese (as Mn), mg/l, Max	0.1mg/l	0.3mg/l	Brownish color, black stains on laundry and fixtures at .2 mg/l, bitter taste, altered taste of water-mixed beverages	Landfills Deposits in rock and soil	Ion Exchange, Chlorination, Oxidizing Filter, Green-sand Mechanical Filter
11.	Sulphate (as SO_4), mg/l, Max	200mg/l	400mg/l	Bitter, medicinal taste, scaly deposits, corrosion, laxative effects, "rotten-egg" odour from hydrogen sulfide gas formation	Animal sewage, septic system, sewage By-product of coal mining, industrial waste Natural deposits or salt	Ion Exchange, Distillation, Reverse Osmosis
12.	Nitrate (as NO_3), mg/l, Max	45 mg/l	No relaxation	Methemo-globinemia or blue baby disease in infants	Livestock facilities, septic systems, manure lagoons, fertilizers Household waste water, fertilizers Fertilizers Natural Deposits	Ion Exchange, Distillation, Reverse Osmosis
13.	Chloride (as Cl), mg/l, Max	250 mg/l	1000mg/l	High blood pressure, salty taste, corroded pipes, fixtures and appliances, blackening and pitting of stainless steel	Fertilizers, Industrial wastes Minerals, seawater	Reverse Osmosis, Distillation, Activated Carbon

14.	Fluoride (as F), mg/l, Max	**1.0mg/l**	1.5mg/l	Brownish discoloration of teeth, bone damage, skeletal damage	Industrial waste Geological	Activated Alumina, Distillation, Reverse Osmosis, Ion Exchange
15.	Total arsenic (as As), mg/l, Max	0.01mg/l	0.05 mg/l	Weight loss; Depression; Lack of energy; Skin and nervous system toxicity	Previously used in pesticides (orchards) Improper waste disposal or product storage of glass or electronics, Mining Rocks	Activated Alumina Filtration, Reverse Osmosis, Distillation, Chemical Precipitation, Ion exchange, lime softening
16.	Total chromium (as Cr), mg/l, Max	0.05	No relaxation	Skin irritation, skin and nasal ulcers, lung tumors, gastrointestinal effects, damage to the nervous system and circulatory system, accumulates in the spleen, bones, kidney and liver	Septic systems Industrial discharge, mining sites Geological	Ion Exchange, Reverse Osmosis, Distillation
17.	Copper (as Cu), mg/l, Max	0.05	1.5	Anaemia, digestive disturbances, liver and kidney damage, gastrointestinal irritations, bitter or metallic taste; Blue-green stains on plumbing fixtures	Leaching from copper water pipes and tubing, algae treatment Industrial and mining waste, wood preservatives Natural deposits	Ion Exchange, Reverse Osmosis, Distillation
18.	Cyanide (as CN), mg/l, Max	0.05mg/l	No relaxation	Thyroid, nervous system damage	Fertiliser, Electronics, steel, plastics mining	Ion Exchange, Reverse Osmosis, Chlorination

19.	Lead(as Pb), mg/l, Max	0.01 mg/l	No relaxation	Reduces mental capacity (mental retardation), interference with kidney and neurological functions, hearing loss, blood disorders, hypertension, death at high levels	Paint, diesel fuel combustion Pipes and solder, discarded batteries, paint, leaded gasoline Natural deposits	Ion Exchange, Activated Carbon, Reverse Osmosis, Distillation
20.	Mercury (as Hg), mg/l, Max	0.001 mg/l	No relaxation	Loss of vision and hearing, intellectual deterioration, kidney and nervous system disorders, death at high levels	Fungicides Batteries, fungicides Mining, electrical equipment, plant, paper and vinyl chloride Natural deposits	Reverse Osmosis, Distillation
21.	Zinc (as Zn), mg/l, Max	5 mg/l	15 mg/l	Metallic taste	Leaching of galvanized pipes and fittings, paints, dyes Natural deposits	Ion Exchange Water Softeners, Reverse Osmosis, Distillation
22.	Total coliform bacteria	Nil/100ml	100ml	Gastrointestinal illness	Livestock facilities, septic systems, manure lagoons Household waste water Naturally occurring	Chlorination, Ultraviolet, Distillation, Iodination
23.	E. coliform bacteria	Nil/100ml	100ml	Gastrointestinal illness	Livestock facilities, septic systems, manure lagoons Household waste water Naturally occurring	Chlorination, Ultraviolet, Distillation, Iodination

Sources : IS10500-2012

Health Effects of Chemical Parameters

S. N.	Parameter	BIS Guideline value (maximum allowable)	General and Health effect
1.	Total dissolved solids	2000 mg/l	Undesirable taste; gastro intestinal irritations; corrosion or incrustation
2.	pH	6.5-8.5	Affects mucous membrane; bitter taste; corrosion; affects aquatic life
3.	Alkalinity	600	Boiled rice turns yellowish
4.	Hardness	600	Poor lathering with soap; deterioration of the quality of clothes; scale forming; skin irritation; boiled meat and food become poor in quality
5.	Calcium	200 mg/l	Poor lathering and deterioration of the quality of clothes; incrustation in pipes; scale formation
6.	Magnesium	100 mg/l	Poor lathering and deterioration of clothes; with sulphate laxative
7.	Iron.	0.3mg/l	Poor or sometimes bitter taste, color and turbidity; staining of clothes materials; iron bacteria causing slime
8.	Manganese	0.3 mg/l	Poor lathering and deterioration of clothes; with sulphate laxative
9.	Aluminium	0.2 mg/l	Neurological disorders; Alzheimer's disease
10.	Copper	1.5 mg/l	Liver damage; mucosal irritation, renal damage and depression; restricts growth of aquatic plants
11.	Zinc	15 mg/l	Astringent taste; opalescence in water; gastro intestinal irritation; vomiting, dehydration, abdominal pain, nausea and dizziness
12.	Ammonia	0.5 mg/l	Indicates pollution; growth of algae
13.	Nitrite		Forms nitrosoamines which are carcinogenic
14.	Nitrate		Blue baby disease (methemoglobineamia); algae growth
15.	Sulphate		Taste affected; laxative effect; gastro intestinal irritation
16.	Chloride		Taste affected; corrosive

17.	Fluoride		Dental and skeletal fluorosis; non-skeletal
18.	Phosphate		Algae growth
19.	Arsenic	0.05 mg/l	Toxic; bio-accumulation; central nervous system affected; carcinogenic
20.	Mercury	0.001 mg/l	Highly toxic; causes 'minamata' disease- neurological impairment and renal disturbances; mutagenic
21.	Cadmium	0.003 mg/l	Highly toxic; causes 'itai-itai' disease- painful rheumatic condition; cardio vascular system affected; gastro intestinal upsets and hyper tension
22.	Lead	0.01 mg/l	Causes plumbism-tiredness, lassitudes, abdominal discomfort, irritability, anaemia; bio-accumulation; impaired neurological and motor development, and damage to kidneys
23.	Chromium	0.05 mg/l	Carcinogenic; ulcerations, respiratory problems and skin complaints
24.	Pesticide		Affects central nervous system
25.	Detergent		Undesirable foaming

4.9.1. Water Quality Index (WQI)

A Water Quality Index provides a single number (like a grade) that expresses overall water quality at a certain location and time based on certain quality parameters. The objective of an *index* is to turn complex water quality data into information that is easily understandable and useable by the public.

The use of such an index to *grade* water quality is a controversial issue among water quality scientists. This is because a single number cannot tell the whole story of water quality. However, a water index based on some very important parameters can provide a simple indicator of water quality. It gives the public a general idea of the possible problems with water in the region.

- The purpose of finding Water Quality Index (WQI) is to decide whether a water body is healthy or not. The parameters are selected on the basis of their impact on overall quality of water and health effects. Guidance is taken from IS 10500 : 2012.

Calculating Water Quality Index

For calculating WQI, Nine (9) important parameters are choosen and *weight* (importance) was attached to each, as mentioned below :

S.N.	Parameter	Unit	Weight
1.	Dissolved Oxygen (DO)	% Saturation	0.17
2.	Faecal coliform	Colonies per 100 ml	0.16
3.	pH	—	0.11
4.	Biological Oxygen Demand (BOD)	mg/L	0.11
5.	Temperature		0.10
6.	Total Phosphate	mg/L	0.10
7.	Nitrates	mg/L	0.10
8.	Turbidity	NTU	0.08
9.	Total Solids	mg/L	0.07
	Total		**1.00**

* NTU- Nephelometric Turbidity Unit.

After completion of water quality tests for all the nine parameters, the results are recorded and transferred to a weighting chart or curve where a numerical value called Q-value is obtained for each parameter.

Now the Q-value is multiplied by the weighting factor. The *nine* resulting values are then added to arrive at an overall WQI-value.

The Water Quality Index is evaluated on a scale of 0 to 100 (zero to hundred). The 100-point index is divided into several ranges corresponding to the general descriptive terms shown in the following table.

Water Quality Index Legend

Range	Quality
90 to 100	Excellent
70 to 90	Good
50 to 70	Medium
25 to 50	Bad
0 to 25	Very Bad

The Q-Values : A Q-value is a way of standardizing all the different water quality test results so that they can be combined and used in calculating WQI.

Q-value Analysis

A Q-value is a way of standardizing all the different water quality test results so that they can be combined and used to find an overall water quality value for the river. You can think of the Q-value like a score on a test. Less than 50 is like a failing grade, whereas 90 or more is like an "A."

Use the following charts to determine the Q-value for each of the water quality tests you perform. Find your water quality value on the x-axis. Determine where that value intersects the curved line on the graph, then follow it horizontally across to the y-axis and read the Q-value.

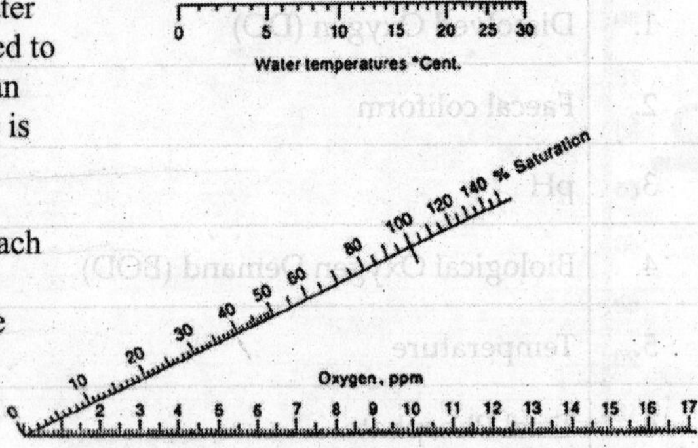

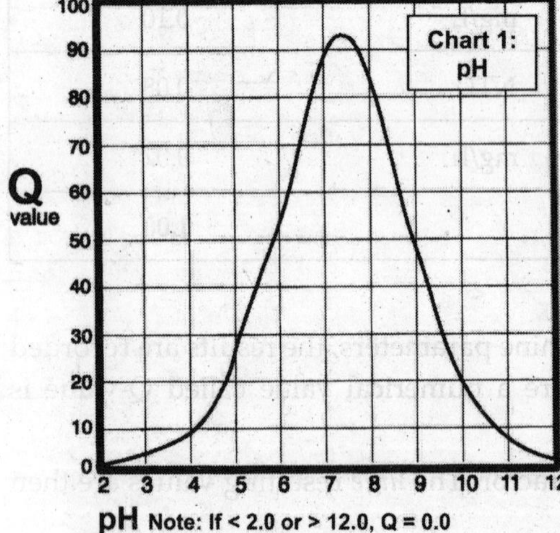

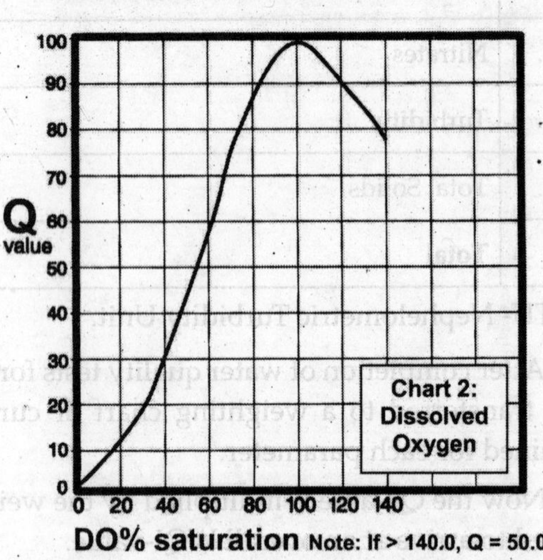

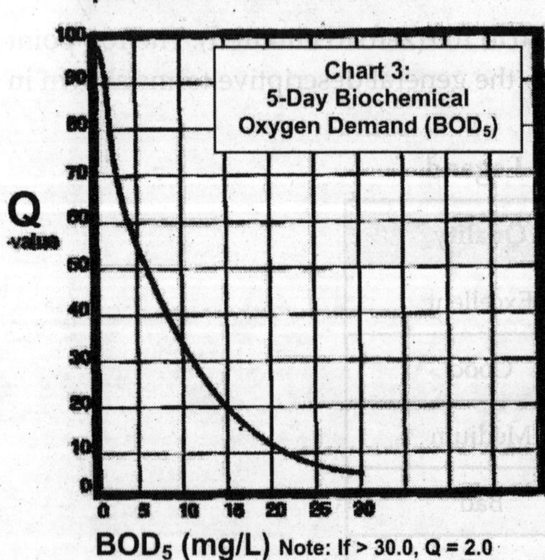

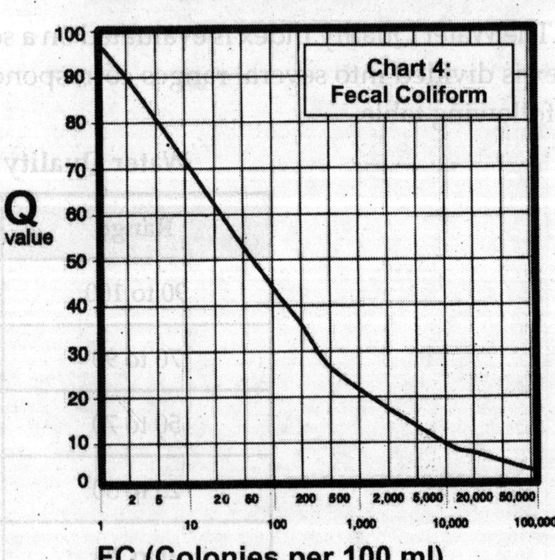

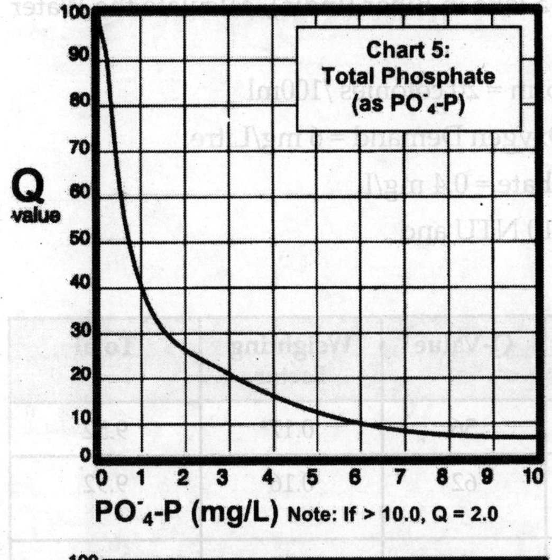

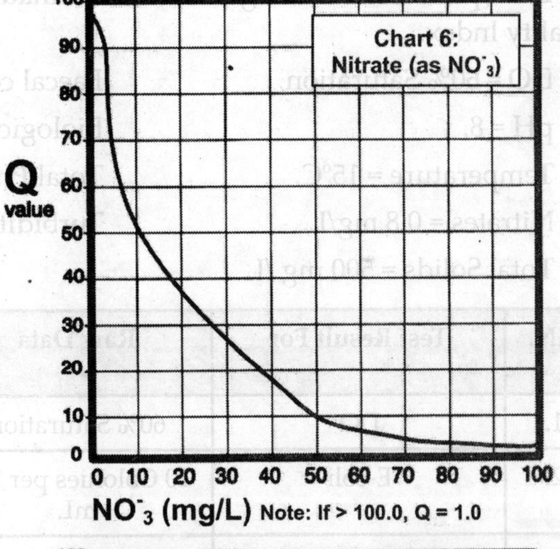

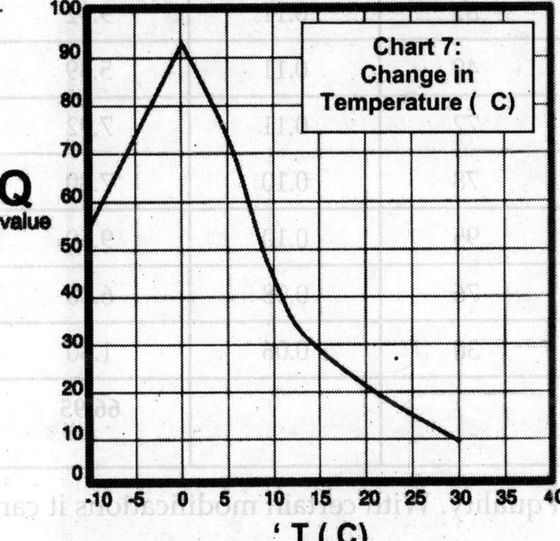

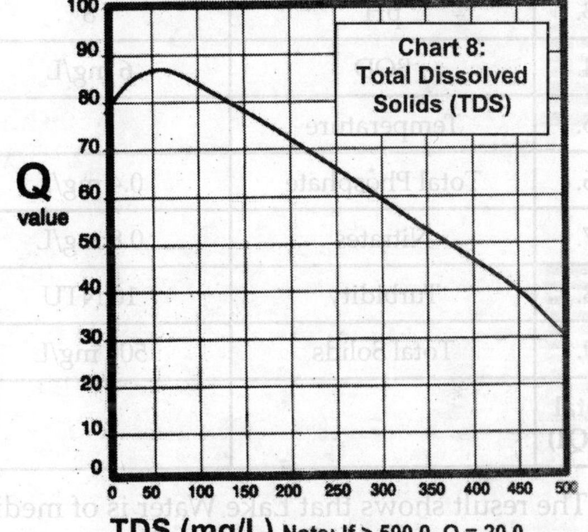

Turbidity Conversion Chart

Cm	NTU	Cm	NTU	Cm	NTU
< 7	> 240	21 to 24	35	44 to 46	13
7 to 8	185	24 to 26	30	46 to 49	12
8 to 9	150	26 to 29	27	49 to 51	11
9 to 10	120	29 to 31	24	51 to 54	10
10 to 12	100	31 to 34	21	54 to 57	9
12 to 14	84	34 to 36	19	57 to 60	8
14 to 16	60	36 to 39	17	60 to 70	7
16 to 19	48	39 to 41	15	70 to 85	6
19 to 21	40	41 to 44	14	> 85	< 5

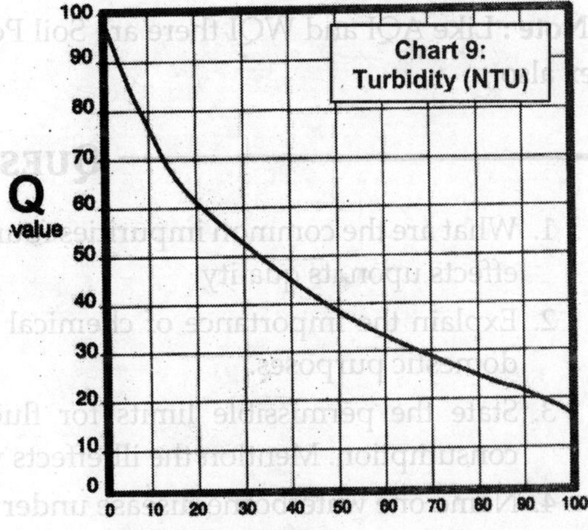

Friends of the Chicago River

CHICAGO RIVER SCHOOLS NETWORK

Example : For following observations made at a lake in Bihar (India), calculate the Water Quality Index.

DO = 60% Saturation,
Faecal coliform = 20 colonies /100ml
pH = 8,
Biological Oxygen Demand = 6 mg/Litre
Temperature = 15°C,
Total Phosphate = 0.4 mg/L
Nitrates = 0.8 mg/L,
Turbidity = 10 NTU and
Total Solids = 500 mg /L

S.N.	Test Result For	Raw Data	Q-Value	Weighting Factor	Total
1.	DO	60% Saturation	56	0.17	9.52
2.	E-coli	20 Colonies per 100 mL	62	0.16	9.92
3.	pH	8	82	0.11	9.02
4.	BOD	6 mg/L	49	0.11	5.39
5.	Temperature		72	0.11	7.92
6.	Total Phosphate	0.4 mg/L	78	0.10	7.80
7.	Nitrates	0.8 mg/L	95	0.10	9.50
8.	Turbidity	10 NTU	76	0.08	6.08
9.	Total Solids	500 mg/L	30	0.06	1.80
Total (WQI)					**66.95**

The result shows that Lake Water is of medium quality. With certain modifications it can be raised to good category.

Note : Like AQI and WQI there are Soil Pollution Index (SPI), Nose Pollution Index, U-V Index also.

QUESTIONS

1. What are the common impurities found in natural sources of water, and explain their effects upon its quality.
2. Explain the importance of chemical and bacteriological analysis of water used for domestic purposes.
3. State the permissible limits for fluorides in water to be supplied for domestic consumption. Mention the ill-effects when they are not in the permissible limits.
4. Name one waterborne disease under each of bacterial, viral and protozal origin that can be controlled by proper treatment of water.

5. Enumerate the important waterborne diseases in india, and explain how you can prevent the incidence of these diseases.
6. What are the causes of taste and odour in water supplies?
7. Enumerate and discuss in brief the various physical, chemical, and bacterial characteristics of testing of raw-water supplies. What steps would you take in order to make them fit for drinking.
8. Enumerate the various impurities of physical chemical, micro-organic, and radiological origin in raw-water supplies, and state their maximum permissible limiting values for obtaining municipal supplies.
9. How does water quality criteria differs for industrial supplies from those for domestic municipal supplies?
10. Write short notes on :
    (*i*) pH value of water
    (*ii*) Hardness
    (*iii*) Algae
    (*iv*) Waterborne diseases
    (*v*) BOD
11. Name organic and inorganic impurities in water.
12. State the standard of good quality drinking water.
13. What are the objectives of treatment of drinking water?
14. Compare the surface water and ground water for their quality, quantity and collection aspects.
15. What do you understand by water quality index?
16. What is meant by 'Water Stress Condition'? Explain.

Transmission of Water

5.1. INTRODUCTION

The water supply scheme consists of collecting water from the source(s) carrying to the treatment plant, and finally distributing the treated water among the consumers. The water is collected by means of an intake structure, and then transported to the city either raw in case the treatment plant is located within the city; or transported after treatment in case the treatment plant is located away from the city near the source. Generally, the treatment plant is located within the city limits.

When the source of supply is near to the city limits, the problem of transmission is less, as the length of the conduits required for carrying water from the intake to the treatment plant or city is less. However, when water is brought from far off sources, the length of conduits required and the resultant expenditure on transportation and maintenance may be very high.

In case of underground sources, such as, wells and tube wells (where no separate intake is required but lift is reqd.), the problem of conveying water from the source to the city does not normally arise as these sources are generally located within the cities.

If the city is situated at an elevation higher than the elevation of the source, the water will have to be pumped and carried through pressure pipes. A low lift is generally required when the water is withdrawn from the rivers. On the other hand, if the source is available at higher elevation than the city, the water can be conveyed without lifting, through open or closed conduits either by simple gravity or under pressure. In such a case the water is withdrawn from storage reservoirs or lakes.

5.2. CONDUITS (WATERWAYS)

A conduit is used to carry water from the reservoir to the consumers. Conduits include canal, pipe or tunnel. The tunnel is made by cutting the mountains where topography prevents the use of a canal or pipeline. The tunnel is made to save the distance. Various appliances used to control the flow of water are called 'head works' or 'control works'. Head works include gates, valves and trash rack etc.

'Head works' are the **structures** at the intake of conduits. They have booms, trash racks, sluices and gates or valves for controlling water flow. Booms divert ice and floating logs from the intake. The gates are used to control discharge through sluices. Trash racks are provided for preventing the debris from getting entry into the intakes from reservoirs. Trash rack is made up of steel bars and is placed across the intake to prevent the debris from going into the intake.

As discussed earlier, **intake structure** is provided at the entrance of the canals or tube-wells or pipes through which water is conveyed to the water supply scheme. Following are the functions of intakes :

(*i*) Control of the flow of water into the conveyance system by providing a gate or valve.

(*ii*) Smooth and turbulence free entry of water into the conveyance system.

(*iii*) Stopping entry of boulders, logs of wood etc., into the conveyance passage by providing trash racks or steel grids.

(*iv*) Not allowing heavy sediment load into the conveyance passage.

5.2.1. Types of Conduits

Depending upon the conditions and characteristics of flow, the conduits may be classified as under :

(1) *Gravity conduits*; and

(2) *Pressure conduits*.

1. Gravity Conduits

In these conduits the water flows under the action of gravity. In such flow the water all along is at atmospheric pressure. Gravity conduits can be of the form of canals, flumes, or aqueducts as given below.

Assuming Uniform flow *i.e.* constant depth throughout the length of the channel and referring to the following Fig. 5.1.

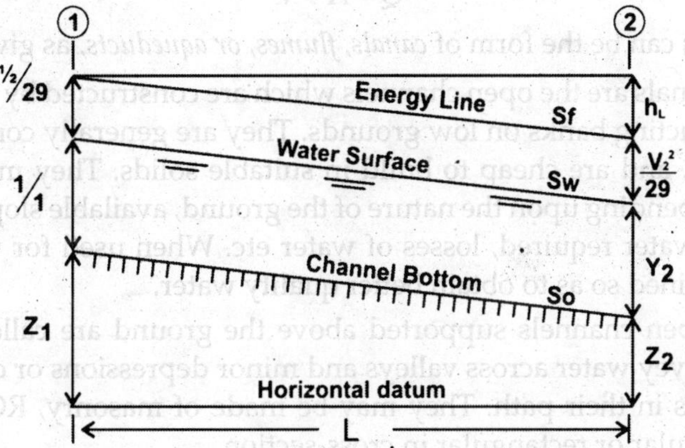

Fig. 5.1. Flow illustration in a gravity conduct

The Energy Equation between section (1) and section (2) can be written as,

$$z_1 + y_1 + \frac{v_1^2}{2g} = z_2 + y_2 + \frac{v_2^2}{2g} + h_L$$

For uniform flow $y_1 = y_2$

and $v_1 = v_2$.

$\therefore \quad s_o = s_w = s_f$

where, s_o is bottom slope

s_w is water surface slope

and s_f is energy line slope.

Now, considering all the forces at section (1) and (2)

Following equation known as Chezy's Equation can be derived,

$$V = C\sqrt{RS}$$

where, V is average velocity through the conduit

R is Hydraulic Radius = A/P

A is Area of cross section

P is Wetted Perimeter and

C is Chezy's Coefficient (having units $m^{½}/s$)

Manning defined C as equal to $\frac{1}{n}R^{1/6}$. Hence getting Manning's Equation as,

$$V = \frac{1}{n}R^{2/3}S_0^{1/2}$$

where, *n* is called roughness or rugosity coefficient. *n* has been found to vary from 0.01 to 0.035 for smooth to rough surfaces.

The discharge through the channel is calculated by the equation.

$$Q = A \times V$$

Gravity conduits can be the form of *canals, flumes, or aqueducts,* as given below :

1. **Canals:** Canals are the open channels which are constructed by cutting high grounds and constructing banks on low grounds. They are generally constructed in balanced cut and fill, and are cheap to build in suitable solids. They may be either lined or unlined depending upon the nature of the ground, available slopes, design velocities, quality of water required, losses of water etc. When used for water supplies, these have to be lined so as to obtain better quality water.
2. **Flumes:** Open channels supported above the ground are called flumes. These are used to convey water across valleys and minor depressions or over drains and other obstructions in their path. They may be made of masonry, RCC or metal, and are usually circular or rectangular in cross-section.

3. **Aqueducts:** These are closed, rectangular, circular or horse-shoe sections built of masonry or RCC. Although they are covered or closed, yet water does not flow under pressure as they are not allowed to run full.

2. Pressure Conduits

Pressure conduits are closed conduits and as such no air can enter into them. In these conduits, the water flows under pressure above the atmospheric pressure. The pressure pipes can follow the natural available ground surface and can freely go up and down hills or can dip beneath valleys or mountains and thus requiring lesser length of conduit.

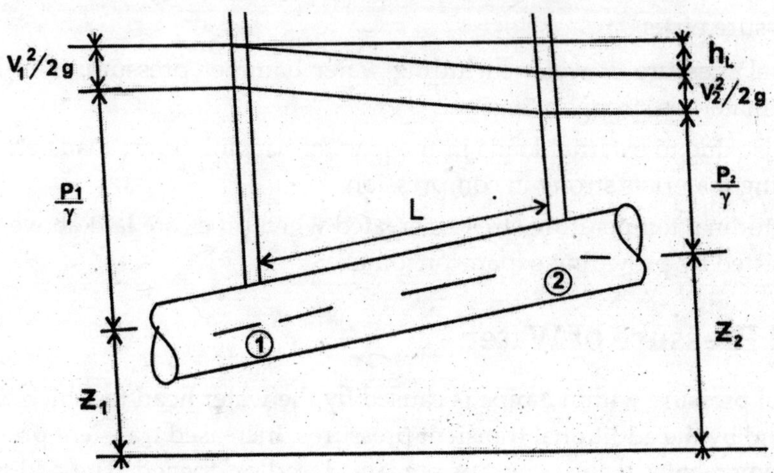

The total energy equation is the Bernoulli's Equation which when referred to points (1) and (2) will be as follows,

$$\frac{p_1}{\gamma}+\frac{v_1^2}{2g}+z_1=\frac{p_2}{\gamma}+\frac{v_2^2}{2g}+z_2+h_L$$

where, p_1 and p_2 are pressures at points 1 and 2
v_1 and v_2 are average velocities at points 1 and 2
h_L is the head loss between points 1 and 2
γ is unit weight of water $1000 \times g = 9810$ N/m^3

Equating the Forces at points 1 and 2, following equation is derived,

$$h_L = \frac{fLv^2}{2gD} \text{ (Daray-Weisbach Equation).}$$

where, L = Length of conduits between points 1 and 2
D = Diameter of the pipe
h_L = head loss = $\frac{p_1 - p_2}{\gamma}$
f = Friction factor (Varies between 0.01 to 0.1)

The pressure pipes prove economical than canals or flumes because they can generally follow shorter routes. The main advantage in these conduits is that, the water moving through such a conduit is not exposed anywhere, and hence, there are little chance of its getting polluted. The pressure pipes are, therefore, invariably used for drinking water supply. In the pressure conduits, because of no evaporation losses, are also used when the water is scarce.

5.3. FORCES ACTING ON PRESSURE CONDUITS

The structural design of the pressure pipes should be carried out, so as to enable them to withstand the various forces likely to come on them. The following forces generally come into play in the pressure pipes :

1. Internal pressure of water including water hammer pressure, is to be resisted by using materials strong in tension.
2. Pressure due to external loads in the form of backfill, traffic loads etc, to be resisted by using materials strong in compression.
3. Longitudinal temperature stresses created when pipes are laid above the ground, to be resisted by providing expansion joints.

1. Internal Pressure of Water

The internal pressure within a pipe is caused by the water head to which the conduit can be subjected, and by the additional transient pressures. Increased transient pressure is caused momentarily, when gates, valves, pumps etc. are closed or opened. The additional pressure created by such a transient phenomenon to known as **water hammer pressure.**

When a liquid flowing in a pipe is abruptly stopped by closing of a valve etc., the velocity of the water column behind is retarded, and its momentum gets dissipated due to the conversion of kinetic energy into elastic energy. A series of positive and negative pressure waves are produced, which travel back and forth in the pipe, till they are dumped out by friction. This phenomenon is known as **water hammer.**

2. Pressure due to External Loads

Pipes are often buried under the ground and placed in trenches which are backfilled. The weight of the pipe, the weight of the bakfill and the superimposed traffic loads, if any, will then be transferred to the pipe. This will produce compressive stresses in the pipe material (when the pipe is empty and there is no internal pressure to counteract these loads) and the material may fail in compression, if it exceeds the allowable compressive stress of the pipe material.

However, in water supply pipelines it is not a big problem, because such pipes are generally not laid very deep. Moreover, the steel pipes, and the cast iron pipes, which are commonly used for transporting water, are quite good in compression.

3. Temperature Stresses

When pipes are laid above the ground, they are exposed to atmosphere and are, therefore, subjected to temperature changes. They expand during day time and contract at night. If this expansion or contraction is prevented due to fixation or friction over the supports, longitudinal stresses are produced in the pipe materials.

Special types of expansion joints at suitable intervals (say 20 m to 30 m or so) must be provided to counteract these stresses.

5.4. TYPES OF PRESSURE PIPES

Depending upon the construction material, the pressure pipes are of the following types. The selection of a particular types of material for a pipe depends mainly upon the relative economy , the pressures likely to come and the working pressures, maximum permissible sizes and capacities etc.

1. Cast Iron Pipes

These are widely used for city water supplies, because they are comparatively cheaper, sufficiently resistant to corrosion and may last as long as 100 years or so. They are generally manufactured in lengths of about 3.5 metres, but on special orders may be manufactured in lengths upto 6 m or so.

The **advantages** of cast iron pipes are :

(*i*) Moderate in cost

(*ii*) Corrosive resistant

(*iii*) Easy to join

(*iv*) Strong and durable

(*v*) Long life upto 100 years or so.

(*vi*) Service connections can be easily made.

The **disadvantages** of cast iron pipes are :

(*i*) Cannot be used for high pressures (above 700 kN/m^2)

(*ii*) Water carrying capacity decreases with time, as the value of friction factor increases.

(*iii*) When large, they are very heavy, and uneconomical.

(*iv*) These are likely to break during transportation or while making connections.

2. Steel Pipes

Steel plates of varying thickness for withstanding different pressures are generally bent and welded or riveted so as to manufacture steel pipes. Welded pipes are smoother and stronger than riveted pipes, and are, therefore commonly used. Since steel is strong in tension, even large size diameter pipes (say upto 6 m or so) can be made of thin shells, and these pipes can withstand high internal pressures. These are much lighter than cast iron (C.I.) pipes and can be more easily cut and threaded and look neat but are less corrosion resistant.

3. Cement Concrete Pipes

These may be plain, reinforced, and prestressed category and may be pre-cast or cast-insitu type. Plain cement concrete pipes are manufactured in small sizes (upto 0.6 m diameter) and pressure upto 0.7 kg/q\sq cm. Reinforced cement concrete (RCC) pipes are used for sizes upto 4.5 metre and pressure upto 6 kg/sq cm. For pressure higher than this prestressed cement concrete pipes are used. These pipes may either be prepared at site by transporting various ingredients or can be manufactured in factories and then transported to the site (known as cast-in-situ pipes).

4. Plastic Pipes

Plastic pipes are rigid PVC pipes and are light in weight tough, exceptionally resistant to the chemical attack and available in large lengths, which reduce the cost of handling, transportation and installation. These are immune to corrosion and has lesser chances of leakage. These also have good elastic properties. Smooth internal surface of the pipes offers less friction which results in saving of energy in the conveyance of water.

Since at higher temperatures, the strength of pipe reduces and therefore working pressure is suitably reduced. These pipes, are therefore recommended for water temperature ranging from 1 to 45°C. The deterioration of these pipes starts with discolouration, surface cracking and ultimately ends with brittleness.

Selection of Pipe Material

The factors which should be considered in selection of pipes are :

(*i*) Strength of pipe.

(*ii*) Water carrying capacity

(*iii*) Life and durability of pipe.

(*iv*) Ease of transportation and handling, and

(*v*) Ease of joining, maintenance and repairs.

(*vi*) Hydraulic properties: The material should be smooth so as to produce lesser friction losses and have higher hydraulic efficiency.

(*vii*) Resistant to corrosion.

5.5. LAYING OF WATER SUPPLY PIPES

Pipes are used for conveying water from the source to the city, and also for distributing the same within the city among the consumers (Fig. 5.2). When pipes are used for bringing water from the source to the city, only one or two lines of the large sizes are generally laid. When the pipes are used for distributing water, they are generally of varying smaller sizes, having many connections and branches. In the former case, pipes used are generally made of steel or R.C.C., whereas in the latter case, cast iron pipes (with bituminous coatings) are generally used. Galvanised iron pipes are used for service connections to the various houses.

Pipes are laid on one side of road either above the ground or below the ground. Generally, the pipes bringing water from the source to the city, are laid on the ground, whereas the distributing mains taking the water within the localities, are laid below the roads and streets. The pipelines, should follow the profile of the ground, and that location is choosen which is most economical.

When pipes are laid on the ground (or above the ground), they must be laid on a well compacted formation of suitable width so as to avoid future settlements. They may be laid directly over the compacted soil formation or may be laid over small masonry or cement concrete supports at 6 to 12 metres apart. This arrangement, through costly, facilities inspection, maintenance, repairs, etc.

When pipes are buried under the ground, they are laid in trenches excavated upto the required depths. The top of the pipe is generally kept about 1 metre below the surface, so as to minimise the impact and traffic loads transmitted to the pipes.

Utmost precaution is to be taken to prevent cross connections between water mains and sewers. Leakage in water mains causes not only wastage of water but also gives opportunity for sewage to enter water mains. After laying pipes, they are jointed and tested for withstanding pressures and leakage.

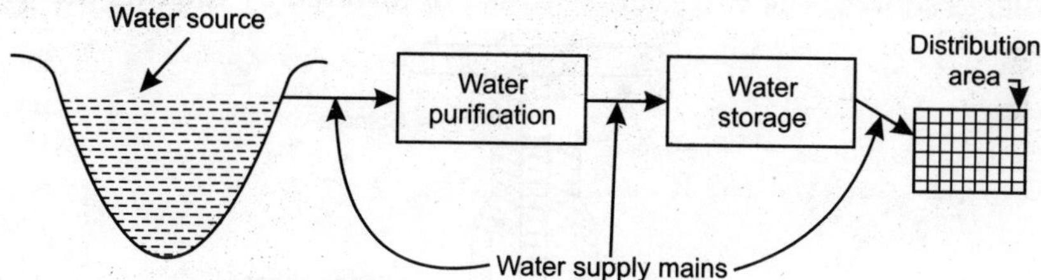

Fig. 5.2. Layout of a water supply scheme.

5.6. PIPE APPURTENANCES

In order to isolate and drain the pipe line sections for test, inspections, cleaning and repairs; a number of appurtenances such as gates, valves, manholes, insulation joints, expansion joints, anchorages, etc., are provided at various suitable places along the pipe lines, as described below.

5.6.1. Gates and Valves in Pipelines

A large number of different types of valves are required for the proper functioning of the pipe line, as described below and shown in Fig. 5.3.

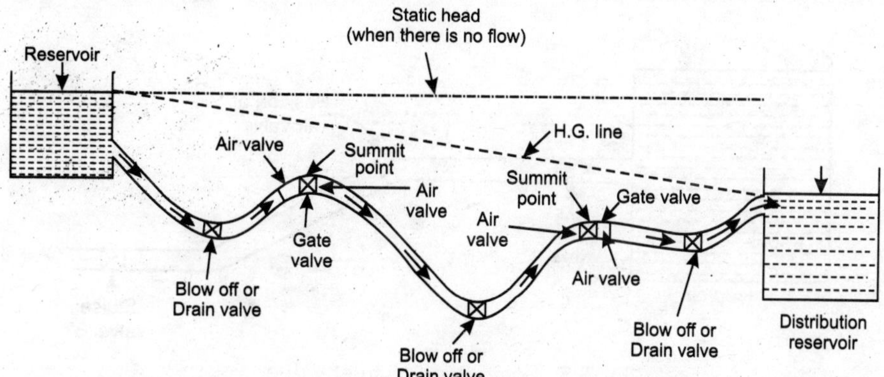

Fig. 5.3. Layout of the pipe showing the locations of gates and valves.

1. Gate Valves (or Sluice Valves)

Gate valves, also known as sluice valves, are used to regulate the flow of water through the pipes. They are similar to gate valve. In large pipe lines, bringing water from the source to the city, they are generally located along the pipe line at intervals of about 3 to 5 kilometres, so as to divide the pipe line into different sections. Thus, during repairs, only one section can be cut off at a time, by closing the gate at both ends of the section.

The most commonly used type of a gate valve or a sluice valve is shown in Fig. 5.4. The valve is made of cast iron with brass, bronze or stainless steel mountings. The ends of the valve are properly jointed on both sides of the pipe lengths. In its simplest form, the valve consists of a wedge shaped circular disc fitted closely in a recess against the opening in the valve. This is connected to a wheel by means of a threaded spindle, as shown. The valve can be raised or lowered by rotating the wheel.

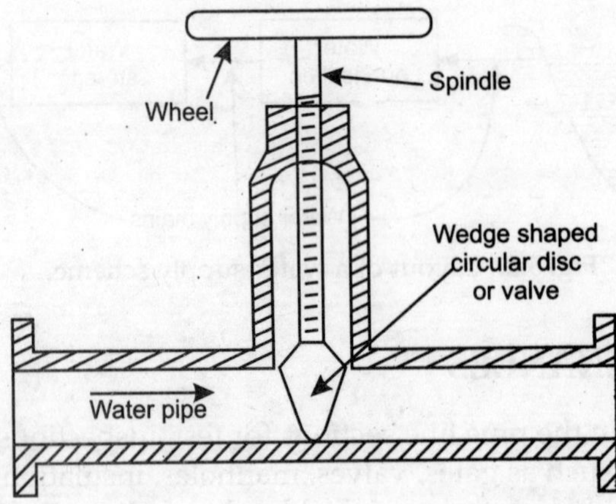

Fig. 5.4. Vertical section of a typical sluice valve.

2. Air Valves

Air relief valves are provided to remove accumulated air at elevated or higher (summits) points in flow through pipes (Fig. 5.5.). This valve takes care of problem arising due to sudden change in pressure in this water pipe.

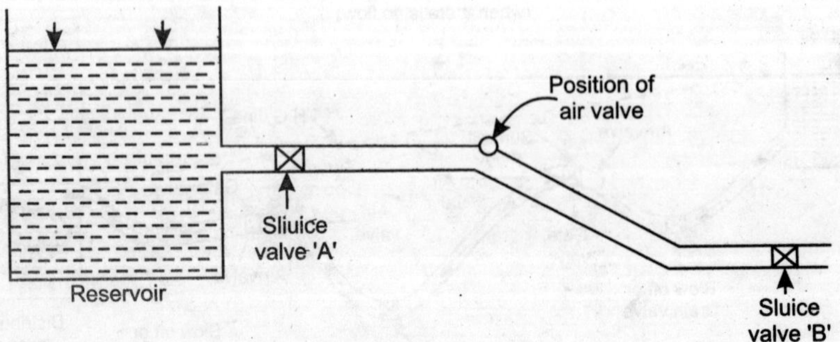

Fig. 5.5. Position of air inlet valve.

These valve open out automatically as soon as the pressure in the pipe falls below a certain fixed predetermined value, and thus allowing air to enter the pipe. Such a valve is, therefore, known as an ***air inlet valve.*** Similarly sometimes air gets accumulated at high points when the supply is restored and the pipe is refilled after repairs. The accumulated air obstruct the free flow of water and the pipe may get air locked. ***Air relief valves*** are, therefore, required to be provided at all summits to remove the accumulated air (Fig. 5.6).

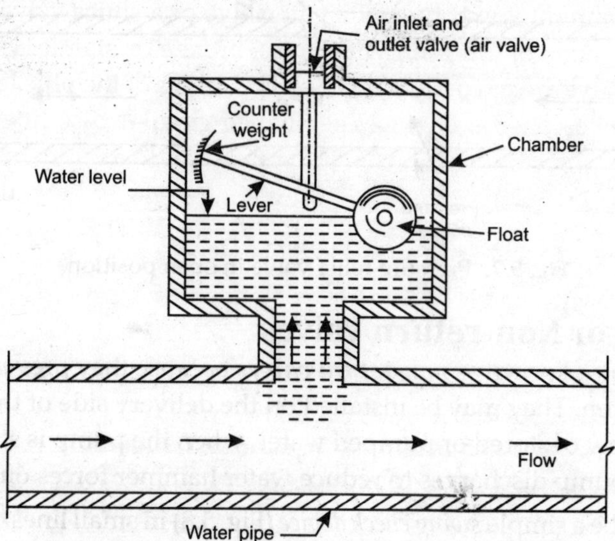

Fig. 5.6. Vertical section of a poppet type air-valve.

3. Blow off Valves or Drain Valves or Scour Valves

In order to empty the entire pipe after closing the supply, small gated off-takes are provided at low points. These valves are known as *blow off valves* or drain valves or *scour valves*. These valves, are necessary at low level points for completely emptying the pipe for inspection, repairs, etc. When opened, water comes out of these valves quickly under gravity and they are made to discharge water into some natural drainage channel or into a sump from which the water can be pumped out.

4. Pressure Relief Valves

Water hammer pressure can be reduced in pressure pipes by using *pressure relief valves*. Such a valve is adjusted to open out automatically as soon as the pressure in the pipe exceeds a certain fixed predetermined value. Due to the opening of this valve, certain water will get out of the pipe, and thus, reducing the pressure in the pipe. As soon as the water hammer pressure reduces, and the pressure in the pipe falls upto the fixed value, the valve will close automatically. A pressure relief valve is shown in Fig. 5.7.

Such valves are often placed especially at low points where the pressure are high, so as to function during emergencies, when pressure rises in the pipe above the design valve, to help in protecting the pipe from getting burst.

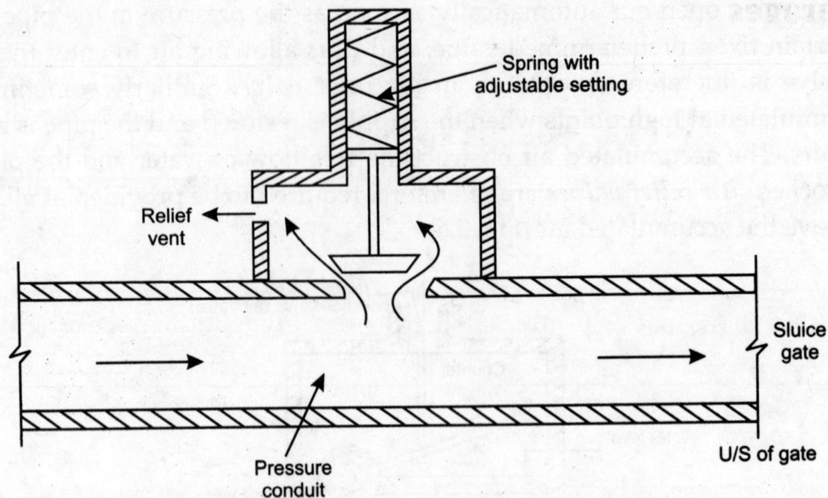

Fig. 5.7. Pressure relief valve in open position.

5. Check Valves or Non-return Valves

Check valves are also known as ***non-return valves*** because they prevent water to flow back in the opposite direction. They may be installed on the delivery side of the pumping set, so as to prevent the back flow of stored or pumped water, when the pump is stopped. Check valves are also installed on pump discharges to reduce water hammer forces on the pump.

Such a valve may be a simple *swing check device* (Fig. 5.8) in small lines. In large installations, they should be designed to close slowly, usually with discharge of some water through a bypass.

A *swing type* of check valve consists of a disc covering the inside area of the pipe, which opens only in one direction, as shown in Fig. 5.8. So, when water tends to flow in the opposite direction, the disc closes the opening, and the flow is prevented.

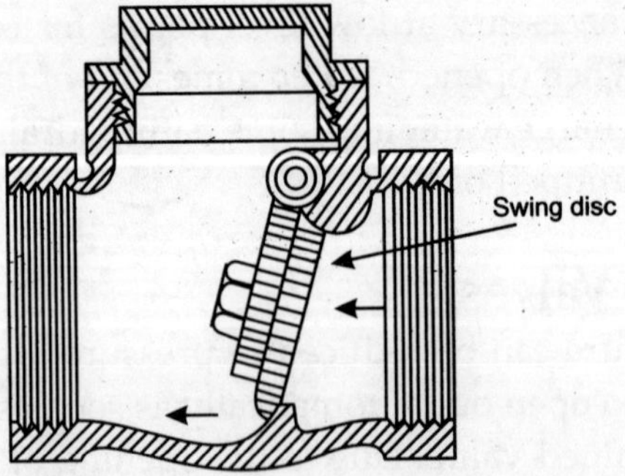

Fig. 5.8. Swing type of check.

6. Insulation Joints

Insulation joints are provided along the pipe lines at suitable intervals, so as to insulate the pipe against the flow of stray electric currents. Rubber gaskets or rings can be provided as insulators, in between the pipe length, so as to prevent the flow of electric current between them.

7. Anchorages

The pipes try to pull apart and get out of the alignment at bends and other points of unbalanced pressures. At such places, the forces exerted on the joints due to longitudinal shearing stresses caused by these unbalanced pressures are enormous, and the joints may get loosened, ultimately leading to excessive leakage or failure of the pipe. In all such circumstances, in order to prevent the pipes from pulling apart, pipes are *anchored* by firmly embedding these portions in massive blocks of concrete or masonry, which absorbs the side thrusts.

Similarly, when pipes are laid on steep, they try to slip and thereby pull apart, and the resistance of their joints may be insufficient to balance the longitudinal shearing stresses. Pipes are, therefore, anchored under such circumstances also.

5.7. PIPE JOINTS

Pipes are manufactured in small lengths for convenience in handling and transporting operations. These small pieces of 2 to 6 m lengths are joined together after placing them in proper position. Following are the common types of joints in use depending upon material of the pipe and condition of support :

1. Socket and spigot joint.
2. Flanged joint.
3. Mechanical joints (or known as dresser coupling).
4. Flexible joint.
5. Expansion joint.

1. **Socket and Spigot Joint:** These are mostly used for cast iron pipes. These pipes are made in such a way that one of their end is enlarged, whereas the other end is normal. The enlarged end is called "socket" or "ball", while the normal end is called "spigot". The spigot is filled into the socket. A few strands of jutes are wrapped around the spigot before inserting it into the socket and then, more jute is packed into the joint. The remaining space between the socket and the spigot is finally filled with molten lead, which gets solidified and tightly caulked into the joint after cooling, and thus making a water tight joint.
2. **Flanged Joint:** Flanged joints are used where it may be necessary to occasionally disjoint the pipes, such as pumping stations, filter plants etc. In these joints, two flanges are brought together, keeping a rubber washer (called gasket) in between them, so as to make them water tight. They are them fixed by means of nuts and bolts.

   These joints are strong but rigid and hence cannot be used where deflections or vibrations are excepted. They are mostly used for indoor works.
3. **Mechanical Joints:** These joints are used to join the plain ends of cast iron pipes. A spherical type of metallic collar is then fitted and tightened over the abutting ends, Thus forming a mechanical joint.

   These joints are strong and rigid. These can withstand vibrations and are, therefore, useful for pipes to be carried over bridge or below bridges in hangers.
4. **Flexible Joint:** These are used where large scale flexibilities are required, for examples, when pipes are laid in rivets with uneven beds, large scale settlements.

In such circumstances ordinary joints may break. Similarly, while laying pipes on curves, the provided joints must be flexible, otherwise they may break.

The pipes to be provided with such a joint are cast with special types of ends. The socket is spherical, and the spigot, though plain, is having a bead at the end. A retainer ring is placed over the bead which keeps the special rubber gasket in position. A spilt cast iron gland ring is then placed over it. They are then tightened by means of nuts and bolts.

5. **Expansion Joint:** These are provided at suitable intervals in the pipelines, so as to counteract the thermal stresses produced due to temperature variations. For providing expansion joints in CI pipes, the socket end is cast flanged and the spigot end is plain. The socket end is connected rigidly to an annular ring which can slide freely over the spigot end. While making this joint, a small space is kept between the face of the spigot and the inner face of the socket, and the spigot is filled up by means of a rubber gasket. The flanges are then tightened by means of nuts and bolts.

   When the pipes expand, the socket-end moves forward and the gap left just gets closed. Similarly, when the pipes contract the socket moves backward creating the gap. All the time, the annular ring follows the movements of the socket and maintains the gasket in position, thus keeping the joint water tight.

5.8. TESTING AND CLEANING OF PIPELINES

After a pipeline has been laid, fitted with all appurtenances and accessories, painted both from inside as well as outside by means of protective paints, etc., the pipe line will be tested for the soundness in its construction. The soundness of the construction is examined by performing the **pressure test** on the pipeline. The step by step procedure adopted for performing this test is described below :

1. The pipeline is tested from section to section. Thus at a time, only one particular section is taken up for testing.
2. The downstream sluice valve is closed, and water is admitted into the pipe through upstream sluice valve. The air valves well be properly operated during filling up of the pipe.
3. The upstream valve, through which water was admitted, is closed so as to completely isolate the pipe section from the rest of the pipe.
4. Pressure gauges are then fitted along the length of pipe section at suitable intervals on the crown, through holes left for this purpose.
5. The pressure in the pipeline is raised by means of a hydraulic pressure pump, till the test pressure (to be measured on the pressure gauge fixed on the pipe) is namely 25 to 50% above the highest working pressure.
6. The pipe and the joints are then visualised for water tightness for a period of atleast 4 hours.
7. The pipe is finally emptied through drain valves, and the observed defects are rectified, so as to make the line fit for use.

Leakage Test: After the satisfactory completion of pressure test, a leakage test for a duration of about 2 hours is required to be performed by maintaining the specified leakage test pressure after the pipe has been filled with water and the air is expelled.

Chlorination (Disinflection) of Pipelines before Use

After the pipeline has been tested and corrected for defects, it is ready for transporting untreated water to the city from the sources. However, when the pipelines are carrying treated water, they must be disinfected before use. The pipes are disinfected by keeping them full with water and adding chlorine in amounts, so as to maintain a residue of 50 mg/L (*i.e.* 50 ppm). This residue is maintained for 12 hours and the pipe is emptied and flushed with fresh treated water, thus making the pipe ready for carrying potable water to the consumers or to the storage tanks.

QUESTIONS

1. Why are the pressure pipes most commonly used for conveying water from distant sources to the towns of supply? What are the drawbacks of open channels and masonry aqueducts in this respect?
2. Discuss as to why the raw-water is carried from its source to the city treatment plant, through pressure conduits as against the gravity canals.
3. What is meant by "water hammer" and how is it produced in pipes conveying water under pressure? What precautions should be taken and arrangements made to reduce its effects? What provisions are normally made in the design of pipe lines on this account?
4. Enumerate the various forces which may act on pressure conduits carrying water supplies. Also discuss briefly, any two of these forces.
5. What are the different materials, which are commonly used for water supply pipes? Discuss their comparative merits and demerits.
6. Discuss the use of cast iron, steel and RCC as materials for water supply pipes. How are the first two types of pipes get corroded?
7. Discuss the procedure followed in laying and testing the water supply mains.
8. Enumerate the various appurtenances which are generally provided while laying water supply mains bringing water from a source to the city.
9. Sketch and explain importance and functioning of gate valves, air valves, pressure relief valves, and manholes
10. Write short note on :
    (*i*) Testing and disinfection of water mains.
    (*ii*) Pipe appurtenances in laying water supply mains.
    (*iii*) Jointing of cast iron water supply mains.
11. State Chezy's equation and discuss Chezy's coefficient.
12. Discuss friction factor and its role.

Pumps and Pumping Stations

6.1. INTRODUCTION

Pumps are required, in a water supply scheme, at one or more of the following stages :

1. To lift the water at the source, when the water cannot flow by gravity into the mains.
2. To lift the water at the treatment plant, if sufficient natural slope is not available.
3. To lift the water after the treatment, so as to force the water into the distributing mains.
4. When the pressure in the distributing mains has to be increased or boosted at some intermediate points within the distribution system.

Pump is a mechanical device to increase the pressure energy of a fluid. Generally the pump is used for raising the fluids from a lower level to a higher level.

Pump can also be defined as the hydraulic machines which converts the mechanical energy into hydraulic energy which is mainly in the form of pressure energy.

Almost all the pumps increase the pressure energy of the liquid which is subsequently converted into potential energy as the liquid is lifted from a lower level to a higher level.

This is achieved by creating a vacuum pressure at the inlet or suction end and high pressure at the outlet or delivery end of the pump. Vacuum pressure raises the fluid from a depth where it is available and the high pressure at outlet forces it upto the height where it is required.

Pumps are used for large variety of applications, like supply of drinking water, irrigation purposes, mine water drainage, handling mixtures of water and solids like pebbles, course gravel, clinker, boulder, ash, rock chips and silt; and for chemical liquids etc., Reciprocating pumps are used for developing very high pressures or heads, while centrifugal pumps are used for supplying large quantities at low and medium pressures with high discharges.

6.2. TYPES OF PUMPS

Following two types of hydraulic pumps are common :

1. **Roto-dynamic pumps, and**

2. Displacement pumps

1. **Roto-dynamic Pumps:** These pumps have a wheel or a rotating element (known as impeller) which rotates the water in a casing, and thus imparting energy to the water. The shape of the impeller may be such, as to force the water outward in a direction at right angles to its axis **(radial flow)**, or to give water, an axial as well as a radial flow **(mixed flow)**, or to force the water in the axial direction alone **(axial flow)**. Radial flow and mixed flow pumps are known as **centrifugal pumps**, whereas the axial flow pumps are called **axial flow pumps**. Centrifugal pumps are the most common and widely used.
2. **Displacement Pumps:** The displacement pumps, also known as positive displacement pumps, work on the principle of mechanically inducing vacuum in a chamber, and thus, sucking in a certain volume of water, which is then mechanically displaced and forced out of the chamber. The two common types of displacement pumps are :
   1. Reciprocating pumps, and
   2. Rotary pumps.

In addition to these two major types of pumps, other common types are: air lift pumps, jet pumps, hydraulic rams etc.,, and are used under special conditions.

6.3. CENTRIFUGAL PUMPS

As discussed above, hydraulic machines which convert the mechanical energy into hydraulic energy are called pumps. This hydraulic energy is in the form of pressure energy.

A centrifugal pump is a hydraulic machine used to raise the liquids form a lower to a higher level by creating a required pressure by means of centrifugal action. These pumps increase the pressure energy of fluids. These may either lift the fluid or just boost the pressure in a pipe line.

These pumps have impellers in the form of a circular disc with radial vanes. The impeller rotates fast drawing water into the centre and discharging out through a chamber at the periphery.

Principle: The basic principle on which a centrifugal pump works is that, when a certain mass of liquid is made to rotate by an external force, it is thrown away from the central axis of rotation and a centrifugal head is impressed which enables it to rise to a higher level. Now if more liquid is constantly made available at the centre of rotation, a continuous supply of liquid at a higher level may be ensured.

6.3.1. Component Parts

The most important component parts of a centrifugal pump are discussed hereunder :

1. Impeller

A series of curved vanes are fitted on a impeller. Impeller is mounted on a shaft, which is driven by an electric motor. The function of impeller is to force the water into rotating machine and the shroud around the impeller is to direct the water through the impeller. The water coming out of impeller is the load through pump causing under high pressure.

The impellers may be classified as under:

(*a*) Closed or shrouded impeller

(*b*) Semi-open impeller

(*c*) Open impeller

(*a*) ***Closed or Shrouded Impeller:*** Fig. 6.1 shows this type of Impeller. It contains two shrouds in which curved inserted. This type is suitable for pumping liquids which are pure and comparatively free from foreign matter or debris, for example, ordinary water, oils, chemicals like acids etc., The material of the impeller should be such that it can resist the reaction of chemical to be handled.

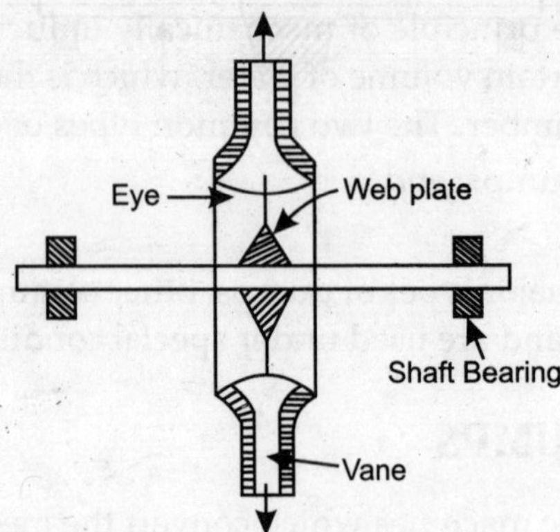

Fig. 6.1. Closed impeller.

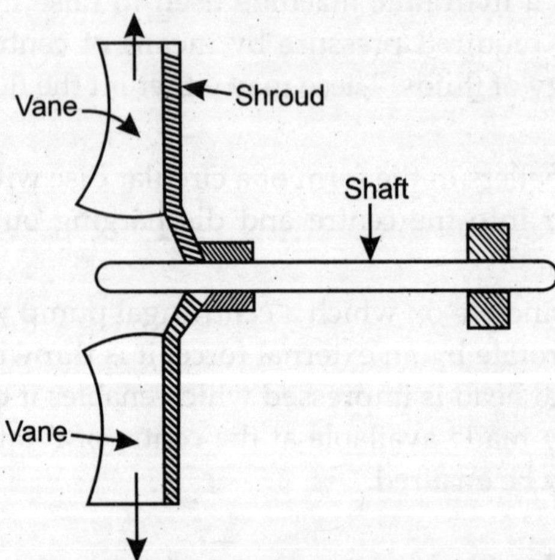

Fig. 6.2. Semi-open impeller.

(*b*) ***Semi-open Impeller:*** Fig. 6.2 shows this type. In this, vanes are fixed on one shroud plate only. This type of Impeller is used for pumping liquids containing debris upto some extent only, such as sewage water, paper pulp etc., These Impellers have less number of vanes but have more length to avoid the clogging of the Impeller.

(*c*) ***Open Impeller:*** In this, vanes are fixed from the centre with web plate as shown in Fig. 6.3. No crown and shroud plates are used. This is used for pumping liquids having coarse debris of large size such as mixture of water, sand, clay and pebbles in which the solid contents are in very large amount. As these Impellers are required to perform very rough duty, these are generally made of forged steel. These Impellers have less life, as they perform very rough duty.

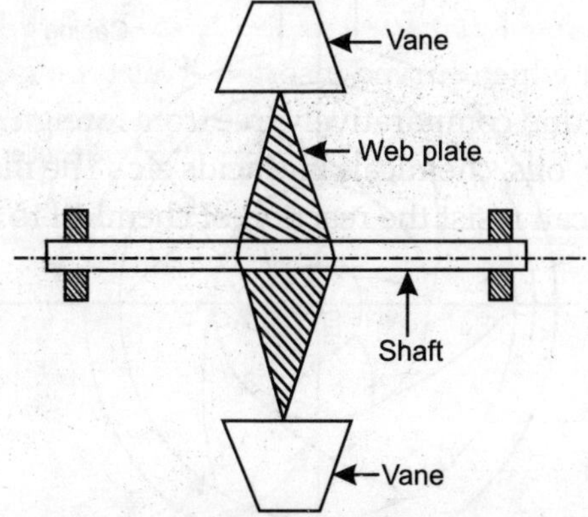

Fig. 6.3. Open impeller.

2. Casing

A casing is provided for housing the impeller and supporting the bearing provided for the shaft carrying the impeller. Two openings are provided in the casing for suction and delivery of water. As it handles the water at a higher pressure, therefore, a stuffing box is provided to prevent the leakage from the gap between the pump casing and shaft. Casing is made of cast iron or cast steel.

As the impeller increases the kinetic energy of water, hence this casing should be such thatleast amount of energy should be lost due to eddy formation. Therefore, the efficiency of a pump also depends on the type of casing used. Following are the there common types of casing, which are generally used :

(*a*) Diffuser Ring
(*b*) Volute Casing
(*c*) Vortex or Whirlpool Casing

(*a*) ***Diffuser Ring:*** Fig. 6.4 (*a*) shows diffuser ring type casing. In this, a diffuser ring having guide vanes surrounds the impeller. These guide vanes are such that they provide gradually enlarged passage for the flow of water. These guide vanes are placed at such an angle that water enters without shock. It has been found that by using this type of casing, large amount of kinetic energy (upto 75%) is covered into pressure energy.

The pumps having diffuser casing are known as diffuser pumps or guide blade pumps. These are commonly known as "Turbine Pumps" because in principle, they resemble to that of a reversed radial flow reaction turbine.

These pumps are more costly than that of the volute pumps, therefore, diffuser pumps are mostly used for developing high heads (60 m or more) by using number of impellers in series (*i.e.* multistage pump). For single stage, volute or whirlpool casing pumps are preferred. Therefore, diffuser pumps are mostly used as vertical pumps and are very suitable for installations in deep wells and mines etc.,

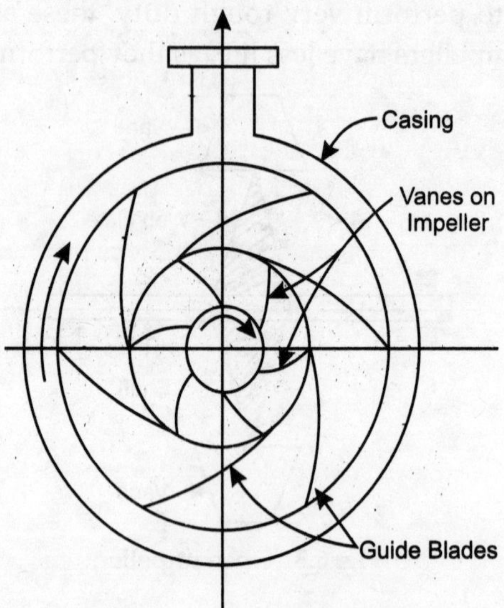

Fig. 6.4. (*a*) Diffuser casing.

(*b*) ***Volute Casing:*** This is a spiral shaped casing which surroundings the impeller as shown in Fig. 6.4 (*b*). The shape of the casing is such that its area of cross-sectional gradually increases towards outlet of the pump. As the flowing water progresses towards the delivery pipe, more and more water is being added from the outlet periphery of the impeller. Increasing cross-sectional area helps firstly to provide the space for the water coming out of the impeller and secondly to reduce the velocity, thus converting kinetic energy into pressure energy

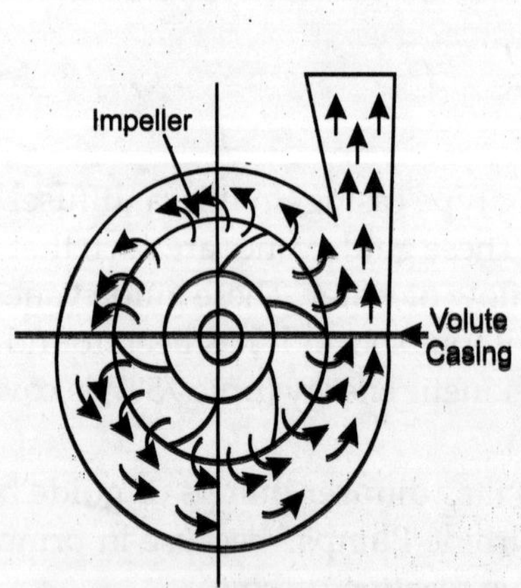

Fig. 6.4. (*b*) Volute casing.

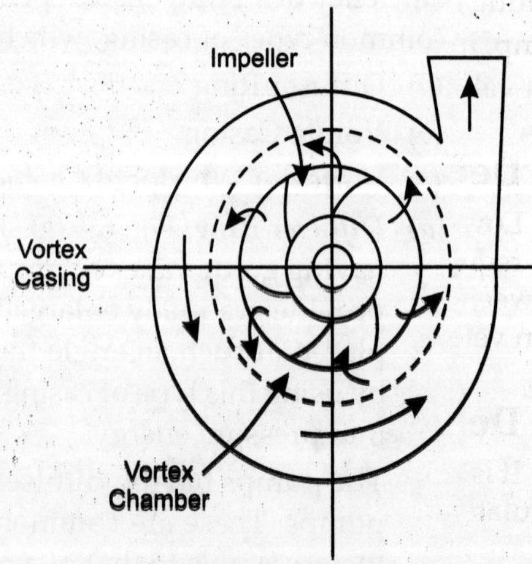

Fig. 6.4. (*c*) Vortex casing.

These casings can convert only a small amount of kinetic energy into pressure energy and a large amount of kinetic energy is lost in eddies. Therefore, these pumps can produce comparatively low heads. These pumps are known as "Volute Pumps".

(c) ***Vortex or Whirlpool Casing:*** This is the improvement on volute pump (Fig. 6.4c). In such pumps an annular space known as vortex or whirlpool chamber is provided between the impeller and volute casing. Thus the water enters the vortex chamber from the impeller with a whirling motion and, therefore, water particles are having spiralling path. In the vortex chamber these water particles have constant energy because no work is done on the water in this chamber. Thus, when the water flows in this vortex chamber, the torque of the whirling motion also remains the same *i.e.* free vortex is formed.

Since for a free vortex velocity of whirl varies inversely as its radial distance from the centre; velocity of flow of liquid reduces as it passes through the vortex chamber. Thus, in this way reduction of flow of liquid reduction in velocity increases the presence of water leaving the vortex. The water after leaving the vortex chamber passes through the volute chamber surrounding it. Thus it increases the efficiency of the volute chamber by converting the kinetic energy into pressure energy in the vortex chamber.

3. Suction Pipe

It is a pipe, whose upper end is connected with the pump on suction side and lower end is submerged in the source of water, *i.e.* sump. The lower end is fitted with a foot valve (one way valve) and a strainer. When water enters in the suction pipe, strainer keeps the debris (*e.g.*, leaves, wooden pieces and other rubbish materials) away from entrance to the pipe. Thus only clear water can enter the suction pipe through foot valve. The velocity of water in the suction pipe should be between 1.5 to 3.0 m/sec.

Strainer and Foot Valve. These are fitted at the end of the suction pipe and submerged in the water (or liquid to be pumped) in such a way that it is only a few cm above the bottom of the water. The water from the sump or well first enters the strainer which is meant to keep the floating bodies (such as leaves, wooden pieces and other kinds of rubbish etc.,) away from the pump. In the absence of strainer, the foreign material will chock the pump. With the strainer a foot value, of non-return or one way type, opening upwards is provided.

4. Delivery Pipe

Lower end of this pipe is connected to the outlet of the pump and takes the water upto the required head and then delivers it. The velocity of water in the delivery pipe should be between 1.5 to 3.5 m/sec The velocity of water in the delivery is kept equal or slightly higher than velocity in suction pipe.

5. Delivery Valve

It is connected between the outlet of the pump and lower end of delivery pipe. It is used to regulate the quantity of water to be lifted by controlling the opening of the valve.

6. Prime Mover

It is used to rotate the impeller of pump. Usually an electric motor is provided for this purpose. This motor is coupled with impeller shaft.

6.4. WORKING OF CENTRIFUGAL PUMPS

Fig. 6.5 explains in detail the working of a centrifugal pump. Before starting a pump, priming is done. In priming, water is filled in the suction pipe, casing and into a portion of delivery pipe upto delivery valve. This is done for the removal of air. With the presence of air, it is not possible to create pressure because the pressure generated by the impeller is directly proportional to the density of the liquid to be handled. Thus, if impeller is run in air, negligible pressure will be produced because of very less density of air and hence no liquid will be lifted by the pump.

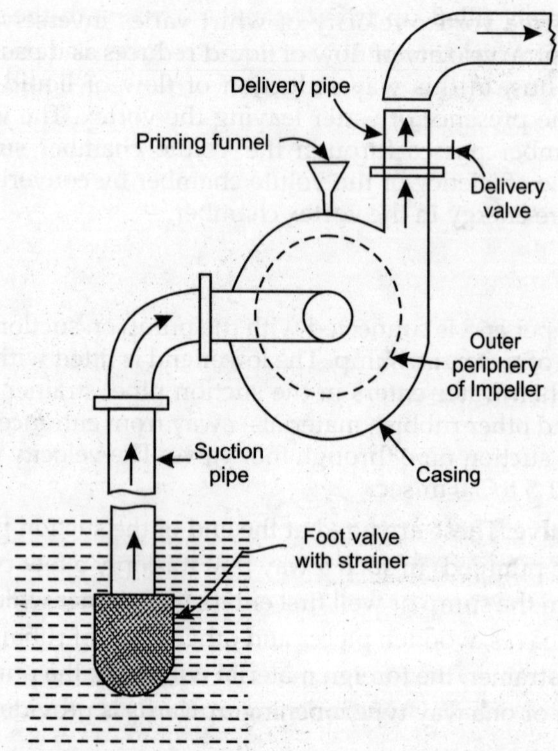

Fig. 6.5. Working of centrifugal pump.

Thus, it is very essential to prime the pump properly and it should be noted that during priming delivery valve is kept closed.

After priming, the delivery valve is kept still closed and prime mover (generally electric motor) is started to rotate the impeller. The rotation of impeller in the casing full of liquid produces centrifugal head in the liquid and thus presence in whole of the liquid is increased.

After a short period, when pump attains a constant speed, delivery valve is gradually opened and thus liquid is allowed to flow in a radially outward direction through the impeller vanes and attains a higher velocity and pressure at the outer periphery. With the help of special shape of casing (Volute, Vortex or Diffuser). This high velocity energy is converted into pressure energy and reduces velocity. Now, this water at higher pressure rises through the delivery pipe upto a desired height.

At the sometime centrifugal action creates partial vacuum at the centre of impeller known as "EYE". The partial vacuum causes the water to rush through the suction pipe towards the impeller eye, to take the place of water which has left the impeller vanes from the sump.

6.5. PRIMING

As stated earlier, a reaction machine must run fill, as the pressure of liquid in the machine is different at all points from atmospheric pressure. In pumps, the pressure has to be developed by the machine and this can be done unless the machine (whole suction side and machine passages) is first completely filled up from outside source with the liquid, to be pumped. This is known as priming. Arrangement is also required to be made such that the pump once started, should remain full, the passage of water from sump to machine is, therefore, provided with a not-return value to prevent the water from going back to the sump, when the pump is stopped.

When priming is being done before starting a pump, air must be allowed to escape by providing an air-vent value on pump casing. When all the air is exhausted, priming liquid itself will flow out of the pump casing, indicating thereby that priming is complete.

As explained earlier priming is the filling of water in the casing and suction pipe of a pump for the removal of air before starting it. If any air remains inside the casing, discontinuity of flow may be caused. Therefore, priming is very essential before starting a pump.

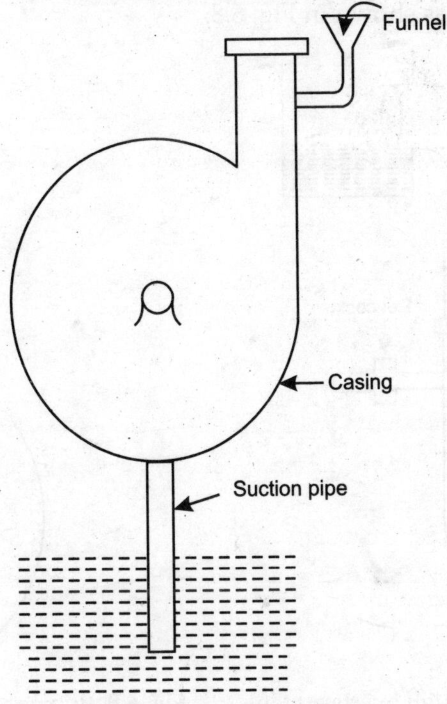

Fig. 6.6. Priming (manual).

Reciprocating pumps do not require priming, if they are in good condition. In those, priming is done only when the suction pipe is long. When the water sump is available the pump level then priming is automatically done by simply opening the fitted value in between the sump and pump suction pipe. If the pump is installed above the source of water then a foot value is fitted at the bottom of the suction pipe submerged in water and priming is done by the following methods:

Pumps can be primed in several ways. These methods can be divide into the following three general categories :

1. Manual Priming

Water is admitted to replace the air from the suction pipe and the pump casing. It may be done by hand or automatic means.

Some of the most common methods in the category are :

(*i*) *Pouring water with funnel* : Fig. 6.6 shows this method. This is an ordinary method, in which water is poured through a funnel into the casing and thus replaces the air.

(*ii*) When the pump is discharging into an overhead tank, then for priming purposes delivery pipe is connected with the casing so that when priming is to be done liquid from overhead tank can be used as shown in Fig. 6.7. If city water at some pressure is available then priming can be done by city water by connecting it with the pump as shown in Fig. 6.8.

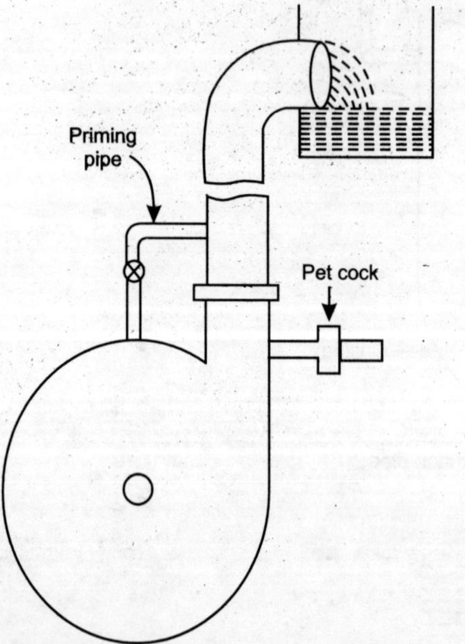

Fig. 6.7. Priming through overhead tank.

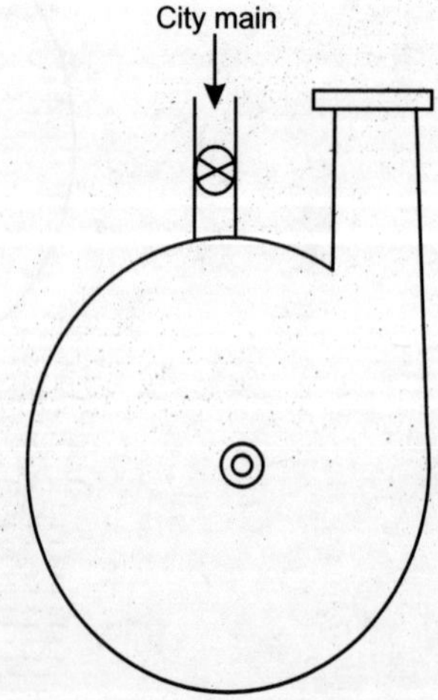

Fig. 6.8. Primary through city main.

Pet cock is provided at the top of the casing to allow air escape from the casting while priming. After priming pet cock is closed.

2. Priming by Vacuum

In these methods, the air from the suction pipe and pump is sucked by any of the devices, namely :

(*i*) Domestic water piston pump connected with the discharge line.

(*ii*) A diaphragm air pump operated by a handle connected with the top of centrifugal pump through a valve.

(*iii*) By an ejector using compressed air or exhaust steam of an engine to create vacuum at the top of casing.

When air is sucked from the pump, and suction pipe, atmospheric pressure in the sump forces the water to rush into the suction pipe and casing. Modern pumps are mostly equipped with vacuum pumps (electric driven) based on the principle of rotary compressors.

3. Self-priming

In these methods pumps are fitted with self-priming devices to recirculate the water in the pump for the removal of air. It is the method in which priming is automatically done. Therefore, self priming centrifugal pumps are very useful. As self-priming devices require large expenses, hence these are used with big size pumps only.

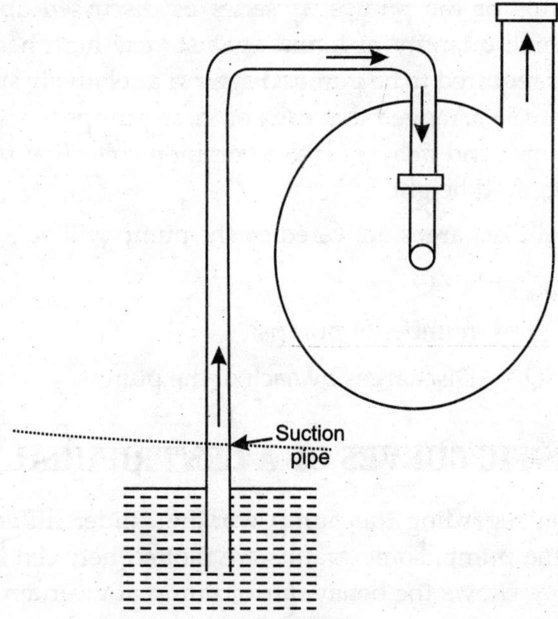

Fig. 6.9.

Each manufacturer use his own design and patent for self-priming pumps. The device used in "Sini Pump" is very simple, in which the suction pipe and delivery pipe are connected at the top of the pump as shown in Fig. 6.9. When the pump comes to rest, it will always have some water for priming purposes.

6.6. MULTI-STAGE PUMPS

For delivering the liquid at high heads, two or more pumps were earlier used is series. But this method is not common, and multi-stage pumps are popular these days. A multi-stage pump consists of two or more identical impellers mounted on the same shaft, and are enclosed in the same casing. All the impellers are connected in series, and the liquid is discharged from one impeller to another with increased pressure. A pump is designated by the number of impellers fitted in the casing, as two stage, three-stage pump,

Total head developed by a multistage pump,

$$H_m = n \cdot H$$

where n = Number of stages

H = Head developed in each stage.

Since capacity of a single impeller is limited (against a maximum head of about 100 m), therefore the lifting still larger head, two or more identical implellers are used in series. The number of stages to be adopted depends on the limitation of speed of the driving motor and also upon the discharge and total head.

Pumps in Parallel

The multi-stage pumps or the pumps as series as discussed above are employed for delivering a relatively small quantity of liquid against very high heads. However, when a large quantity of liquid is required to be pumped against a relatively small head, two or more pumps are used, which are so arranged that each of these pumps working separately lifts the liquid from a common sump and delivers it to a common collecting pipe through which the liquid is carried to the required height.

In such cases, the total discharges delivered by the pump will be

$$Q_{total} = n \times Q$$

where, n = number of pumps;

Q = Discharges by each of the pumps.

6.7. CHARACTERISTIC CURVES OF A CENTRIFUGAL PUMP

To know the information regarding the pump working under different conditions and to study the behaviour of the pump, some graphs are plotted between head, Shaft power and η_{man} against Q. Each curve shows the behaviour of pump at a certain constant speed. These curves are known as "main and operating" characteristic curves.

In order to draw these curves, the tests are conducted, when the rate of flow Q is varied by means of delivery value and at several different values of Q the quantities of H, HP and η are measured or calculated. By plotting these sets of values a set of three curves (as shown in Figs. 6.10, 6.11 and 6.12) is obtained. The pump is normally designed to run at the same speed as the driving unit which is generally an electric motor. The speed of such motors is either 1450 or 2900 r.p.m. for a frequency of 50 cycles per second. When electric power is not available, the pump may be driven by diesel engine.

In actual practice the operation may be at some other condition of head or flow rate, and for changed condition the behaviour of the pump may be quite different than the designed one. Therefore, in order to predict the behaviour and performance of a pump under varying conditions, tests are performed and the results of the test are plotted. The curves thus obtained are known as the characteristics curves of the pump.

The pump is designed for maximum efficiency conditions at a particular speed. From the characteristics curves corresponding value of Q, H_m and P are obtained.

Fig. 6.10 shows the characteristics curves between η_{man} and Q.

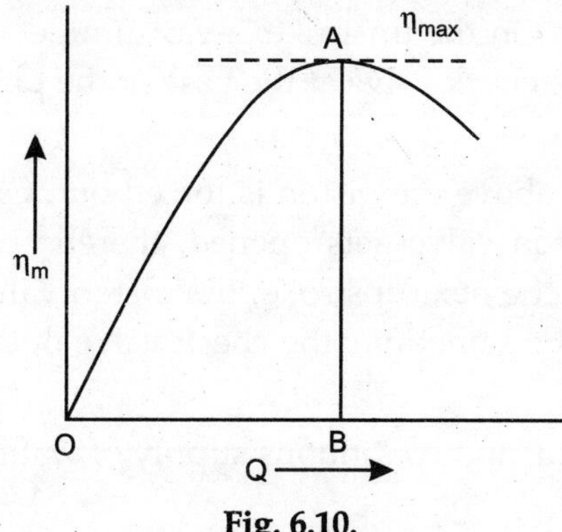

Fig. 6.10.

In the Fig. 6.10, point A shows the maximum efficiency. Then the value of OB gives the discharge for maximum efficiency at which pump should be designed.

Fig. 6.11 shows the characteristics curve between H_m and Q. From this curve the value of $H_m(BC)$ gives the designed H_m for the designed discharges (OB) to give maximum efficiency. The value of H_m on the curve, when $Q = 0$ is known as shut off head.

Fig. 6.12 shows the characteristics curve between P and Q. From this curve the power required for any discharge can be known. BD gives power required for designed discharge (OB) at maximum efficiency.

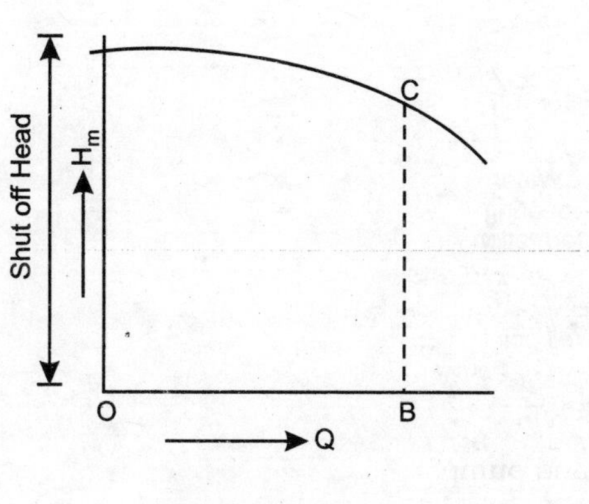

Fig. 6.11.

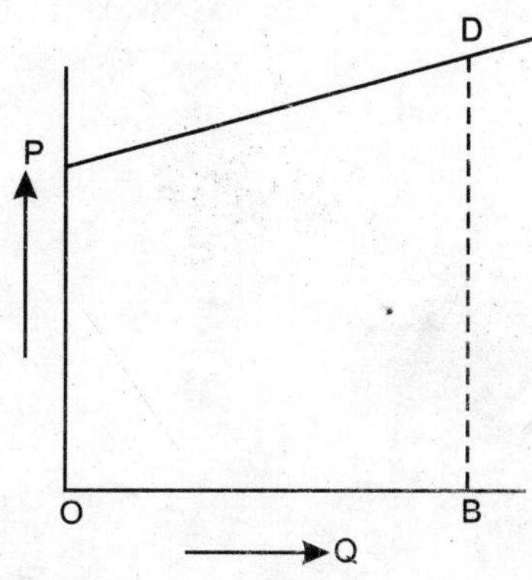

Fig. 6.12.

6.8. POSITIVE DISPLACEMENT PUMPS

As discussed earlier, positive displacement pumps are of two types; namely, reciprocating pumps, are rotary pumps.

Reciprocating Pumps

These pumps have now almost become obsolete because of their high capital cost, high maintenance cost, lower efficiency, and small capacity. However, these can be seen in villages where there is no electricity. These are hand operated well pumps, and commonly known as **hand pump** (Fig. 6.13).

In such a pump, a piston or a plunger reciprocates in a closed vertical cylinder. This is moved up and down by hand. On the upstroke, a vacuum gets created into cylinder, below the piston, thereby opening the check valve at the base of the pump, thus withdrawing meter into the space below the piston.

Simultaneously, the water above the piston is forced out. On the down stroke, the check valve gets closed and the piston valve gets opened, thereby permitting water to enter the cylinder above the piston. On the next upstroke, the piston valve close and the water above the piston comes out, and at the sametime, the check valve opens, permitting water to enter the space above the piston.

The cycles is thus continued, and continuous supply of water can be obtained.

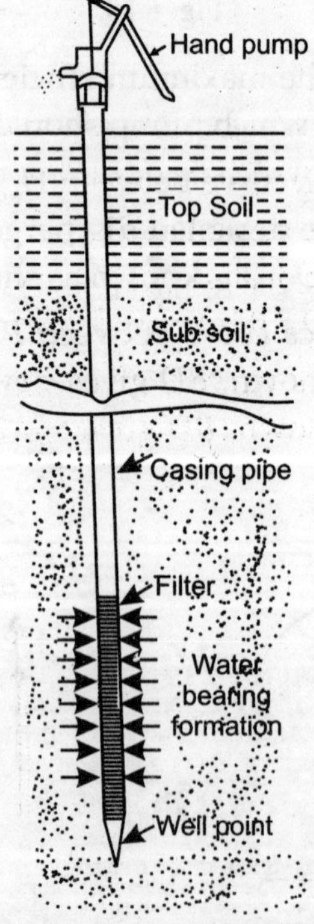

Fig. 6.13. Hand pump.

6.9. ROTARY PUMPS

Rotary pumps are the positive displacement type pumps having circular motion, and are without valves and can deliver a constant capacity against variable pressure. There are generally used for producing pressure upto 25 kg/cm^2. These are suited for handling viscous fluids, maintain high volumetric efficiency, require very less clearance between rubbing surfaces. Although used successfully for clean water and gasoline, their field of application is primarily in pumping oils or other liquids having some lubricating qualities and sufficient viscosity to prevent excessive leakages. Air leakage into the oil pump should be prevented, as enterance of air causes foaming which present a difficult problem when these lubricating oils are for bearings, gears, and similar other services. Oil foaming can be prevented by (*i*) low velocities in suction and discharge (*ii*) low suction lifts and (*iii*) preventing air leakage.

Classification. Mainly these are of two types :

1. *Constant Delivery Pumps*. These type include :
   (*a*) Gear pump. (*b*) Screw pump.
   (*c*) Vane pump. (*d*) Radial piston pump.
2. *Variable Delivery Pumps*. These type include :
   (*a*) Rotary piston pump (sliding piston pump).
   (*b*) Radial or seated valve piston pump.

"Constant" delivery pump is a pump in which there is continuous discharge of liquid at a uniform rate of flow whereas a "variable" delivery pump is a pump in which there is continuous discharge of liquid at varied rate of flow.

Since these pumps are not used in water supply schemes, hence are not being discussed in detail.

6.10. SELECTION OF A PUMP

Various factors which must be considered while selecting a particular type of a pump for a particular project are:

(*i*) Capacity of pumps.
(*ii*) Importance of water supply scheme.
(*iii*) Initial cost of pumping arrangement.
(*iv*) Maintenance cost.
(*v*) Space requirements for locating the pump.
(*vi*) Number of units required.
(*vii*) Total lift of water required.
(*viii*) Quantity of water to be pumped.

Since reciprocating pumps are outdated these days, therefore the centrifugal types of rotodynamic pumps are used these days, as they provide effective and economic service. However, pumps other than those of centrifugal types may be used under extreme conditions. The choice between various types of pumps depends on following :

Centrifugal pumps are especially useful for pumping waste water and water containing solids, although they are equally suitable for pumping treated waters.

Among the centrifugal pumps, a choice is sometimes made out of the horizontal shaft centrifugal pump and the vertical spindle turbine pump and the submergible pump. The horizontal pump is the cheapest and is used under a wide range of pumping conditions, but vertical spindle and submergible pumps may be preferred for handling large quantities of water under low heads, and for wells and tubewells.

6.11. HEAD, POWER AND EFFICIENCY OF PUMPS

The total head (H) against which a pump has to work consists of

(*i*) the suction lift (H_s);

(*ii*) the delivery head (H_d); and

(*iii*) the head lost due to friction, entrance and exit in the rising main (H_L).

$$\therefore \quad H = H_s + H_d + H_L$$

The work done by the pump in lifting Q cumecs of water by a head H

$$= \gamma_w . Q . H \text{ kN. m/s } (\textit{i.e.} \text{ J/s of Watts})$$

where γ_w = unit wt. of water in kN/m^3; *i.e.* 9.81 kN/m^3

Q = discharge to be pumped in m^3/sec.

The water horse power of the pump is then given as :

$$\text{W.H.P.} = \frac{\gamma_w . Q . H}{0.735} \qquad (\because 1 \text{ HP} = 735 \text{ W} = 0.735 \text{ kW})$$

If η is the efficiency of the pump set, then the brake horse power of the pump (B.H.P.) is given by

$$\text{B.H.P.} = \frac{\text{W.H.P}}{\eta} = \frac{\gamma_w . Q . H}{\eta \times 0.735} .$$

6.12. ECONOMICAL DIAMETER OF PUMPING MAINS

As diameter of a pipe can be reduced (for passing a certain fixed discharge) by increasing the flow of velocity through the pipe. But however, the increased velocity will lead to higher frictional head loss, and thus increased cost of pumping. Hence, although the dia and the cost of pipe can be reduced by choosing a higher flow velocity, the horse power of the pump required will increase, thus increasing the cost of pumping. For optimum conditions, we must choose such a diameter, which together with the pumping cost, will make the total annual expense, as *minimum*. The diameter which provides optimum conditions is known as the **economical diameter** of the pipe used for pumping.

If the diameter is less than the economical diameter, the cost of pipe will be less, but the head loss will be high and the cost of pumping shall be much more than the resultant saving in the pipe cost. Similarly, if the diameter choosen is more than the economical dia, the cost of pumping will be less but the increase in the cost of pipe will be much more than the resultant saving in the pumping cost.

An empirical formula given by Lea, which is commonly used in practice for giving economical diameter is:

$$D = 0.97 \text{ to } 1.22 \sqrt{Q}$$

where D = economical dia in metres

Q = discharge to be pumped in cumecs

This relationship gives optimum flow velocity varying between 1.35 to 0.8 m/sec.

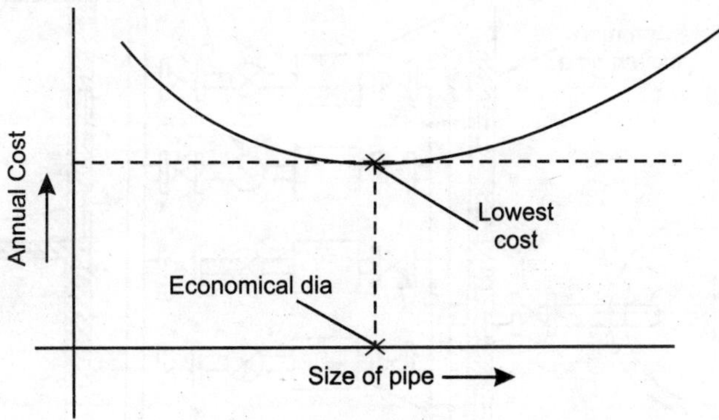

Fig. 6.14.

For a rigorous analysis, the total cost of pipe and pumping should be worked out at different assumed flow velocities (between 0.8 to 1.35 m/sec) and a graph plotted between the yearly cost and the size of the pipe (Fig. 6.14.).

6.13. PUMPING STATIONS

Various points considered for selecting a suitable site for pumping station are:

(*i*) The site should be away from all the sources of contamination or pollution.

(*ii*) The site should be above the highest flood level of the river.

(*iii*) It should be so selected that its future expansion is easily possible.

(*iv*) In case of rivers, the site for the pumping stations should be such that the pumping machinery is able to draw sufficient water so as to meet the peak demand. For this, the dry weather flow in the river, the likely changes in the river course, etc., must be properly studied and considered.

The pumping stations should be properly lighted, ventilated and kept neat and clean. Different pumping units and driving engines (generally motors) should be spaced wide apart so as to provide sufficient space for their operation, repairs, maintenance and cleaning (Fig. 6.15).

Setting of Pumps. The pumps are generally placed *above* the level of the water in the sump well, they can be placed at an elevation which has to be decided by the possible *suction lift* of the pump, the velocity in the suction pipe, the frictional losses, etc.,, and is usually limited to metres or so. To keep the losses to minimum, suction lines should be made as short as possible and with a continuous upward slope from the foot valve to the pump in order to avoid air pockets. The foot valves should be at the ends of the suction pipes, which should remain sufficiently submerged in water, so as to prevent entry of air into the suction pipe, so as to avoid priming difficulties. The motors, in such a setting, can be placed at the same elevation as that of the pumps, and thus horizontally coupled to them, so as to avoid difficulties in alignment.

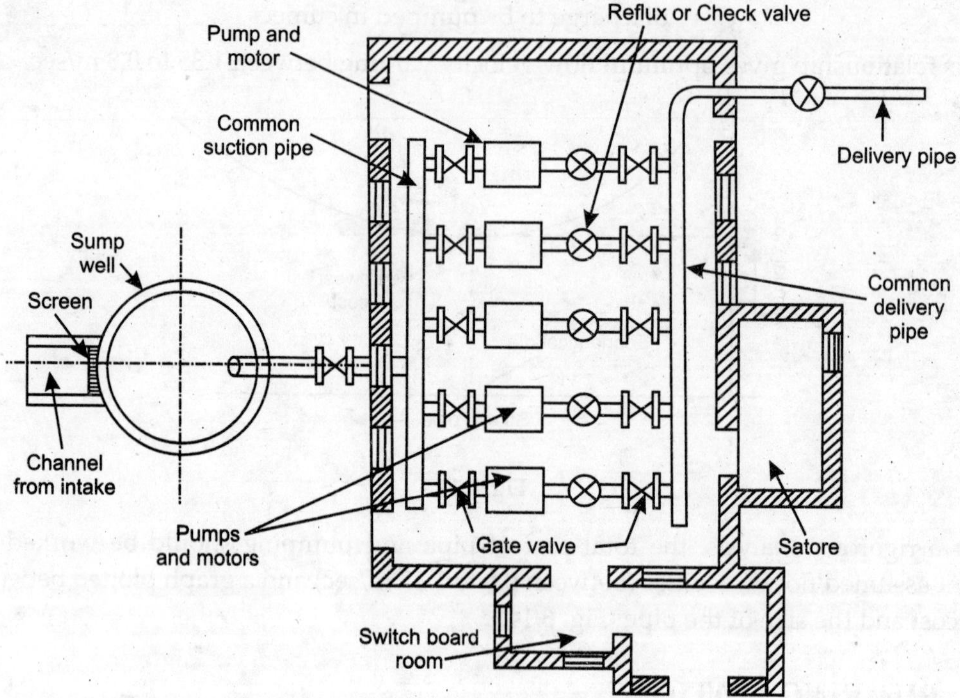

Fig. 6.15. Layout plan of a pumping station.

Valves at the Pumping Station

The various valves which are used in pumping stations with their common positions are as under :

(*i*) *Sluice or Gate valves*—one on suction side, and the other on the delivery side;

(*ii*) *Check valves or Reflux valves* on the delivery side in between the pump and the gate valve, and

(*iii*) *Pressure relief and Air valves* on the side (not shown).

Sizing of the Pump Units and the Stand-by Capacity

A waterworks pumping station is generally not operated at full capacity, all of the times. Since the efficiency of a pumping unit varies with the load, it is a usual practice to so design a station, that some of the pump units can be operated at full capacity, all of the time. Hence, two, three, or four pumps may have to be installed. Say for example, if there are four pumps, the two largest may supply the maximum demand; one of these and the smallest may handle the normal demand, and the other pump may care for the minimum requirement. Thus, there will always exist some standby capacity to take care of the repairs, breakdown etc., Generally 100% standby capacity against average demand, and 33.5% to 50% standby capacity against the peak demand is considered sufficient and may, therefore, be provided at the pumping stations.

QUESTIONS

1. Name the different type of pumps used generally in water supply scheme. What are the factors on which their selection depends.
2. List the appurtenances necessary to be installed on rising main between the pumps and an overhead tank. Indicate their relative positions on a line diagram and mention the function of each.
3. Enumerate the advantages and disadvantages of various types of pumps generally used in water supply.
4. Discuss the construction and working of a centrifugal pump used for water supply.
5. Write short note on :
   (*a*) Priming.
   (*b*) Centrifugal pump used in water supply.
   (*c*) Hand operated well pumps.
6. What is meant by economical diameter of a rising main. On what factors does it depend?

Purification of Water Supplies

7.1. INTRODUCTION

Water whether from underground or surface sources found in nature are generally polluted. The pollution may be due to various reasons namely from sewage, industrial wastes or from natural contaminants. Waterborne epidemic may spread by supply of such water. Natural water may also contain high amount of turbidity, organic load from decaying leaves, vegetations or animals. Materials in solution or suspension forms may also contaminate water.

All the above contaminants are required to be removed so that the water is fit for domestic use as well as for a wide variety of industrial uses. To make water consumable, the following factors should be taken into consideration:

(*i*) Safety for preventing waterborne diseases.

(*ii*) Palatability (mineral content, temperature) from aesthetic considerations.

(*iii*) Wholesomeness of water (*i.e.* Safe and potent for drinking to human health) taking the dissolved salts into account (in desirable concentrations).

(*iv*) Ensuring water quality so as to attain the permissible standards till the water reaches the consumer.

(*v*) Removing turbidity (presence of suspended material such as clay, silt finely divided organic material, plankton and other particulate material in water), colour, taste and odour which cause the water to be unacceptable.

The available raw waters must be treated and purified before they can be supplied to the general public for their domestic, industrial or any other uses. The extent of treatment required to be given to a particular water depends upon the characteristics and quality of the available water, and also upon the quality requirements for the intended use.

In general, the public water supplies are treated as per the quality requirements of drinking water. The available water must be made safe, good in appearance, and attractive to human taste . Various methods which are used to make the raw water safe and attractive to the consumers are described below.

7.2. METHODS (TECHNIQUES) OF PURIFICATION OF WATER

A conventional water treatment plant (Flow diagram shown in Fig. 7.1) consists of three main units, viz. sedimentation (with or without coagulation), filtration, and disinfection (chlorification). The dissolved salts are removed only when necessary.

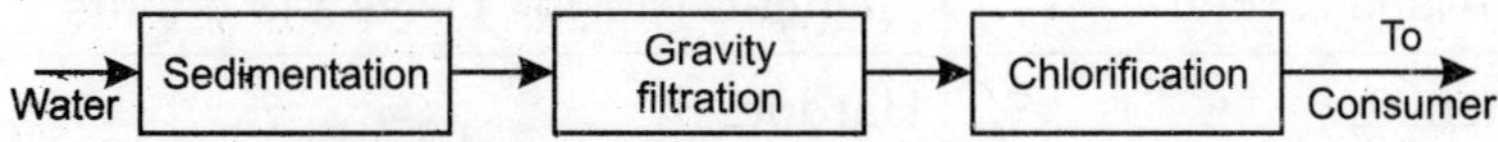

Fig. 7.1. Flow diagram of conventional water treatment plant.

Screens are provided at the intake to exclude coarse material from source of collection. Plain sedimentation removes about 35% of suspended solids, while coagulation process of sedimentation removes fine suspended and colloidal matter. Filtration removes more than 90% of suspended solids, turbidity. Disinfection kills all the bacteria. Besides above aeration helps in controlling the tastes and odours of water.

The various methods or the techniques which are now-a-days adopted for purifying the public water supplies are:

1. Screening
2. Plain sedimentation
3. Sedimentation aided with Coagulation
4. Filtration
5. Disinfection
6. Aeration
7. Softening
8. Miscellaneous treatments, such as fluoridation, recarbonation liming, desalination, etc.

Degree and methods of treatment, depend on the nature of source quality of water of the source, purpose for which it is to be supplied. Quality of domestic water may require different standards than that of an industry. Further, quality requirements of one industry may differ from the other.

All these above mentioned factors determine which treatment process to be employed and which may not be necessary. The various treatment processes and the nature of impurities removed by them are given in the Table 7.1

Table. 7.1: Water Treatment Techniques

S.N.	Process	Impurity removed
1.	Screening	Floating matter
2.	Plain sedimentation	Large suspended solids
3.	Coagulation	Fine particles

4.	Filtration	Colloidal particles, micro-organisms
5.	Disinfection	Living organisms including pathogens
6.	Aeration	Tastes and odour removal, oxygen deficiency
7.	Activated carbon	Elements causing tastes and odours
8	Softening	Hardness

It is pointed out that all the above techniques may or may not be used for treating a particular raw water. The necessity of a few or all of these treatments depends solely upon the quality of the available raw water. The functions served by each of these techniques in effecting purification of raw water are discussed in brief as under.

Large and visible objects, such as trees, branches, sticks, vegetations, fish, animal life, etc., present in raw waters of surface sources can be removed by **screening**. The coarser suspended material can then be removed by letting the water settle in **sedimentation** basins. This process is called **plain sedimentation**. The sedimentation effectiveness can be increased by mixing certain chemicals with the water, so as to from flocculent precipitate, which carries the suspended particles as it settles. This process is called **chemical coagulation**.

The finer particles in suspension, which could not settle even after using chemical coagulation, may then be removed by filtering the water through filters. This process is called **filtration**.

The filtered water which may still contain pathogenic bacteria, is then made bacteria – proof by adding certain chemicals such as chlorine etc.

This process of killing of germs is called disinfecting. The resulting water, though by now becomes safe yet may not be attractive to the tongue of the consumers. Unpleasant tastes and odours may then be removed by adding certain chemical compounds such as activated carbon or by using ozone. This process is called **aeration**.

The resulting water may sometimes be much harder than permissible and therefore, have to be softened by a process called **softening**. Sometimes, the resulting water may be given further treatment, such as **fluoridation** (addition of soluble fluoride for controlling dental caries), liming (addition of lime so as to control acidity and reduce corrosive action), **desalination** (removal of excess salt) etc. All these techniques/processes shall now be discussed in detail.

7.3. SCREENING

Screens are generally provided infront of the intake works, so as to exclude the large sized particles such as debris, animals, trees, branches, bushes, ice etc. **Coarse screens** (generally called **trash rack**) are sometimes placed infront of the **fine screens**. Coarse screens consist of parallel iron placed vertically or slight slope, at about 2.5 to 5 cm apart. The fine screens are usually made of woven wire mesh with openings not more than 6 mm square. The coarse screens first remove the bigger floating bodies and the organic solid; and the fine screens then remove the fine suspended solids.

The fine screens normally get clogged, and are to be cleaned frequently. The fine screens are, therefore, avoided these days, and the finer particles are separated in 'Sedimentation' rather than in 'screening", The coarse screens are also now normally kept inclined at about 45° to 60° to the horizontal, so as to make the screening more effective.

The material which is collected on the upstream side of the screens is removed either manually or mechanically. In mechanically cleaned screens, a rake traverses infront of the screen either continuously or intermittently. In mechanically cleaned screens, the cross bars, if provided, will obstruct the raking, and should, therefore, be avoided as for as possible.

A fixed bar type screen is shown in Fig. 7.2.

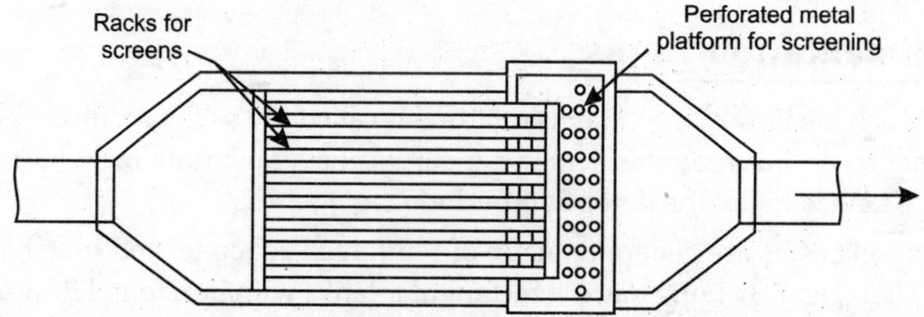

Fig. 7.2. Fixed bar type hand cleaned screen.

A stationary find screen is cleaned by raising it out of the water and washing off the collected material. Moveable fine screens are endless belts of fine screening or a series of panels of fine wire mesh mounted on a chain. These are driven slowly by a motor. As the screen is raised above the water level, the debris collected on it are discharged into a trough, usually with the aid of a water spray or a jet.

7.4. PLAIN SEDIMENTATION

It is a process of getting the particles, heavier than water, settles due to gravitational forces under still water. Particles settle as discrete (single, unattached) particles in a plain sedimentation tank. Particles combine together and settle quickly in coagulation – sedimentation tank. In this process, two forces simultaneously act on a particle. A downward pull or a driving force and a frictional force, *i.e.*, drag force acting upward on the particle. The resultant force decides whether a particle settles, floats or is carried away with flow.

The settling velocity of the particle depends upon:

(*a*) Horizontal flow velocity of water,

(*b*) Shape and size of the particle,

(*c*) Relative density of the particle, and

(*d*) Temperature of the water.

7.4.1. Theory of Sedimentation

The settlement of a particle in water brought to rest, is opposed by the following factors:

1. *The velocity of flow.* The velocity of flow carries the particle horizontally. The greater the flow area, the lesser is the velocity, and hence more easily the particle will settle down.
2. *The viscosity of water.* The viscosity varies inversely with temperature. The warm water is less viscous and, therefore, offers less resistance to settlement. Since the temperature of water cannot be controlled in 'water purification process', hence, this factor is generally ignored.
3. *The size, shape and specific gravity of the particle.* The greater is the specific gravity, more readily the particle will settle. The size and shape of the particle also affect the settling rate. The small sized particle will settle very slowly.

7.4.2. Sedimentation Tanks

The basin in which the flow of the water is retarded is called the settling tank or sedimentation tank or sedimentation basin or clarifier, and theoretical average time for which the water is detained in the tank is called the **detention period.**

Sedimentation basins are generally made of reinforced concrete, and may be rectangular or circular in plan (Fig. 7.3). Long narrow rectangular tanks with horizontal flow are generally preferred to the circular tanks. The capacity and other dimensions of the tank should be properly designed, so as to effect a fairly high percentage of removal of the suspended materials. A plain sedimentation tank under normal conditions may remove as much as 70% of the suspended impurities present in water.

Ideal inlet for a sedimentation tank should,

(*i*) Reduce entering velocity to prevent pronounced currents towards the outlet,

(*ii*) Distributes the water uniformly over the cross-section of the tanks, and

(*iii*) Mixes entering water with already in the tank and prevent it from sinking to the bottom and moving along the bottom towards the outlet.

The outlet of the tank may be a free falling weir which skims the water from the top surface.

Regardless of the type of basin, the design can be divided into four zones: inlet, settling, outlet, and sludge storage.

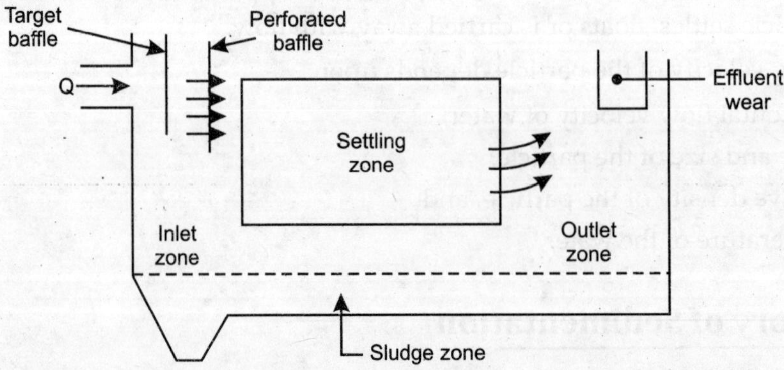

Fig. 7.3. Schematic diagram showing four zones of sedimentation tank.

7.4.3. Tube Settlers

Tube settlers are a bunch of plasic tubes 5 cm × 5 cm; 1 m long kept in an inclined position as settling tank (Fig. 7.4). In these settlers, water flows upward and the settled material moves downwards. In this sludge storage capacity is small. The main advantage of the tube settler is its small size and low cost of construction.

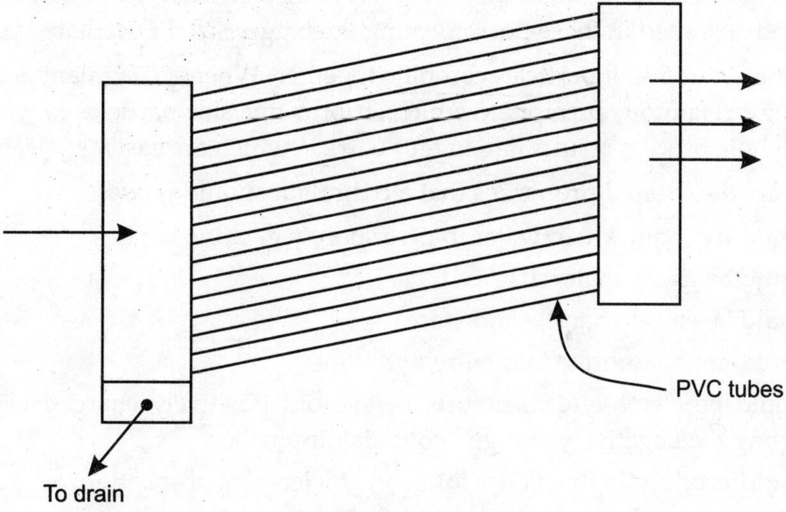

Fig. 7.4. Tube settlers.

The overall dimension of a tube settlers may be say 5 m × 5 m, or so. The height of this module may be around 1 m. Because of the inclined tubes, continuous gravity drainage of the settleable material (sludge) can be obtained.

7.4.4. Cleaning and Sludge Removal

The suspended materials contained in the raw waters settle down at the bottom of the sedimentation tank, and are removed periodically. The removed of the deposited sludge and sediments is necessary because it reduces the capacity of the tank and also it leads to the formation and evolution of certain foul gases due to the decomposition of the settled organic matter. The sedimentation tanks are cleaned either manually or they are provided with mechanical arrangements for cleaning.

For manual cleaning, the tank is, first of all, put of service and the supply of raw water is stopped. The already contained water is drained off till the depth remains about 30 cm or so. The sludge is now stirred and removed as a slurry, through a separate pipe provided with a gate value at the bottom of the tank. Pick axes and phaorahs may be used for displacing hard deposits.

In modern water treatment plants (WTPs), sedimentation tanks are generally fitted with mechanical cleaning devices in which the sludge is scrapped and brought to the hopper at the outlet end, and is removed daily or oftener.

7.5. SEDIMENTATION AIDED WITH COAGULATION

As very fine suspended mud particles and the colloidal matter present in water cannot settle down in plain sedimentation tank. Such particles can, however, be removed easily by increasing their size by changing them into flocculated particles. For this purpose, certain chemical compounds, called coagulants, are added to the water, which on through mixing, form a gelatenous precipitate called 'floc'. The very fine colloidal particles present in water, get attracted and absorbed in these flocs, forming the bigger sized flocculated particles.

In coagulation individual particles combine together. When a coagulant is used in water, it forms a spongy gelatinous precipitate which absorbs fine size particles in water bind them together. The whole process results into bigger particle which are easily settleable.

Following are the desired properties that a coagulant should possess:

(*i*) It should react quickly in water to give spongy gelatinous flocs.

(*ii*) It should be cheap material.

(*iii*) It should be easy to handle and store.

(*iv*) It should not deteriorate in quality with time.

(*v*) It should be electrolyte in nature to give out positively charged electric ions for attracting the negatively charged colloidal impurities.

(*vi*) It should produce high valence ions for efficiency of absorption.

(*vii*) It should react in the long range of pH.

It may also be noted that overdosing of coagulants may reverse the electric charges on colloids and which retard the coagulation.

For all practical purposes, the entire process of addition of chemicals (coagulants) and mixing (flocculation) is usually referred to as coagulation. The coagulated water is finally made to pass through the sedimentation tank, where the flocculated particles settle down, and are thus removed.

7.5.1. Chemicals used for coagulation

Various chemicals such as alums; iron salts like ferrous sulphate, ferric chloride, ferric sulphate etc., are generally used as coagulants. These chemicals, are most effective when water is slightly alkaline. In the absence of such an alkalinity in raw supplies, external alkalies like sodium carbonate, or lime, etc., are added to the water, so as to make it slightly alkaline, and thus to increase the effectiveness off the coagulants. The important coagulants and the chemical reactions associated with them are described below:

1. **Use of Alum as Coagulant.** The chemical name of 'Alum' is aluminium sulphate with its chemical formula as $Al_2(SO_4)_3 . 18\ H_2O$. The alum when added to raw water, reacts with the bicarbonate alkalinities, which are generally present in raw supplies, so as to form a gelatinous precipitate (floc) of aluminium hydroxide. This floc attracts other fine particles and suspended matter (colloids), and thus grows in size, and finally settles down to the bottom of the tank.

The amount of alum required for coagulation depends on the turbidity and colour of the raw water. The use of the optimum amount of a coagulant can be approximately determined by laboratory testing. The dose of alum may vary from 5 mg/l for relatively clear waters to 85 mg/l for highly turbid waters. The average normal dose is about 17 mg/L.

Alum a very effective coagulant, and is now extensively used. It is quite cheap, forms an excellent stable floc, and does not require any skilled supervision for handling. The water so obtained is quite clear, it also helps in reducing taste and colour of raw water.

2. **Use of Copperas (Iron Salt) as Coagulant.** Copperas is the name given to ferrous sulphate with its chemical formula as $FeSO_4 . 7H_2O$. Copperas is generally added to raw water in conjunction with lime. When lime is added, the following reaction takes place:

$$\underset{\text{Copperas}}{FeSO_4} + 7H_2O + \underset{\text{Hydrated lime}}{Ca(OH)_2} \longrightarrow CaSO_4 + \underset{\text{Ferrous hydroxide}}{Fe(OH)_2} + 7H_2O$$

The ferrous hydroxide formed further gets oxidised, forming ferric hydroxide, as given below:

$$4Fe(OH)_2 + O_2 + 2H_2O \longrightarrow \underset{\text{Ferric hydroxide ppt.}}{4F(OH)3 \downarrow}$$

The ferric hydroxide forms the floc, and thus, helps in sedimentation.

Copperas is extensively used as a coagulant for raw waters that are not coloured. It is generally cheaper than alum, and functions effectively in the pH range of 8.5 and above.

3. **Use of Chlorinated Copperas as Coagulant.** When chlorine is added to a solution of copperas (*i.e.*, ferrous sulphate), the two react chemically, so as to form ferric sulphate and ferric chloride. The chemical equation is as follows:

$$\underset{\text{Copperas}}{6(FeSO_4 . 7H_2O)} + 3Cl_2 \longrightarrow \underset{\text{Ferric sulphate}}{2Fe_2 (SO_4)_3} + \underset{\text{Ferric chloride}}{2FeCl_3} + 42H_2O$$

The resultant combination of ferric sulphate and ferric chloride is known as chlorinated copperas, and is a valuable coagulant for removing colours, especially where the raw water has a low pH value.

Comparison of Alum and Iron Salts (as Coagulants). The alum and the iron salts are having their own advantages and disadvantages, as summarised below:

(*i*) Iron salts produce heavy floc and can, therefore, remove much more suspended matter than the alum.

(*ii*) Iron salts, being good oxidising agents, can remove hydrogen sulphide and its corresponding tastes and odours from water.

(*iii*) Iron salts can be used over a wider range of pH values.

(*iv*) Iron salts cause staining and promote the growth of iron bacteria in the distribution system.

(*v*) Iron salts impart more corrosiveness to water than that which is imparted by alum.

(*vi*) The handling and storing of iron salts require more skill and control, as they are corrosive and deliquescent. Whereas, no such skilled supervision is required for handling alum.

Note. Iron salts are used as coagulants more frequently for treating sewage, and alum is used more frequently for treating raw waters.

7.6. FILTRATION

Screening and sedimentation as discussed above, removes a large percentage of the suspended solids and organic matter present in raw supplies. The percentage of removal of the fine colloidal matter increases when coagulants are also used before sedimentation. But even the resultant water will not be free of impurities, and may contain some very fine suspended particles (discrete, or flocculated when coagulation is used) and bacteria present in it. To remove or to reduce the remaining impurities, and to produce potable and palatable water, the water is filtered through the beds of fine granular material, such as sands, etc. The process of passing the water through the beds of such granular materials (called **filters**) is known as **filtration**. Filtration may help in removing colour, odour, turbidity, and some pathogenic bacteria from the water. The two types of filters are commonly used for treating municipal water supplies. They are

(*i*) The slow sand gravity filters; and

(*ii*) The rapid sand gravity filters.

A third type of a rapid sand filter works under pressure and is known as a pressure filter. This type of filters are generally used for small plants, such as for individual industrial supplies, or for swimming pools; and are generally not adopted for treating large scale municipal supplies.

The slow sand gravity filters often called slow sand filters are useful in the sense that they can remove much larger percentage of impurities and bacteria from the water, as compared to what can be removed by rapid sand gravity filters. However, slow sand filters yield a very slow rate of filtration (about $\frac{1}{30}$ th of that given by rapid gravity filters and require large areas, and are costly.

In the modern treatment plants, rapid gravity filters are now-a-days almost universally adopted. The water from the coagulation-sedimentation plant is directly fed into the rapid gravity filter, and the resultant supplies are disinfected for complete killing of germs and colour removal.

The filters, purify the water under following **processes.** These processes or action are:

1. **Mechanical Straining.** The suspended particles present in water, and which are of bigger size than the size of the voids in the sand layers of the filter, cannot pass through these voids and get arrested in them. The resultant water will, therefore, be free from them. The upper sand layers arrest particles including the coagulated

flocs, which forms a mat on the top of the bed, which further helps in removal of the impurities.

2. **Flocculation and Sedimentation.** It has been found that the filters are able to remove even particles of size smaller than the size of the voids present in the filter. The colloidal matter arrested in these voids is a gelatinous mass and, therefore, attract other finer particle. These finer particles thus settle down in the voids and get removed.
3. **Biological Metabolism.** Certain micro-organisms and bacteria are generally present in the voids of the filters. These organisms require organic impurities (such as algae, plankton, etc.) as their food for their survival. These organisms, therefore, utilise such organic impurities and convert them into harmless compounds by the process of biological metabolism.

7.6.1. Filter Materials

Either fine or coarse sand is generally used as filter media. The layers of sand may be supported on gravel, which permits the filtered water to move freely to the under-drains, and allows the wash water to move uniformly upward.

1. **Sand.** The filter sand should generally be obtained from rocks like quartzite, and should contain the following properties:
   (*i*) It should be free from dirt and other impurities.
   (*ii*) It should be uniform in nature and size.
   (*iii*) It should be hard and resistant.
2. **Gravel.** The gravel which may be used below the sand should be hard, durable, free from impurities and properly rounded. Graved of different sizes are usually placed in 3-4 layers, each of 15-20 cm depth, with the coarsest size (20 to 60 mm) placed in the bottom-most layer, and the finest size (3 to 6 mm) in the top most layer.
3. **Other materials.** Instead of using sand, sometimes, anthrafilt is used as filter media. Anthrafilt is made from anthracite, which is a type of coal-stone that burns without smoke or flames. The use of anthrafilt as a filter media in two or multi layers have proved very successful in some countries. It is cheaper and has been able to give a high rate of filtration. However, this type of coal is not available in India. Research is, therefore, being conducted at Central Public health Research Institute, Nagpur, to find out the suitability and usefulness of the bituminous coal as a substitute to anthracite.

Types of Filters

As pointed out earlier, there are mainly three types of filters, *viz.*

(*i*) Slow sand filters:
(*ii*) Rapid gravity filters; and
(*iii*) Pressure filters.

(*i*) and (*ii*) above can be categorised as gravity filters, while (*ii*) and (*iii*) above are rapid sand filters.

In gravity flow fitters, two flow rates are adopted, one is a low rate of filtration (100 – 150 lit/hr/m^2) known as slow sand filter, the other is a high rate of filtration (5000 – 6000 lit/hr/m^2) known as rapid sand filter. Pressure filter maintains high rate (more than 8000 lit/hr/m^2).

7.6.2. Slow Sand Filters

Slow sand filters were first devised when the process of filtration was invented in England in 1829. They were widely used since then, till the last decade of the 19th century, when the repaid gravity filters were invented. They are becoming obsolete these days.

Slow sand filters (shown in Fig. 7.5.) require sand filter and even number of units are preferred in sizes of 60 m × 40 m (approximately). The effective size of sand used is 0.25 to 0.35 mm and uniformity co-efficient is 2 to 3. When top layer is clogged (generally in 2 to 3 months), top few centimetres of sand is removed for cleaning. Now sand is replaced for starting the filter again. The dirty sand is cleaned with jets of water, dried and again used later. As the flow rate, in the system is less and sand size is small, filter efficiency is very high. This method has no maintenance and operational difficulties. Filtered water is chlorinated and then supplied directly to consumers.

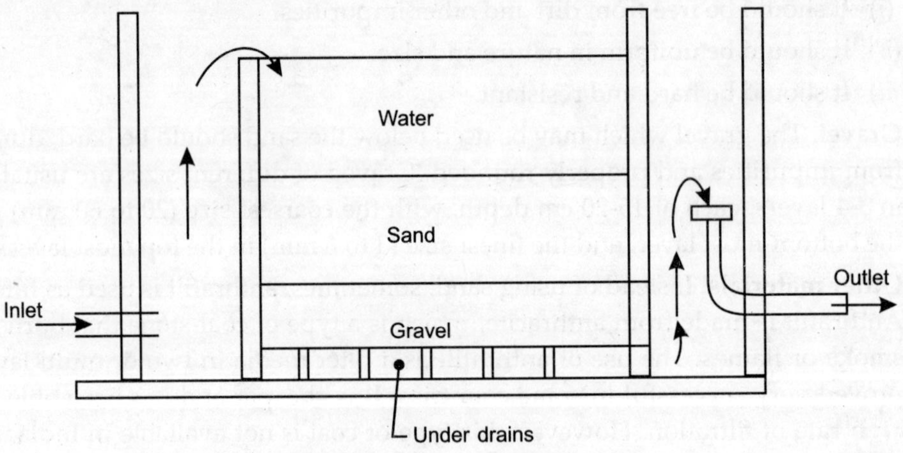

Fig. 7.5. Slow sand filter.

7.6.3. Rapid Sand Filters

The slow sand filters require huge quantities of filtering materials, and require huge space which may prove to be very costly, particularly in congested cities and industrial towns. In order to reduce the requirement of space and to increase the rate of filtration, rapid sand filters, or sometimes called mechanical sand filters were developed (Fig. 7.6). The rapid sand filters are of two types : *viz.*

(*a*) **Rapid gravity filters.** These utilise comparatively large sized sand particles, and allow greater rate of filtration as compared to that of slow sand filters.

(b) Pressure filters. These utilise the development of pressure over the filtering water and thereby increasing the rate of filtration. Rapid gravity filters are used for large municipal supplies, whereas pressure filters, being more handy, may be used for small installations, such as for industrial plants and swimming pools.

1. Rapid Gravity Filters

These filters employ coarser sand, with effective size as 0.5 mm or so. On an average, these filters may yield as high as 30 times the yield given by the slow sand filters. Waters from the coagulation-sedimentation tanks are used in these filters, and filtered water is treated with disinfectants, so as to obtain potable supplies.

The filtering media consists of sand layers about 60 + 90 cm in depth and placed over a gravel support. The effective size of the sand varies from 0.35 to 0.55 mm. The sand must be laid in layers. The finer variety should be used towards the top and coarser variety towards the bottom.

Like slow sand filter, the rapid gravity filters have the base material as gravel and supports the sand. But in a rapid gravity filter, in addition to supporting the sand, the gravel distributes the wash water. It consists of 60 to 90 cm thick gravels of different sizes, placed in 5 to 6 layers. The coarsest gravel (upto about 40 mm size) is used in the bottom-most layer and finest gravel (about 3 mm in size is used in the top - most layer).

The size of the gravel in the bottom-most layer is generally kept between 20 to 40 mm; in the intermediate layers, between 12 to 20 mm, and 6 to 12 mm (when two intermediate layers are used); and in the topmost layer, between 3 to 6 mm.

In slow sand fitters, the under - drainage system was provided only to receive and deliver the filtered water, whereas, in rapid gravity filters, the under - drainage system serves two purposes, *viz.*

(*i*) To receive and collect the filtered water, and

(*ii*) To allow the back - washing for cleaning of filter.

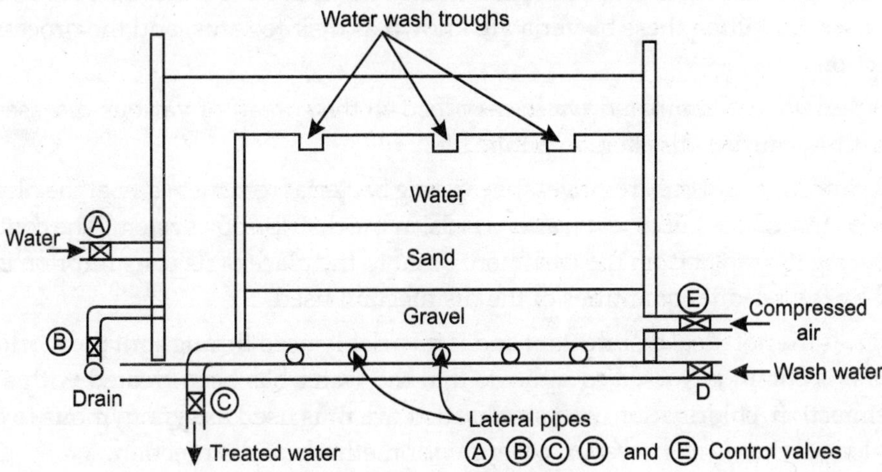

Fig. 7.6. Rapid sand filter.

In these rapid sand filters, water is distributed uniformly over the filter surface area, with a constant flow rate. The impurities penetrate into deeper layers which results into clogging of filter bed. The filter are cleaned by backwashing with compressed air and (or) backwashing with water. The sand must be sufficiently agitated or scrubbed to enable the release of impurities clogged. Wash water gutters are situated at a height over the filter bed such that all the impurities are collected towards a drain avoiding loss of sand.

A rapid sand filter requires constant and skilled maintenance as it contains a number of control devices, such as flow rate controller, loss–head indicators, back washing system and various valves.

2. Pressure Filters

Pressure filters are just like small rapid gravity filters placed in closed vessels, and through which water to be treated is passed under pressure. Since water is forced through such filters at a pressure greater than the atmospheric pressure, it is necessary that these filters are located in air tight vessels. The raw water is pumped into the vessels by means of pumps. The pressure developed may normally vary between 30 to 70 metres head of water, *i.e.* 300 to 700 N/m^2.

The filter vessels are steel cylinders of diameters between 1.5 to 3 m and their length/height may vary between 3.5 to 8 m.

The pressure filters can yield filtered water at rates much higher, *i.e.* 2 to 5 times than what can be obtained from rapid gravity filters. Their rate of filtration normally ranges between 6000 to 15000 litres/hr/sq m of filter area. The lower rates are used for domestic supplies for colonies of few houses, individual industries; and higher rates for recirculating swimming pool supplies.

7.7. DISINFECTION

The filtered water obtained as above may, normally contain some harmful disease producing bacteria in it. These bacteria must be killed in order to make the water safe for drinking. The chemicals used for killing these bacteria are known as disinfectants, and the process is known as **disinfection.**

The bacterially contaminated water may lead to the spread of various diseases and their epidemics, thus, causing disasters to public life.

The 'disinfection' not only removes the existing bacteria from the water at the plant, but also ensures their immediate killing even afterwards, in the distribution system. The contamination of water during its transit from the treatment plant to the place of its consumption is also thus, prevented by the residual quantities of the disinfectant used.

Chlorine is the not ideal disinfectant and is invariably used throughout the world. The term Chlorination is commonly used to indicate that the water has been treated with a sterilising agent. Disinfection, chlorination, and sterilisation are thus used as synonymous to each other. Besides chlorine, there are, however, other minor methods of disinfection.

The commonly used minor methods of disinfection are:

1. Boiling of water.

2. Treatment with line.
3. Treatment with ozone.
4. Treatment with iodine and bromine.
5. Treatment with ultraviolet rays.
6. Treatment with potassium par magnet.

7.7.1. Chlorination

Chlorine is used for disinfecting public water supplies. It is cheap, reliable, easy to handle, easily measurable, and above all, it is capable of providing residual disinfecting effects for long periods. Thus, it provides complete protection against future recontamination of water in the distribution system.

Its only disadvantage is that when used in greater amounts, it imparts bitter and bad taste to the water, which may not be liked by certain consumers.

When chlorine is added to water, it forms hypochlorous acid or hypochlorine ions, which have an immediate and disastrous effect on most forms of microscopic organisms. The reaction that took place, in this process are:

$$Cl_2 + H_2O \xrightarrow{pH > 5} \underset{\text{Hypochlorous acid.}}{HOCl} + HCl$$

The hypochlorous acid is unstable and may break into hydrogen ions and hypochlorite ions

$$HOCl \xrightleftharpoons{pH > 8} \underset{\text{Hydrogen ions}}{H^+} + \underset{\text{Hypochlorite ions}}{OCl^-}$$

The above reaction is reversible and depends upon the pH value of water. The dissociation of hypochlorous acid into ions is more effective at high pH values and vice versa

The amount of chlorine required for a water depends upon the inorganic an organic impurities in it, because when chlorine is added to a water, it first of all reacts with the inorganic impurities that converts the chlorine into chlorides, which has no residual oxidising power. The chlorine will also react with organic impurities present in water. The chlorine consumed in all the reactions represents the chlorine demand of water, when once it gets satisfied, the chlorine will appear as free chlorine.

The chlorine dose must be increased during the rainy season and epidemics. Moreover, the supply of water, to reach the first consumer should take atleast about 30 minutes after the chlorine is mixed in water, so as to ensure sufficient disinfecting action on harmful bacteria and viruses present in water.

The requirement of chlorine dose depends upon the following:

(1) The organic load including microbial like.

(2) The pH value of water.

(3) The amount of carbon dioxide present.

(4) The temperature of water.

(5) The available line of contact.

Direct Feed Chlorinator

A simple method of disinfection for piped water supplies is to feed bleaching powder solution direction directly with the suction line of a centrifugal pump used for drinking water (Fig. 7.7). Bleaching powder is fed from an storage tank. Flow into the intermediate container is regulated such that sufficient solution is always present to prevent air bubbles from being sucked in. The strength of the chlorine solution is 1%.

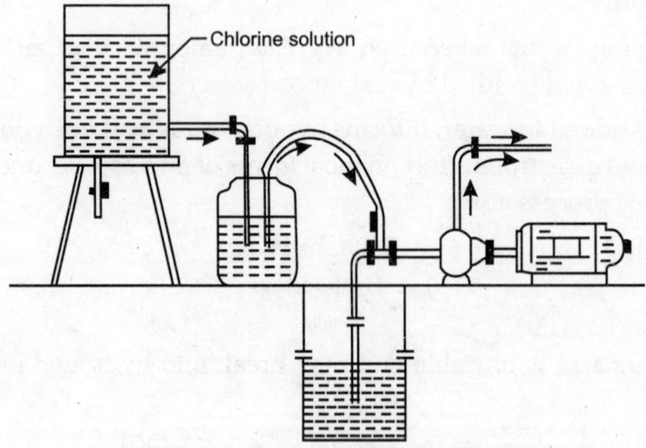

Fig. 7.7. Direct feed type chlorinators

7.8. WATER SOFTENING (REMOVAL OF HARDNESS)

The reduction or removal of hardness from water is known as water softening. Normally prevalent hardness in raw waters are not too high to cause any major harm to human health. The advantage of softening, however, lies chiefly in the reduction of soap consumption, lowered cost in maintaining plumbing fixtures, and improved taste of food preparations and drinking of water.

In view of the above, whether or not the hardness of a water supply should be reduced, depends on the relation between the cost of treatment and the obtained resultant saving and satisfaction to the consumers. However, for industrial supplies, the softening, the softening is more important, because the hard water are likely to cause scaling troubles in boilers and interfere in the working of dyeing systems.

The temporary carbonate or carbonate hardness, which is caused by the carbonates and bicarbonates of calcium and magnesium; can be removed by boiling, or by adding lime; whereas, the **permanent or non-carbonate hardness** which us caused by the sulphates, chlorides and nitrates of calcium and magnesium, has to be removed by special methods of water softening. The permissible hardness for public supplies normally ranges between 75 to 115 mg/L (where 14.25 mg/L is equivalent to one degree of hardness). Normally the hardness for all practical purposes is attributed to Ca^{++} and Mg^{++} only.

7.8.1. Removing Temporary Hardness

The temporary hardness can be removed either by boiling or by adding lime to the water, as explained below:

1. Boiling

Calcium carbonate, being only slightly soluble in water, will however, usually exist in water as. calcium bicarbonate because it easily dissolves in natural water containing carbon dioxide. When such a water is boiled, the carbon dioxide gas will get out, leading to the precipitation of $CaCO_3$, which can be sedimented out in settling tank. The reaction can be represented as:

$$\underset{\text{Cal. bicarbonate}}{Ca(HCO_3)_2} + \text{Heat} \longrightarrow \underset{\substack{\text{Cal. carbonate}\\\text{(insoluble)}}}{CaCO_3\downarrow} + CO_2\uparrow + H_2O$$

The magnesium bicarbonate and magnesium carbonate, however, cannot be satisfactorily removed under the above type of reaction, since $MgCO_3$ is fairly soluble in water.

On the other hand, magnesium carbonate and magnesium bicarbonate are removed by precipitating them as insoluble $Mg(OH)_2$, by treating the hard water with lime

2. Addition of lime

Lime (CaO), generally hydrated lime $[Ca(OH)_2]$ is added to the water. The following reactions take place:

$$(ia)\quad \underset{\substack{\text{Mag. carbonate}\\\text{(soluble)}}}{MgCO_3} + \underset{\substack{\text{Hydrated}\\\text{lime}}}{Ca(OH)_2} \longrightarrow \underset{\substack{\text{Mg. hydroxide}\\\text{(insoluble)}}}{Mg(OH)_2\downarrow} + \underset{\substack{\text{Cal. carbonate}\\\text{(insoluble)}}}{CaCO_3\downarrow}$$

$$\underset{\text{Mag. bicarbonate}}{Mg(HCO_3)_2} + \underset{\text{Hydrated lime}}{Ca(OH)_2} \longrightarrow \underset{\text{Mag. hydroxide}}{Ca(HCO_3)} + Mg(OH)_2\downarrow$$

$$\underset{\text{Cal. bicarbonate}}{Ca(HCO_3)_2} + Ca(OH)_2 \longrightarrow 2CaCO_3\downarrow + 2H_2O$$

The calcium carbonate and magnesium hydroxide are precipitated, and can be removed in the sedimentation tank.

This method is generally adopted for softening waters which contain only temporary hardness.

7.8.2. Removing Permanent Hardness

The permanent hardness is more difficult to remove. The three methods, which are commonly adopted for softening waters containing either permanent hardness, or permanent as permanent as well as temporary hardness both, are

(1) Lime-Sada process;

(2) Base-Exchange process, generally called Zeolite process; and

(3) Demineralisation process.

1. Lime-Soda Process for Removing Hardness.

In this process, lime [$Ca(OH)_2$] and soda ash [Na_2CO_3] are added to the hard water; which react with the calcium and magnesium salts, so as to form insoluble precipitates of calcium carbonate and magnesium hydroxide [$Mg(OH)_2$]. These precipitates can be sedimentation tank. The chemical reaction which may be involved are:

(*i*) $\underset{\text{Carbonate hardness of Ca}}{Ca(HCO_3)_2} + \underset{\text{Lime}}{Ca(OH)_2} \longrightarrow 2CaCO_3\downarrow + 2H_2O$

(*iia*) $\underset{\text{Carbonate hardness of Mg}}{Mg(HCO_3)_2} + \underset{\text{Lime}}{Ca(OH)_2} \longrightarrow \underset{\text{Carbonate hardness of Ca}}{Ca(HCO_3)_2} + Mg(OH)_2\downarrow$

(*iib*) $\underset{\text{Carbonate hardness of Mg}}{MgCO_3} + \underset{\text{Lime}}{Ca(OH)_2} \longrightarrow Mg(OH)_2 + CaCO_3\downarrow$

(*iii*) $\underset{\text{Non-Carbonate hardness of Mg}}{MgCl_2} + \underset{\text{Lime}}{Ca\ (OH)_2} \longrightarrow Mg\ (OH)_2\downarrow + \underset{\text{Non-Carbonate hardness of Ca}}{CaCl_2}$

(*iv*) $\underset{\text{Non-Carbonate hardness of Ca}}{CaCl_2} + \underset{\text{Soda}}{Na_2CO_3} \longrightarrow CaCO_3\downarrow + 2NaCl.$

(*v*) $\underset{\text{Non-Carbonate hardness of Ca}}{CaSO_4} + \underset{\text{Soda}}{Na_2CO_3} \longrightarrow CaCO_3 + Na_2SO_4$

2. Zeolite or Cation - Exchange Process

Zeolite are the natural salts or clays, which are hydrated silicates sodium and aluminium, having the general. formula as $Na_2\,O\,Al_2\,O_3x.\,Si\,O_2y.\,H_2O$. The usual value of x is 2 or more, and that of y is ranging amount. The zeolites or Rasins have the excellent property of exchanging their cations. Hence, during softening operation, the sodium ion of the zeolite get replaced calcium and magnesium ions present in hard waters.

The chemical reactions which may be involved are given in the following equation, where Z stands for zeolite.

$$Na_2Z + \begin{matrix}Ca \\ Mg\end{matrix}\left\{\begin{matrix}(HCO_3)_2 \\ So_4 \\ Cl_2\end{matrix}\right. \rightarrow Na_2\left\{\begin{matrix}(HCO_3)_2 \\ So_4 \\ Cl_2\end{matrix}\right. + \begin{matrix}Ca \\ Mg\end{matrix}\Big\}Z.$$

The calcium and magnesium zeolite can be regenerated into active sodium zeolite by treating it with 5 – 10 per cent solution of sodium chloride. The exchange reaction takes place during regeneration can be represented as under:

$$\left.\begin{matrix}Ca\\Mg\end{matrix}\right\}Z + 2\,NaCl \longrightarrow Na_2Z + \left.\begin{matrix}Ca\\Mg\end{matrix}\right\}Cl_2$$

Used zeolite — Sodium chloride solution — Regenerated zeolite

A zeolite softener (or a cation exchanger unit) resembles a sand filter in which the filtering medium is a zeolite rather than sand. The hard water enters through the top, and is evenly distributed on the entire zeolite bed. The softened water is collected through trainers at the base.

When a significant portion of the sodium in the zeolite has been replaced by calcium and magnesium, it is regenerated by, first washing it with water by reversing the flow, and then treating it with 10 per cent solution of brine (NaCl). The zeolite softeners may be either gravity or pressure filters.

Advantages of Zeolite Method.

1. Water of zero hardness can be obtained and hence useful for specific use in textile industries and boilers.
2. The plant is compact, automatic, and easy to operate.
3. No sludge is formed, and hence there is no problem of sludge disposal.
4. The operation and maintenance cost is quite less.
5. It also remove ferrous iron and manganese from water.
6. These is no difficulty in treating water of varying quality.

7.9. MISCELLANEOUS TREATMENTS

Other than the normal treatment processes, such as, coagulation sedimentation, filtration and disinfection (or chlorination); certain other special treatments are sometimes required to remove the special minerals, tastes, odours, colours, etc. from water. Minerals such a iron and manganese are not removed in the normal treatment processes, and hence need to be removed by some special treatment method. Similarly, dissolved gases such as carbon di-oxide (CO_2) or hydrogen sulphide (H_2S), etc., may sometimes be present in water, and may impart certain tastes and odours, which are not removed under the conventional treatments.

Dissolved carbon di-oxide gas will also make the water corrosive. Organic matter derived from algae and other micro-organism may also impart tastes and odours which may not be easily removed by the ordinary conventional methods. Similarly, the industrial waste waters may contain organic matter, such as phenols, etc., which may create bad tastes and odours after chlorination. All these aspects must be looked into, and steps taken to remove all the taste and odours from water, so as to make it attractive to the consumer's tongue.

7.9.1. Removal of Colours, Odour and Tastes from Waters

The special treatments given to the water for removing colours, tastes and odours from it, are discussed hereunder.

1. Aeration

In this process, water is brought in intimate contact with air, so as to absorb oxygen and to remove carbon dioxide gas. It may also help in killing bacteria to a certain extant. It also helps in removing H_2S gas, and iron and manganese to a certain extent, from the treated water. The aeration of water can be carried out in one of the following ways:

(*i*) **By Using Spray Nozzles.** In this method, water is sprinkled in air or atmosphere through special nozzles which breaks the water into droplets, thus permitting the escape of dissolved gases. Carbon di-oxide is thus considerably removed (upto 90% or so) in this method. However, 10 to 14 m head of water is required for the working of these nozzles, to function efficiently.

(*ii*) **By Permitting Water to Trickle Over Cascades.** In this method, the water is made to fall through a certain height (1 to 3 m) over a series of steps (three to ten) with a fall of about 0.15 to 0.3 m in each step. The structure so formed is known as a **freefall aerator.** The simplest type of a freefall aerator is known as a **cascade aerator.** They will take large quantities of water in a comparatively small area at low head, are simple to be kept clean and can be made of robust and durable materials with a long life. These cascade aerators are efficient in raising dissolved oxygen (DO) content of water.

(*iii*) **By Air Diffusion.** In this method, compressed air is bubbled through the water, so as to thoroughly mix it with water. Perforated pipes are, therefore, installed at the bottom of the settling tanks, and the compressed air is blown through them. During its upward movement through the water body, it gets thoroughly mixed up with the water contained in the tank, thereby completing the aeration process.

2. Treatment with Activated Carbon

Activated carbon is a specially treated carbon, which possesses the property of absorbing and attracting impurities, such a gases, and finely divided solids. Because of its excellent property of absorbing impurities, it is widely used for removing tastes and odours from public supplies.

It can be manufactured by heating or charring wood or saw dust or some other similar carbonaceous materials at about 500 C in a closed vessel and then slowly burning it (or oxidising or activating it) under very closed controlled conditions at 800 °C, thus removing hydrocarbons (which may interfere with the absorption of organic matter) from it. It is available in granular as well as powder forms. It is highly porous and absorptive, and is very useful for removing impurities.

The activated carbon is mostly used in the powdered from, and may be added to the water either before or after the coagulation, but before filtration.

Activated carbon may also be used in the granular form as a filter media, instead of using sand in the rapid gravity filters or preferably in pressure filters. With the passage of time, it looses its absorptive powers and requires to be rejuvenated. This regeneration is carried out by forcing the live steam upward through the perforated pipes placed in the gravel bed of filters. This process is to be carried out at intervals of 1 month to 1 year, depending upon the quality of water being treated.

The use of activated carbon may thus serve the following **advantages,**

1. When used in powdered form before coagulation, it aids in coagulation.
2. It reduces the requirement of chlorine.
3. It removes the organic matter present in water.
4. It removes the tastes, odours, and colours caused by the presence of iron, manganese phenols, excess chlorine, hydrogen sulphide etc.
5. Its overdose is not harmful.

7.10. REMOVAL OF SALT (DESALINATION)

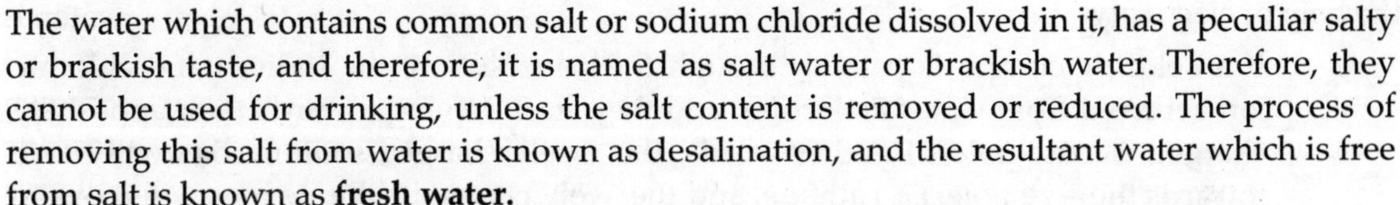

The water which contains common salt or sodium chloride dissolved in it, has a peculiar salty or brackish taste, and therefore, it is named as salt water or brackish water. Therefore, they cannot be used for drinking, unless the salt content is removed or reduced. The process of removing this salt from water is known as desalination, and the resultant water which is free from salt is known as **fresh water.**

Although desalination strictly means removal of salt, yet since common salt (NaCl) cannot be removed in isolation, this term is practically used to describe process, which is used for the reduction of total dissolved solids (TDS) in water, which is measured in mg/L.

Types of salty water

S. No.	Type of water	TDS value in mg/L
1.	Sweet water	0–1000
2.	Brackish waters	1000–5000
3.	Moderately saline waters	5000–10,000
4.	Severaly saline waters	10,000–30,000
5.	Sea water	Above 30,000

Methods of Desalination

Common methods used for the conversion of salt water into fresh water are as under.

1. Evaporation and Distillation Method:

Water obtained by this process is the purest form of water produced on large scale. In modern distillation plants (also called as **evaporators**), water is boiled, and the water vapour is caught and condensed into fresh water. In its simplest form, an evaporator consists of a metal box in which the salt water (generally called as brine) is heated by a nest of pipes, carrying vary hot steam. Heat passes from the steam through the pipe walls and boils the brine. The boiling brine evaporates and the vapour is led into a second box, where another nest of pipes filled with cold brine, condenses it to fresh and pure water. The heat coming out from the vapour during cooling operation, warms the cold brine, which is then sent to the evaporator.

Since the process of converting sea water into drinkable water is costly, efforts are being used to reduce its cost. In view of this, the same the heat is being used again and again. Further, the ways are being found to boil water at lower temperature, so as to reduce the cost of fuel. The water can be boiled at low temperatures, by reducing the pressure.

2. Desalination by Electrodialysis Method

We know that in a salt-water, H_2O molecules are bonded together with the sodium and chlorine ions. These hydrogen–bonds between the H_2O molecules and Na^+ and Cl^- ions must be broken up, in order to separate the salt from water. These bonds were broken by heat in the "method of distillation", while in the "electrodialysis method", these bonds are broken with the help of electricity.

If an electric current is passed through the salt-solution, the sodium and chlorine ion get freed from water molecules, and they start moving towards their appositely charged electric poles. In other words, the + vely charged sodium ions will move towards the – ve pole, *i.e.* cathode, and the –vely charged chlorine ions will move to towards the + ve pole, i.e anode. If these cations (+ ve ions) and anions (–ve ions) are allowed to segregate in different compartment, what is left, is fresh–water.

The segregation is achieved by means of thin plastic sheets called 'membranes'. They are made of peculiar chemical substances called 'ions exchange resins' and they are very very thin, say about 180th part of a cm.

Advantages of Electrodialysis Method.

1. When two much purification is not required, this method is the best of all the available methods.
2. The cost of buying and erecting the plant is small.
3. It is easy to operate.
4. This is an ideal method for purifying water in a small town or at a remote place. This method removes all merols from water.
5. These plants can be assembled for different outputs just by changing the number of units added together.

3. Desalination by Freezing Process

This method is based upon the principle that when salt water freezes, the ice formed in the beginning is almost free from salt. This ice, when melted, can give us good quality water, but the cost of production is high and prohibitive.

4. Desalination by solar Distillation Method

Solar heat is a free heat, and everyday the sun vaporises billions of tonnes of sea water and lifts the vapour into clouds. If this vapour can be caught and distilled to form pure water. This method gives cheaper water.

The solar energy has been used in vapourising the water of a 'huge still' – a pond or shallow basin covered with glass or plastic covers. It is a costlier process and the investigations are still going on, in order to make it cheaper.

7.11. REMOVAL OF IRON AND MANGANESE FROM WATER

Iron and manganese salts are generally found dissolved together in well water or anaerobic reservoir waters, in invisible dissolved state. When exposed to air, these slowly transform to soluble visible oxidized ferriciron and manganic manganese. When their contents exceed about 0.3 mg/L, and 0.05 mg/L respectively, they become objectionable due to the following reasons:

(*i*) They cause discolouration of clothes washed in such waters due to deposition of red/ brown coloured oxides of iron/ manganese;

(*ii*) They cause incrustation of the water-mains due to deposition of ferric hydroxide and manganese oxide;

(*iii*) They make the water unpleasant in taste.

The iron and manganese may be present in water either in combination with organic matter or without such combination. When present without combination with organic matter, they can be easily removed by aeration, followed by coagulation, sedimentation and filtration. During aeration, the soluble ferrous and manganous compounds present in the water may get oxidised into insoluble ferric and manganic compounds, which can be sedimented out easily.

On the other hand, when iron and manganese are present in combination with organic matter, it becomes difficult to break the bond between them and to cause their removal. However, when once this bond is broken, they can be removed as above. This bond may be removed either by adding lime, and thereby increasing the pH value of water to about 8.5 to 9; or by adding chlorine or potassium permanganate.

Manganese zeolite, a natural green sand, coated with manganese dioxide, can also be used for removing soluble iron and manganese from solution. After the zeolite becomes saturated with metal ions, it can be regenerated by back washing with potassium permanganate.

7.12. ADDITION AND REMOVAL OF FLUORIDES

Fluorides in drinking water, must neither be totally absent nor should exceed an optimum value of about 1 mg/L. To ensure this, fluorides are added to waters found deficient in fluoride concentrations, under a process known as Fluoridation. When, however, the fluoride concentration in a given water exceeds the limiting value of 1 to 1.5 mg/L, the fluorides are removed from water under a process known as Defluoridation.

1. Fluoridation

It has been believed in the past that when fluoride concentration in water supply is less than an optimum value of about 1 mg/L, the water proves harmful, as it may lead to formation of weeker tooth enamel leading to early tooth decay. It had also been widely suggested that fluoride also proves beneficial to older people in reducing hardening of the arteries; and as fluoride stimulates bone formation, it is helpful in the treatment of osteoporosis. In public water supplies, the presence of fluorides must be ensured by the addition of extra outside dose of fluoride, wherever the water is found to be deficient in fluoride concentration.

The compounds which may be used for adding fluoride to the water are: sodium fluoride (NaF), sodium silico fluoride (Na_2 Si F_6) etc. Sodium fluoride is most widely used and is generally fed under pressure, as a solution, through a chemical feed equipment,

2. Defluoridation

The dose of fluoride should be carefully worked out because the presence of excess fluoride (above 1 to 1.5 mg/L) is also harmful, as to cause spotting and discolouration. The long term exposure may result in permanent grey to black discolouration of the enamel (dental fluorosis). Still larger and continuous use of excess fluoride (30 to 50 mg/L) may lead to gastroenteritis, skin irritations, deformation of bones and other skeletal abnormalities.

Since the ground waters in countries like India and Bangladesh are found to contain large quantities of fluoride, it becomes imperative to remove fluoride from water, wherever the waters are found to be polluted with excess fluoride, so as to make the water safe for drinking. The technique of removal of fluoride from water is known as **defluoridation**.

Following methods are used for removing fluorides from water:

1. Absorption by Activated Alumina (AA).
2. Iion exchange Adsoption method.
3. Nalgonda Technique.

7.13. REMOVAL OF RADIO ACTIVITY FROM WATER

Dangerous radioactive substances, such as uranium, thorium, stronium, caesium, etc. from the wastes of the atomic reactors or from the atomic explosions, may sometimes pollute the water sources. Such pollutions may occur as more and more use of atomic energy is made in industry, research or medicines. The radioactive pollutions of water sources must be controlled by properly controlling their waste discharge. A lot of research is being undertaken for safely discharging these wastes, and thus, to prevent any major health hazard.

The ***lime soda softening method,*** however, is quite effective in bringing down the radioactivity present in water. With relatively high doses of lime soda, such as 200 mg/l or so, upto 99 per cent of radioactivity can be removed. Similarly, the ***zeolite softeners*** are also very effective in removing radioactivity from water, especially if vermiculite zeolite (which is a complex silicate of magnesium, iron and aluminium) is used.

In fact, the removal of radioactivity is generally not needed for treating ordinarily available supplies, and therefore, the subject remains more or less a topic of research.

7.14. DOMESTIC METHODS OF TREATING WATER

In our homes or offices, we generally get treated municipal water supplies, which usually are safe and disinfected. However, the stored supplies in overhead or underground tanks are liable to get recontaminated and become unsafe. Even the direct piped supplies may often contain dirt and mud, making the water unhealthy and unaesthetic, besides being contaminated on some occasions due to faulty distribution system. In order to obtain better quality safe drinking water, many households are now-a-days using their own water treatment devices, like filters and ultraviolet treatment unit like Aquaguards.

Boiling of water is also sometimes adopted in houses, which needs proper boiling and is thus expensive and time consuming. Such boiled waters are also liable to recontamination during subsequent cooling, storage and handling.

Sometimes, **chlorine tablets** are added in domestic storage tanks to safeguard from the storage contamination. Apart from giving peculiar taste and odour, chlorine is a toxic chemical if consumed in consumed in excess. Hence, the use of chlorine tablets in our houses do require utmost care and caution.

The domestic filters and Aquaguard units are more common and need more elaborate description, and hence are discussed hereunder:

1. **Domestic Filter Containers.** Domestic filters are usually made of stainless steel in the form of two circular containers placed one over the above. The upper container's base fits in the rim of the lower container. The upper container is fitted with vertically standing porous ceramic filter candles, usually two to three in number, and is covered by a lid from the top. The water from the storage tank is filled in the upper container, which slowly gets filtered through the filter candles, and falls down in the lower container, which usually is provided with a water tap (bib cock) at its bottom to take out the filtered water. Such filters, however, take a long time in filtering water. The candles also often get clogged and need to be frequently cleaned (once a weak or so). If not cleaned regularly, the filter candles may serve as the breeding ground for more bacteria.

2. **On-Line Filters.** A variety of domestic filters fixed directly on a water tap (bib cock) are now-a-days available in the market, claiming zero bacteria water at the outlet. These filters (Fig. 7.8.) use granulated filter media, activated carbon, and crystals of silver iodide. The filter media traps the suspended particles, while the activated carbon removes colour, odour and gases from the water. The silver iodide finally causes disinfection, as it is a very effective oxidizing agent and is available in the form of crystals. Silver iodide therefore causes effective disinfection when water passes over these crystals. Silver iodide of course is costly, and may make such filters to be somewhat costlier.

   The filter media provided in such filters may sometimes be in the form of a cartridge having a limited life, and hence needs regular replacement. But it is difficult for the user to know as to when the filter may cease to function efficiently and continued use of water from such filters may add to contamination.

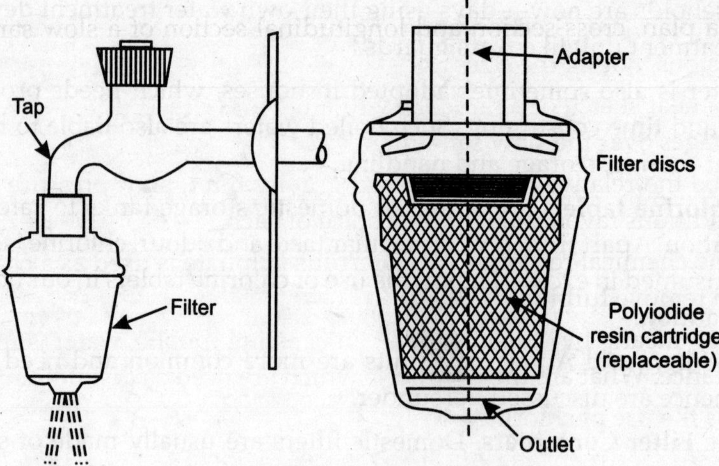

Fig. 7.8. Tap filter.

3. **Domestic Ultra-Violet Purification Systems (like Aquaguards).** When exposed to UV rays, most living cells (bacteria and viruses) either die or loss their capability to reproduce in the same water. However, ultra violet sterilization does not have a residual effect like that of chlorine, and is therefore considered to be an effective disinfectant when installed immediately before the point of use. The water should also be absolutely clear and free from suspended matter, as otherwise the penetration of U.V. rays in water gets affected. That is why, in a U.V. purification system, the water is initially passed through a filter candle, then through a bed of activated carbon (for removing dissolved gases and foul odours), an the resultant clear and sparkling water is finally subjected to ultraviolet rays treatment in a U.V. sterilizer. The rate of flow of water is controlled through the U-tube of the sterilizer (when it is exposed to U.V. radiation) to ensure complete disinfection. To prevent untreated water from passing through the system, an electrically operated automatic shut off valve is provided. Such a system or a unit or a machine requires water under good pressure, such as a supply from an overhead tank; or water lifted by a pump from the underground tank etc. It also requires electricity and regular maintenance, besides being a costly unit.

QUESTIONS

1. Enumerate and discuss briefly the various methods which are adopted collectively for treating public water supplies drawn from a perennial river.
2. Explain the sedimentation process used in a water treatment plant. Draw a neat sketch of a sedimentation tank in which coagulant is used.
3. Describe various constituents of a coagulation sedimentation plant.
4. Enumerate the chemicals which are used for coagulation. Discuss their comparative merits and demerits.
5. Discuss and distinguish between slow sand, and rapid sand filters.

6. Draw a plan, cross-section and longitudinal section of a slow sand filter bed, with dimensions, noting the sizes and depths of filtering materials used.
7. Draw a neat sketch of a rapid gravity filter, and describe how it works. What are its advantages over the slow sand filter.
8. Describe the relative merits of rapid gravity filter; and 'pressure filters', indicating the conditions favourable for the choice of each.
9. Give the chemical reactions when ferrous sulphate is used as a coagulant alongwith lime to remove turbidity.
10. What is meant by "disinfection" in treating public water supply? What is its importance. What are the chemicals which are used as disinfectants.
11. Discuss the use of chlorine a disinfecting agent.
12. Differential between temporary and permanent hardness. Mention any three methods of softening water.
13. Discuss various methods used for treating public supplies of water for removing colour and taste from it.
14. Write short notes on:
    (*i*) Chlorination.
    (*ii*) Water aeration.
    (*iii*) Fluoridation.
    (*iv*) Zeolite method of water hardness removal.
    (*v*) Desalination.
15. Explain the following in brief:
    (*a*) Removal of radioactivity from water.
    (*b*) Chemical coagulation.
    (*c*) Temporary hardness
    (*d*) Permanent hardness.
16. Discuss the following in brief.
    (*a*) Aeration
    (*b*) Screening.
    (*c*) Filtration.
    (*d*) Disinfection.
    (*e*) Activated carbon treatment.
17. List out various impurities associated with water quality resulting in water pollution.
18. Enumerate the common pollutants affecting the quality of water supply. Describe their effects on human body.
19. Draw up specifications regarding standards of purity required for public water supplies.
20. What objections would be there if (*i*) turbidity, (*ii*) chloride, (*iii*) nitrates, and (*iv*) hardness are present in excess in water? State the permissible limit of these in drinking water.
21. Explain the significance of each of the following in water purification:
    (*i*) Turbidity
    (*ii*) Fluorides
    (*iii*) Iron
    (*iv*) Ammonia nitrates

22. Discuss the construction and working,, of "Domestic ultraviolet purification system".
23. Discuss briefly the functions of each unit of treatment plant.
24. Draw a layout plan of the purification works.
25. Draw the flow sheet showing the different treatment units of water purification plants.
26. How is the site of water treatment plant selected.
27. Write a detailed note on the design and working of a continuous flow type sedimentation tank.
28. Write short notes on:
    (*a*) Coagulants,
    (*b*) Alum as coagulant,
    (*c*) Effect of pH on coagulation.
    (*d*) Coagulation and sedimentation.
29. Why softening of water is required? Discuss the procedure of lime soda princess for water softening.
30. What is meant by hardness of water? How is it expressed? Describe the method of softening by Zeolite process.
31. Write briefly on carbonate and non-carbonate hardness in water and their removal.
32. Write short notes on:
    (*i*) Zeolite water softening. (*ii*) Softening.
    (*iii*) Hardness of water. (*iv*) Taste and odour control.

Water Distribution System

8.1. INTRODUCTION

After the treatment of water, it has to be supplied to the consumers in their individual premises. The water is taken from the treatment plant to the roads and street in the city, and finally to the individual houses. This function of carrying the water from the treatment plant to the individual homes is accomplished through a well planned distribution system.

A distribution system consists of pipelines of various sizes for carrying the water to the streets; valves for controlling the flow in the pipes, hydrants for providing connections with the water mains for releasing water during fires; meters for measuring discharges; service connections (called services) to the individual homes; pumps for lifting and forcing the water into the distribution pipes, distribution or service reservoirs for storing the treated water to be fed into the distribution pipes; etc.

The water is supplied to the public either continuously for all the 24 hours of the day, or it may be supplied intermittently during certain fixed hours of the day. Also, the water may either be pumped directly into the distribution pipes, or it may first be stored in a overhead distribution reservoir and then fed into the distribution pipes by gravity. The water should be supplied to consumer with adequate pressure and in required quantity.

8.2. DESIGN (REQUIREMENTS) OF DISTRIBUTION SYSTEMS

The basic requirements for proper functioning of a distribution system are:

(*i*) It should supply water to all the intended consumers within the city with a reasonably sufficient pressure head.

(*ii*) It should supply the requisite amount of water, including that for fire fighting during such needs.

(*iii*) It should be cheap with the least capital construction cost, because the distribution system is the most costly item in the entire water supply scheme.

(*iv*) It should be simple and easy to operate and repair, thereby keeping the minimum O&M cost

(*v*) It should be safe against likely future pollution problems of water. This can be achieved by keeping the water pipelines above and away from the sewerage and drainage lines and also by improving the general sanitary conditions of the area through which the distribution pipes have to pass.

(*vi*) It should be able to keep the "losses due to leakage" to the minimum.

8.3. DISTRIBUTION PIPES AND ACCESSORIES

The distribution pipe system consists of supply mains, submains, branches, and laterals; usually made of cast iron and jointed by means of "spigot and socket joints". The service connections are made of galvanised cast iron pipes. These water mains and submains are usually laid sloping from the high level to the low areas. The distribution system serving consumers consists of:

(*a*) **The primary feeders (also known as arterial feeders).** These carry large quantity of water to and from the storage tanks and to the various parts of area served. Large and long feeders should have drain valves (blow offs) at the low point and air relief valves at high points.

(*b*) **Secondary feeders.** These carry large quantities of water from the primary feeders to the various areas to take care for normal supply and fire fighting. These form small loops.

(*c*) **Distribution mains.** These form a grid over the area to be served and supply water to the fire and service pipes of the residences and other buildings. The velocity in the pipe will depend upon the discharge and the size of the pipe.

(*d*) **Valves.** The importance of the valves is essential from the point of view of design of new systems and extension as well as in maintenance. A properly located valve is a necessity for proper operation and control of the pipe system.

Sluice valves are placed along the length of the pipes at suitable intervals, and also at all the junctions and branching off points, so as to control the flow of water into the different sections.

Drain valves are placed at all the low points in the distribution pipe system, so as to drain off the water from the pipes for carrying out any repairs, etc.

Air valves are placed at all the high points, so as to remove air from the pipe during the filling operations, and also to admit air while emptying the pipe. The description of all these valves have been given earlier in chapter 5.

In smaller cities the distribution mains generally take off from the treatment plant itself, and distribute the water into the different branches and laterals. However, large cities are generally divided into various zones, and each zone is served with a separate distribution system.

The distribution pipes are generally laid on one side of the roads or streets, usually below the footpaths, and at about 2 m above (*i.e.* vertically) and 3 m away (*i.e.* horizontally) from the sewers, so as to avoid any future contamination of water through the leaky joints. It is desirable that the sewer lines run on one side of the streets, and the water pipes run on the other side of the streets.

In wider streets, however, a **subsidiary lateral** may also be provided on the other side of the street, which is cross-connected to the main lateral at suitable intervals. This will avoid the otherwise needed long house connections. The sizes of the distribution pipes mainly depend upon the amount of flow to be carried, and the permissible loss in the pressure head.

System Accessories

These include the following:

1. *Booster stations.* These are used to increase water pressure from storage tanks or low-pressure mains.
2. *Valves.* These control the flow of water is the distribution system by isolating areas for repair or by regulating system flow or pressure.
3. *Hydrants.* These are designed to allow water from the distribution system to be used for fire-fighting purposes.
4. *Meters.* These record the flow of water in a part of the distribution system.
5. *Service connections.* These are used to connect individual buildings.
6. *Air valves.* are provided at all peaks, and wash out valves are provided at low points between peaks.

8.4. LAYOUT OF DISTRIBUTION NETWORKS/TYPES

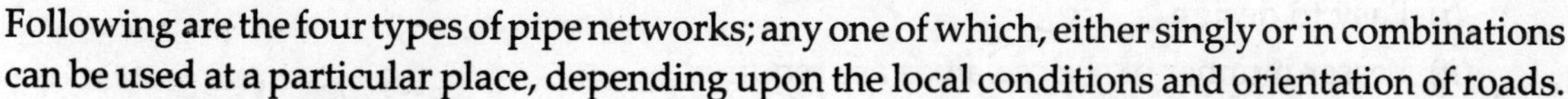

Following are the four types of pipe networks; any one of which, either singly or in combinations can be used at a particular place, depending upon the local conditions and orientation of roads.

1. Dean-end system;
2. Grid iron system;
3. Ring system; and
4. Radial system.

1. Dead-end System.

In the dead-end system, (also called ***Tree system***). there is one main supply pipe (starting from service reservoir), from which originates (generally at right angles) a number of submain pipes. Each submain, then divides into several branch pipes. Called laterals. From the laterals, service connections are given to the consumers. A typical plan of such a network is shown in Fig. 8.1.

This type of layout is adopted for old towns which have developed in a haphazard manner without properly planned roads. The water supply mains are taken along the main roads, and branches taken off wherever needed, thus resulting in the formation of a number of dead ends as shown.

The *advantages* of this system are:

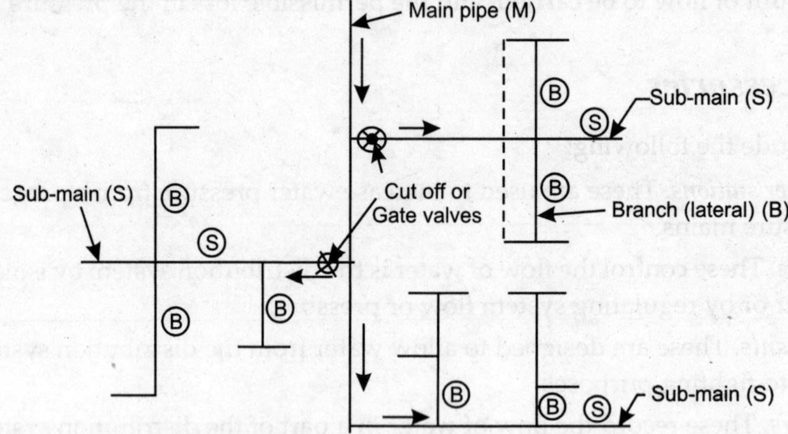

M – Main pipe.

S – Sub-main pipe.

B – branch (lateral)

⊗ – Cut off or gate valve

Fig. 8.1. Dead-end or Tree system.

(*i*) Easy to design.

(*ii*) Lesser number of sluice valves are required.

(*iii*) Shorter pipe lengths are needed, and the laying of pipes is easier.

(*iv*) It is cheap and simple, and can be extended or expanded easily.

The *disadvantages* of this system are:

(*i*) In this method, water can reach a particular point only through one route therefore, any damage or repair in any pipe line will completely stop the water supply in the area being fed by that pipe.

(*ii*) There are numerous dead ends in this system, which prevent the free circulation of water. This stagnation of water may lead to the degradation in its quality.

(*iii*) During fire fighting, supply cannot be increased by diverting any other supplies from any other side.

2. Grid-iron System

In this system, the mains, sub-mains and branches are all inter-connected with each other, as shown in Fig. 8.2. In a well planned city, the roads are generally developed in a grid-iron pattern, and the pipe lines in such places can follow them easily. Hence, this system is more suitable for well planned towns and cities.

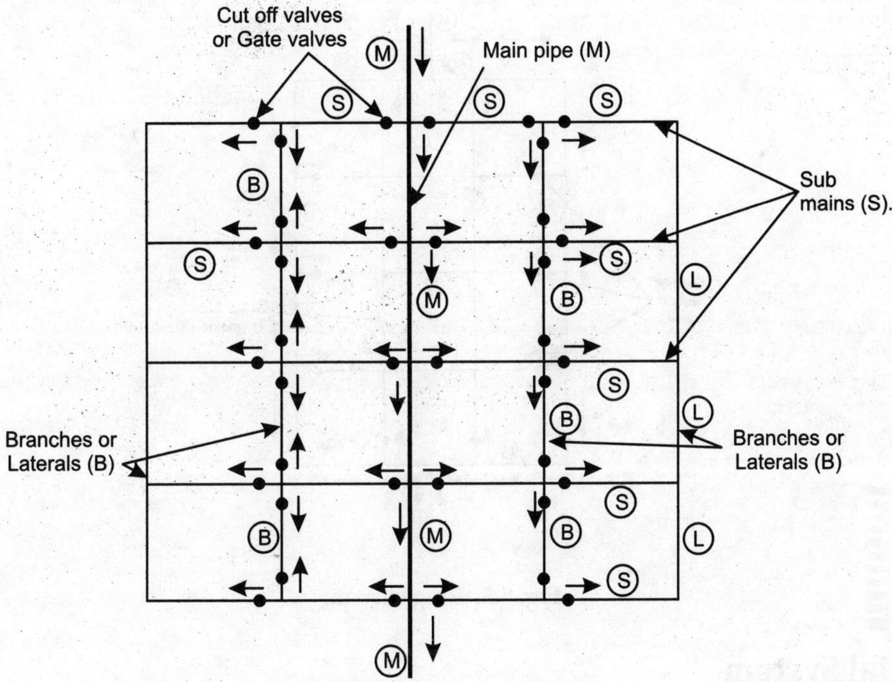

Fig. 8.2. Grid-iron system.

The **advantages** of this system are:

(*i*) Since the water reaches at different places through more than one route, the discharge to be carried by each pipe, and the size of the pipe, therefore, get reduced.

(*ii*) In case of repairs, very small area will be devoid of complete supply.

(*iii*) Because of the different inter-connections, the dead ends are completely eliminated, and hence not liable to pollution due to stagnation.

(*iv*) During fire, more water can be diverted towards the affected point from various directions.

The **disadvantages** of this system are:

(*i*) This system requires more length of pipe lines, and a larger number of sluice valves.

(*ii*) Its construction is costlier.

(*iii*) The design is difficult and costlier, the calculations for determining accurately the sizes of he pipes and the pressures at various key points, is a real tedious job.

3. Ring System

This system (also known as **Circular system)** is a closed ring, either circular or rectangular, of the main pipes, formed around the area to be served, as shown in Fig. 8.3. The distribution area is divided into rectangular or circular blocks, and the main water pipes are laid on the periphery of these blocks. The sub-mains may be placed as shown. The ring system is very suitable for towns and cities having well planned roads.

The advantages and disadvantages of this system are the same as that of the grid-iron system.

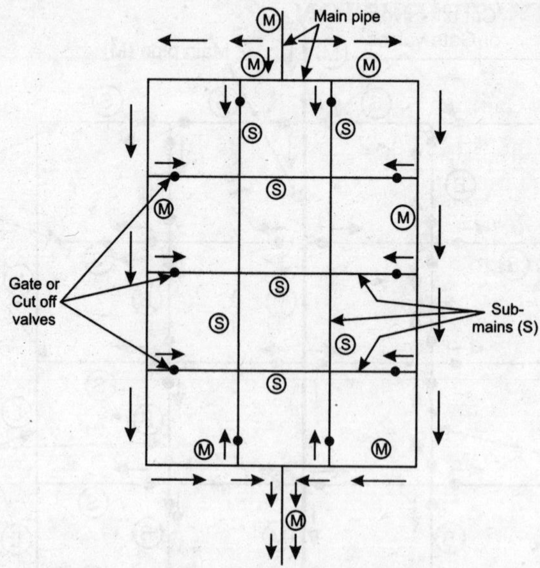

Fig. 8.3. Ring system.

4. Radial System

If a city or a town is having a system of radial roads emerging from different centres, the pipe lines can be best laid in a radial method by placing the distribution reservoirs at these centres. In this system, water is taken from the water mains, and pumped into the distribution reservoirs placed at different centres, as shown in Fig. 8.4. The water is then supplied through radially laid distribution pipes **towards** the outer periphery from inside. This method ensures high pressures and efficient water distribution.

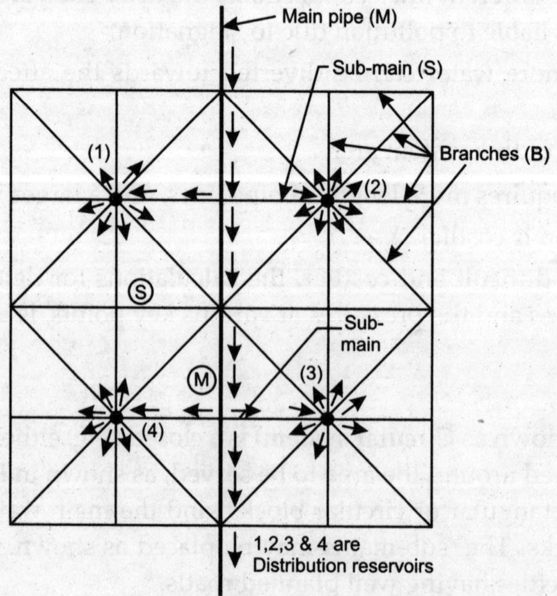

Fig. 8.4. Radial system.

8.5. METHODS OF DISTRIBUTION

The distribution system is required to develop adequate water pressures at various points of the consumer's taps. Depending upon the level of the source of water and that of the city, topography of the area, and other local conditions and considerations, the water may be supplied by following three methods:

1. *By gravitational system;*
2. *By pumping system;* and
3. *By combined gravity and pumping system.*

These systems are described below:

1. Gravitational System

In this system, the water from the high level source is distributed to the consumers at lower levels, due to gravity. For proper functioning of this system, the difference of head available between the distribution reservoir at the treatment plant and the localities, must be sufficient enough, so as to maintain adequate pressure at the consumer's door-steps, after allowing the frictional and other losses in the pipes. This method is the most economical and reliable, since no pumping is involved at any stage. However, it needs a lake or a reservoir as a source of supply.

Such a system can be adopted for cities which are situated at the foot-hills and the source of supply is available in the hills at sufficient elevation in relation to the elevation of the city. Fig. 8.5 shows the gravitational system with hydraulic gradients during maximum and minimum demands.

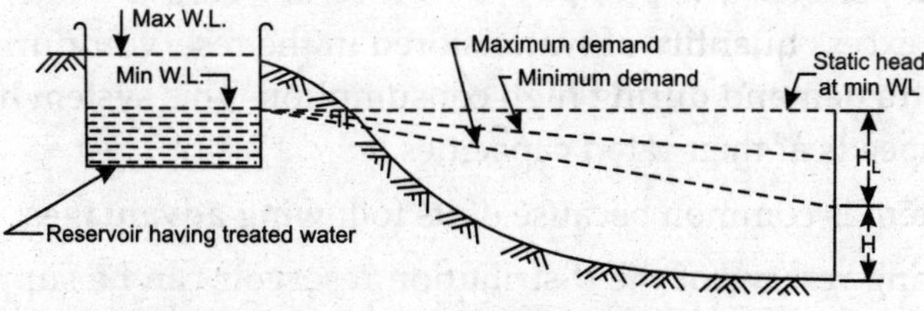

Here, HL = Head loss

H = Head available to consumers.

Fig. 8.5. Gravitational distribution system.

In a case, where the source, such as a lake is situated at a hill, and the treatment plant is also situated almost at the same level on the hill itself, the treated water will be distributed to the low levelled consumers situated at the foothill by mere gravity.

2. Direct Pumping System

In this pumping system, the treated water is directly pumped into the distribution mains without storing it anywhere. High lift pumps are required in this system, which have to operate at variable speeds, so as to meet the variable demand of water. Thus, a continuous attendance is needed at the pumping station so as to ensure the desired flow in the distribution system. In this system, if the power supply fails, there will be complete stoppage of water supply.

Fig. 8.6 shows the pumping system with hydraulic gradients during maximum or minimum demands.

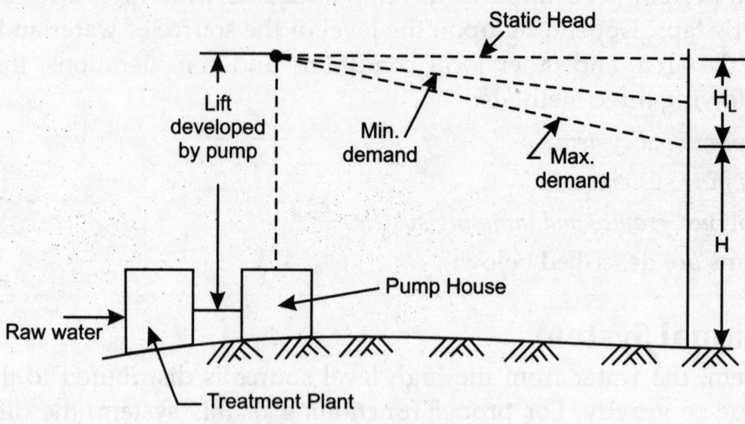

Fig. 8.6. Pumping system.

3. Combined Gravity and Pumping System.

In this system, the treated water is pumped at a constant rate and stored into an elevated distribution reservoir (Overhead tank), from where it is distributed to consumers by the action of gravity. This method thus, combines pumping as well as gravity flow, and is also called as ***pumping with storage system.***

The excess water during low demand periods gets stored in the reservoir and supplied during high demand periods. The pumps are worked at a constant rate, which is adjusted in such a way that the excess quantity of water stored in the reservoir during low consumption nearly equals the extra demand during high consumption. This system helps in operating the pumps at constant speeds at their rated capacities.

This type of system is common because of its following **advantages**:

(*i*) The balancing reserve of the distribution reservoir can be supplied to the places of fire. The necessary pressure for fire fighting can be achieved by closing down the supply of some localities or by using motor pumps.

(*ii*) The pumps are to be worked at uniform rate and, thereby operating them to their rated capacities. In smaller towns, this method of enables us to operate the pumps only for a part of the day.

(*iii*) This method is quite reliable because even during the power failure or pumps failure, certain amount of water can be supplied from the storage.

(*iv*) This system proves to be cheap, efficient and reliable, and hence commonly used everywhere.

8.6. PRESSURES IN THE DISTRIBUTION SYSTEM

From the time the water enters the distribution mains, and upto the time it comes out at a consumer's tap, water-head is constantly lost in overcoming friction of the pipes (H_L), and also in the fittings, meters, valves, etc. The net available heat at a consumer's door step is thus reduced by these losses. This net available head or effective head (*H*) is the most important pressure, because it will make the water rise upto the different storey heights. It is necessary that sufficient pressures are maintained in the distribution system, so as to supply water to the consumers upto the desired heights.

The pressures should be maintained as high as possible, so as to give higher flow and also to serve the upper storey consumers. But from economic considerations, it is not possible to permit very high pressures; because,

(*i*) This will necessitate the use of higher lift pumps and

(*ii*) Thicker and stronger pipes, thereby increasing the operational as well as the initial cost of installation,

(*iii*) It will increase the water lost due to leakage and waste. Therefore, pressure in the distribution system should be so designed that the entire system is economically optimum.

8.7. SYSTEMS OF SUPPLY

The water may be supplied either:

(*i*) Continuously for all the 24 hours of the day, or

(*ii*) Intermittently only for the peak periods during morning and evening. This system may lead to some saving in water consumption due to losses occurring for lesser time and more vigilant use of water by consumers. This method may, therefore, be adopted at places where there is a shortage of water at the source.

The limitations or disadvantages of the intermittent system are:

1. Intermittent system of supplies do not cover fire risks during non-supply periods.
2. This system causes great inconvenience to consumers for receiving and collecting water as soon as the supply is restored.
3. Since the supply of water for the whole day has to be pumped during 6 to 8 hours only, it will necessitate the use of bigger sized distribution mains.
4. When the supply of water is stopped and the water from the pipe is drawn off, a partial vacuum may be created in the pipe. This will contaminate the initial supplies as and when restored.
5. A number of air valves and sluice valves are required to be fitted and operated daily while opening or closing the supply. This will necessitate additional staff and high maintenance cost.

8.8. DISTRIBUTION (OR SERVICE) RESERVOIRS

Distribution reservoirs are the storage reservoirs, which store the treated water for supplying water during emergencies (such as during fires, break-downs, repairs, etc.) and also to help in absorbing the hourly fluctuations in the normal water demand.

The main functions served by these reservoirs are:

1. They absorb the hourly variations in demand, and allow the water treatment units and pumps to operate at a constant rate. This will reduce the operation and maintenance costs and improve the efficiency.
2. These help in maintaining constant pressure in the distribution mains.
3. The pumping of water in shift is made possible by them without affecting the supply.
4. The water stored in these reservoirs can be supplied during emergencies, such as break-down of pumps, heavy fire demand, etc.
5. They lead to an overall economy by reducing the sizes of pumps, pipe lines, and treatment units.

8.9. TYPES OF DISTRIBUTION RESERVOIRS

The distribution reservoirs may be made of steel, R.C.C., or masonry. Depending upon their elevation with respect to the ground. These are of following two types:

1. Surface reservoirs; and
2. Elevated reservoirs.

8.9.1. Surface Reservoirs

Surface reservoirs are circular or rectangular tanks, constructed at ground level or below the ground level (ground reservoirs.) They are generally constructed at higher levels in the city. If a city has more than one high point, more than one reservoir may be provided, by dividing the city into several zones.

In a gravitational type of distribution system, water is stored in the ground service reservoir, and then directly sent from there into the distribution system. However, in a combined gravity and pumping system of distribution, the treated water is first of all stored in ground reservoir, and then pumped into an elevated service reservoir, from where it can be supplied into the distribution mains. This has already been discussed earlier.

8.9.2. Elevated Reservoirs

Elevated reservoirs (overhead water tanks) are the rectangular, circular, or elliptical overhead tanks erected at a certain suitable elevation above the ground level and supported on towers. They are constructed where the pressure requirements necessitate considerable elevation above the ground surface. They are constructed in areas where the combined gravity and pumping system for water distribution is adopted. Water is pumped into these elevated tanks from the filter units or from the surface reservoirs, and then supplied to the consumers.

These overhead tanks are made of steel, R.C.C, or prestressed concrete. However, because of strength RCC or prestressed concrete tanks are preferred. Since these elevated reservoirs are costly, they are designed for the capacities of 6 to 8 hours of the average daily supply of area. These are generally covered at the top by roof slabs so as to avoid contamination of water due to dirt, birds, insects etc. The roof should have manholes/covers for cleaning, and provided with ventilators so as to allow free circulation of air.

The commonly used accessories provided with these tanks are:

1. Inlet pipe for entry of water.
2. Outlet pipe connected to the distribution mains.
3. Overflow pipe discharging into drain gutters and maintaining a constant level.
4. A float gauge or an indicator for indicating the depth of water can be read from outside.
5. A wash - out pipe (or drain pipe) for removing water after cleaning of the reservoir.
6. Automatic devices to stop pumping when the tank is full.
7. Ladders to reach the top of the reservoir and then upto the bottom of the reservoir for inspection
8. Manholes for providing entry into the tank for inspection purposes.
9. Ventilator for fresh air circulation.

8.10. CAPACITY OF DISTRIBUTION RESERVOIR

The total storage capacity of a distribution reservoir is the summation of:

1. Balancing storage (or operating storage);
2. Breakdown storage; and
3. Fire storage.

(1) Balancing Storage or Equalising Storage. The main and primary function of a distribution reservoir is to meet the fluctuating demand with a constant rate of supply from the treatment plant. The quantity of water required to be stored in the reservoir for balancing this variable demand against the constant supply is known as the balancing reserve or the storage capacity of a balancing reservoir.

(2) Breakdown Emergency Storage. The breakdown storage or often called emergency storage is the storage preserved in order to tide over the emergencies posed by the failure of pumps, the electricity or any other mechanism driving the pumps. The amount of provision to be made for this factor is very difficult to assess, therefore, a lump sum provision of about 25 per cent is made for this purpose.

(3) Fire Storage. This provision takes care of the requirement of water for extinguishing fires. In case of fires, sufficient amount of water must remain available in the reservoir for throwing it over the fire, in case the fire coincides with pumps or electricity failure at the waterworks.

Therefore, the total reservoir storage can finally be obtained by adding all the three storages, *viz.*, balancing storage, emergency storage, and fire storage.

8.11. WASTAGE OF WATER IN THE DISTRIBUTION SYSTEM

Considerable amount of water gets lost and wasted in the distribution system, either through leakage or through unauthorised water connections. Even in the best managed distribution system, these losses may amount to about 15% of the total consumption; while in a faulty distribution system, they may be as high as 40% of the total consumption. This wastage of water will increase the per unit cost of the treated water. Hence, for the proper management of the distribution system, it is necessary to detect the places where water is being wasted, and to take immediate steps in order to stop that wastage.

The factors which are responsible for the wastage of the water are:

(1) Watertight joints.

(2) Pressure in the distribution system.

(3) System of supply *i.e.* intermittent supply or continuous supply.

(4) Metering system.

(5) Unathorised connections.

8.11.1. Detection of Leakage in the Distribution Pipes

The following methods are generally used for detecting the leakage of water from the underground water mains:

(*i*) By direct observations

(*ii*) By using sounding rods

(*iii*) By plotting hydraulic gradient line; and

(*iv*) By using waste detecting meters.

These methods are discussed below:

1. **By Direct Observations.** This method consists of observing a wet soft spot on the unpaved ground, or the emergence of a spring at an odd place. All such hints will indicate the presence of the underground water pipe leaking somewhere in the vicinity of the area.
2. **Ey Using Sounding Rods.** In this method, a sharp pointed metal rod is thrust into the ground along the pipe line and pulled up for inspection. Its moist or muddy point will indicate the presence of leakage. The sound of the escaping water can also be heard by placing the ear on the top of the inserted rod. This will confirm the leakage point. Sound produced by leaks can also be magnified by a stethoscope type instrument called an **aquaphone or a sonoscope**.
3. **By Plotting the Hydraulic Gradient line.** In this method, the pressures at various points along a suspected pipe line are measured, and the hydraulic gradient line is plotted. The appearance of any kink or change in the slope of the hydraulic gradient line will indicate the location of a leak in the pipeline.
4. **By using Waste Detection Meters.** These meters are widely used for detecting leakage in distribution pipes. This method is quite scientific and useful, but is costlier. This methods require to install additional valves and cutouts for inserting the meters in the pipelines.

8.12. CONDITIONS FOR PIPE NETWORKS

In any pipe network, the following two conditions must be satisfied:

(i) The algebraic sum of the pressure drop around a closed loop must be *zero, i.e.,* there can be no discontinuity in pressure.

(ii) The flow entering a junction must be equal to the flow leaving the same junction; *i.e.,* the law of continuity must be satisfied.

Based upon these two basic principles, the pipe networks are generally solved by the methods of successive approximation, because any direct analytical solution is not possible, as the same will involve various equations to be solved simultaneously and many of which are non-linear. The important procedures or methods used for such solutions are given below:

8.12.1. Hardy-Cross Method

The procedure suggested by Hardy and Cross requires that the flow in each pipe is assumed by the designer (in magnitude as well as in direction) in such a way that the principle of continuity is satisfied at each junction (*i.e.* the inflow at any junction becomes equal to the outflow at that junction). This method is also known as "quantity balance method",

A correction to these assumed flows is then computed successively for each pipe loop in the network, until the correction is reduced to an acceptable magnitude.

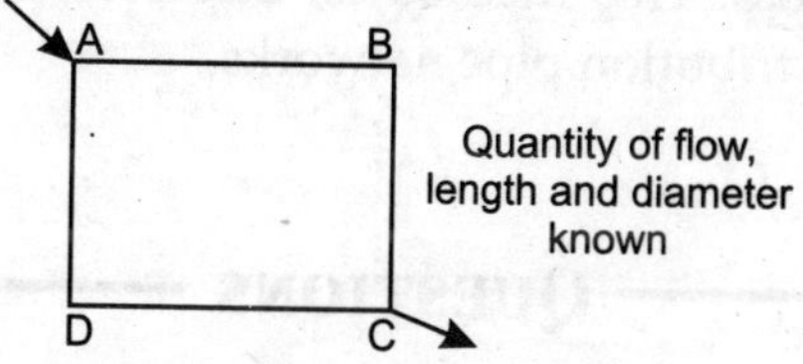

Fig. 8.7. Hardy-Cross method.

In this method, the head loss from one direction (ABC) is kept approximately equal to that from the other direction (ADC). Selection of pipe diameter and quantity of flow through such system is economical and efficient.

In a small loop ABCD (Shown in Fig. 8.7) discharges lengths of pipes along ABC and ADC are known. Pipe diameters are selected reasonably with experience. Loss of heads H_1 and H_2 from two routes ABC and ADC are computed. If they are not the same, some adjustment is to be done for either pipe diameter or flows. If the difference is small, correction to discharge is sufficient. Otherwise pipe diameters are to be revised to balance head losses from two directions.

The value of correction to the flow can be obtained using the formula;

$$q = \frac{H_1 - H_2}{n\left\{\dfrac{H_1}{Q_1} + \dfrac{H_2}{Q_2}\right\}}$$

where, $n = 1.85$

According to this formula, flow from the direction ABC may be increased by an amount equal to 'q' and the flow from other direction may be reduced accordingly. The loss of head is then again calculated in each direction. The process is continued till they are some.

8.12.2. Equivalent Pipe Method

Equivalent pipe is an assumed main to replace a system for equal heat loss under same quantity of flow. AC is an equivalent pipe to replace the system of pipes, *e.g.* ABC and ADC (refer Fig. 8.8).

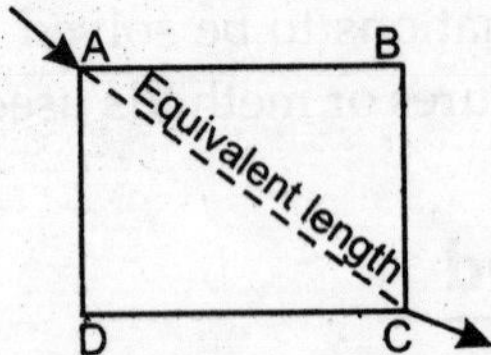

Fig. 8.8. Equivalent pipe

Head loss from A to C is calculated from two routes ABC and ADC. For this heat loss and same discharge, the equivalent length of AC main is calculated.

This method is used in solving large network for pipes, in which it is convenient to replace the different small loops by single **equivalent pipes** having the same discharging capacities and causing the same head loss. This method is, therefore, also called the **head balanced method** of analysing water distribution pipe networks.

QUESTIONS

1. Explain briefly the common methods of distribution of water employed in municipal water supply schemes.
2. Explain with the help of sketches the different types of layouts of pipe systems in distributing the water.
3. Compare the merits and demerits of the 'continuous' and intermittent' systems of water supply. Under what circumstances would you recommend the use of the latter?
4. How would you detect and prevent wastage of water in the distribution system.
5. Draw the systems of conveying the water supply from the source to the centre of gravitational systems or pumping systems or combined systems.
6. State the factors considered and procedure followed in designing a distribution system for the water supply of a city.
7. State the functions of a service Reservoir, and sketch the sectional elevation of the same, showing various appurtenances.
8. Write short notes on:
   (*a*) Distribution of water to consumers.
   (*b*) Courses of water wastes, its detection and prevention.

9. What factors would you consider in selecting a site for the location of a distribution reservoir?
10. Draw a neat sketch of an over-head distribution reservoir showing the various fitting and connections, and explain the function it can serve in public water supply scheme.
11. How will you estimate the quantity of water to be stored in the distribution reservoir.
12. Write short notes on the following with regards to water distribution system.
    (*i*) Hardy – Cross method.
    (*ii*) Different types of distribution networks.
    (*iii*) Fire hydrants.
    (*iv*) Metering in distribution systems.
13. Discuss the procedure commonly adopted for planning and designing of a distribution system in a water supply scheme. What precautions would you take to make the design economical?
14. Write a note on the advantages and disadvantages of various water distribution methods and layouts.
15. Explain the following in brief.
    (*a*) Accessories used in distribution pipe system.
    (*b*) Dead-end system, and radial systems of water distribution networks.
    (*c*) Gravitational system of water distribution.
    (*d*) Pressure in the distribution system.
    (*e*) Capacity of distribution reservoir.

Plumbing Systems in Buildings

9.1. INTRODUCTION

Modern buildings require certain amenities for comfort of the occupants, without which they are incomplete. These amenities (services) include the following:

1. Provision of water supply and their fittings .
2. Provision of sanitary fittings.
3. Provision of adequate drainage.

The buildings require essential services like, drinking, bathing, washing of clothes, washing of floors, flushing of water closets etc. In addition water is also required for fire protection, heating and air-conditioning. The water after use gets polluted and has to be drained off from the building. Hence an elaborate pipe system and storage have to be provided in the building and after use an elaborate drains have to be provided to drain off the contaminated water.

All the bathrooms, lavatory, wash basins and kitchen should be preferably grounds on one side of the building so as to supply water and their subsequent drainage.

Plumbing is a general term which indicates the practice, materials and fixtures used in the installation or maintenance of all pipings, fixtures, appliances, and other appurtenances used in connection with both public or private water supply systems as well as sanitary or drainage systems, within or adjacent to any building or its connection with any point of public disposal.

The entire system of pipings, fixtures, appliances, etc. used in providing water supply and drainage facilities in building is, therefore, called a **plumbing system.** Hence, a ***'plumbing water supply system'*** in a building consists of supply and distribution pipes (called *service pipes), connections, domestic storage tanks, flushing storage tanks,* etc., while a ***'plumbing sanitary or drainage system'*** includes water closets, urinals, and other such appliances like traps, soil waste and vent pipes and stacks, anti-siphonage pipes, house drains and house sewers, flushing cisterns, etc.

The plumbing systems broadly includes the following systems:

1. Entire system of **water supply and distribution pipes** including fitting and fixtures like valves, tanks etc., used in connection with water supply.

2. Entire system of **sanitary drainage** including fittings and fixtures like wash basins, sinks, water closets, urinals, traps, waste pipes, vent pipes, sewer, septic tanks etc.
3. Entire system of storm water including collection and carrying of rain (from roofs, paved areas and ground surface) to a public storm water drain or to a pond or river etc.

Service Connection. It is a connection given by local authority *i.e.*, Municipal Committee or Corporation from the city distribution system (normally from street main to different consumers). This consists or of two parts *viz.*, the ***communication pipe*** which runs from the street mains to boundary of consumer's premises and then ***service pipe*** which runs inside the premises. The responsibility of laying and maintaining the communication pipe if of local authority while the service portion is laid and maintained by consumer at his cost.

Storage Tanks in Buildings. Water supply to a building from city mains could be either continuous or intermittent. Generally, due to shortage of water, the local authority plan distribution of water in shifts (*i.e.*, morning or/and evening). Therefore provision of storage tank is made in building to ensure availability of water during non-supply hours. The minimum **storage capacity** of the tank is equal to 50 per cent of daily demand if the water supply is in two shifts, and it is equal to 100% if it is in one shift per day.

9.2. WATER CONNECTIONS IN BUILDINGS

In installing a 'water supply plumbing system' in a building, the first and main step is to obtain a 'water connection' from the municipal water main. A water connection, connecting the service pipe with the municipal water main, is shown in Fig. 9.1. The water connection consists of : (*i*) a ferrule; (*ii*) a goose neck; (*iii*) a service pipe; (*iv*) a stop cock; and (*v*) a water meter, as discussed hereunder.

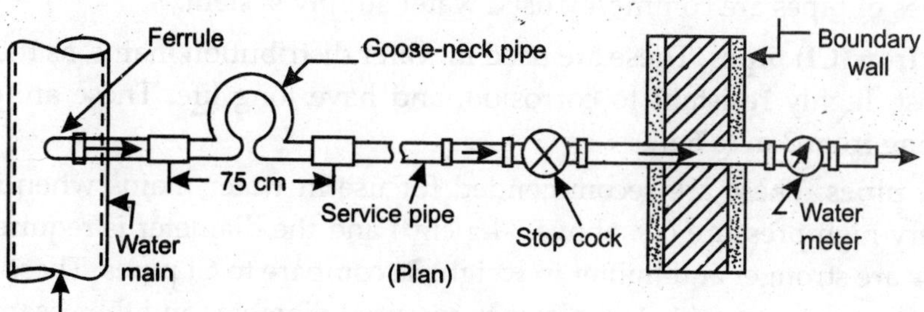

Fig. 9.1. The water connection.

1. **Ferrule.** A ferrule is a right angled sleeve made of brass or gun metal, and is joined to a hole drilled in the water main, to which it is screwed down with a plug. Its size usually varies between 10 to 50 mm dia.
2. **Goose Neck.** Goose neck is a small sized curved pipe made of a flexible material and is about 75 cm in length forming a flexible connection between the water main and the service pipe.
3. **Service Pipe.** Service pipe is a galvanised iron pipe which should be laid underground in a trench in which no sewer or drainage pipe is laid. The service pipe which supplies

water to the building from the municipal main and is to the main through the goose neck and ferrule.

4. **Step cock.** The stop cock is provided before the water enters the water meter in the house. It is boused in a suitable masonry chamber with a removable cover, and is fixed in the street close to the boundary wall in an accessible position. Sometimes, it is provided just before the water meter inside the house, keeping both of them in one chamber.
5. **Water Meter.** Water meter measures and records the quantity of water consumed in the house. The domestic type water meter generally employed for house is fitted into the service pipe with unions, which enables the meter to be changed where necessary. The water meter is generally fixed in an iron box fitted in an opening or cavity made in the boundary wall of the house, and is covered with a movable iron cover.

9.3. PIPES, PIPE FITTINGS AND ACCESSORIES

In water supply system of a building galvanized iron pipe is used. The size of he pipe depends upon various factors like, head of water, number of users etc. Generally, under normal conditions residential buildings are provided with 15 mm diameter pipe. The pipe line is taken to plinth level using bends and small length of vertical pipe on which necessary water taps in bathrooms, kitchen, water closets, wash basins and sinks etc are provided. For laying the pipeline system in a building various attachments are required. These attachments are mainly sockets, bends, elbows, Bib-cock or water tap, stop-cock, check nut, union etc.

9.3.1. Types of Pipes

Following types of pipes are commonly used water supply system:

1. **Cast Iron (CI) pipes.** These are used in water distribution mains, as they are cheaper in cost, highly resistant to corrosion, and have long life. These are cast in length varying from 3 m to 6 m.
2. **Steel pipes.** These are recommended for use in water mains when it is subjected to very high pressure (*i.e.* above 7 kg/cm^2) and the diameter is required to be large. These are stronger and lighter in weight as compare to CI pipes. These are fabricated by rolling flat wrought steel plates to required diameter and then seam welded.
3. **Galvanised Iron pipes.** Galvanised Iron (G.I) pipes are wrought steel pipes provided with zinc coating. These are most commonly used for water supply work inside the buildings. Screw and socket joints are used for G.I. pipe connections.
4. **PVC pipes.** These are rigid PVC pipe and are light in weight, tough, exceptionally resistant to the chemical attack and available in large lengths. These are immune to corrosion and has less chances of leakage. These also have good elastic properties. Smooth internal surface of the pipes offers less friction which results in saving of energy in the conveyance of water.

Selection of Pipe Material

As discussed above, the factors which should be considered in selection of pipes include:

(*i*) Strength of pipe.

(*ii*) Water carrying capacity.

(*iii*) Life and durability of the pipe,

(*iv*) Ease of transportation and handling, and

(*v*) Ease of joining, maintenance and repairs.

9.3.2. Pipe Fittings

The various pipe fittings, such as sockets, bends, crosses, bees elbows, wye unions, caps, plugs, nipples, etc., are frequently used in making service connections and also sometimes in bigger sized mains or sub-mains. Various types of bends and other important pipe fittings are shown in Fig. 9.2.

(*i*) **Sockets are used to join two pieces of pipes.** Sockets have threads inside and they are tightened on the external threads made at the end of the pipes.

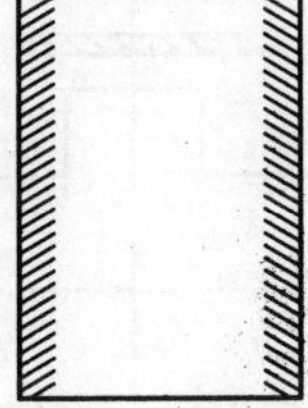

Fig. 9.2. Socket.

(*ii*) **Bends. These are made of galvanised used iron.** Whenever these is need to change the direction (alignment) of pipe the bends are used. The bends are available for different angles, such as $\frac{1}{4}$ bend (90°), $\frac{1}{8}$ bend (45°), $\frac{1}{16}$ bend (22.5°) or $\frac{1}{32}$ bend (11.25°), $\frac{1}{4}$ and $\frac{1}{8}$ bends are shown in Fig. 9.3.

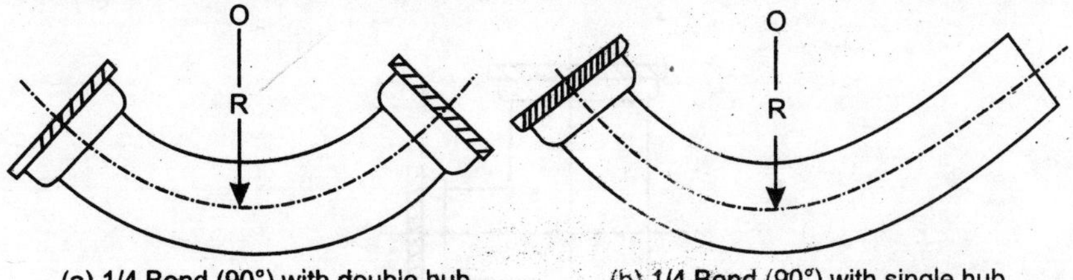

(a) 1/4 Bend (90°) with double hub. (b) 1/4 Bend (90°) with single hub.

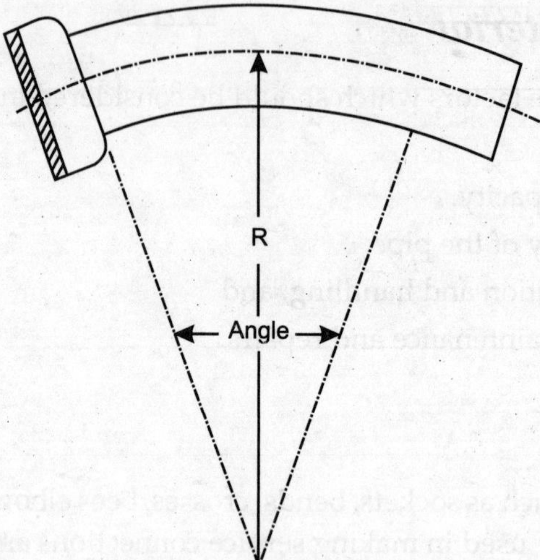

Fig. 9.3. Bends of different angles.

(iii) **Cross.** These are used when pipes crosses each other or flow coming from one direction is diverted to other 3 directions, as shown in Fig. 9.4.

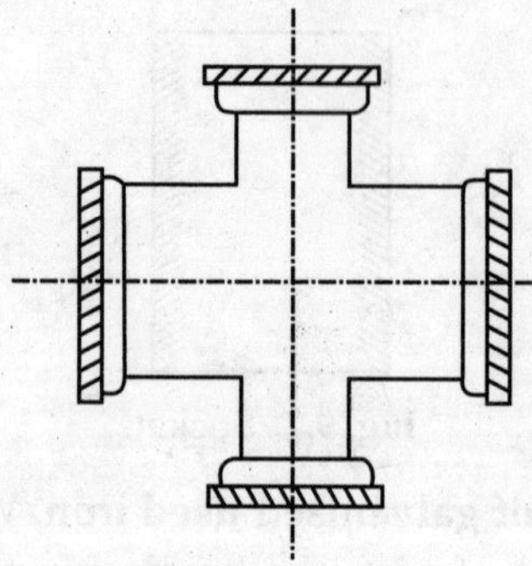

Fig. 9.4. Cross.

(iv) **Tee.** These are provided at 'Tee' points in the piping or when water coming from one direction is diverted to other two other directions, as shown in Fig. 9.5.

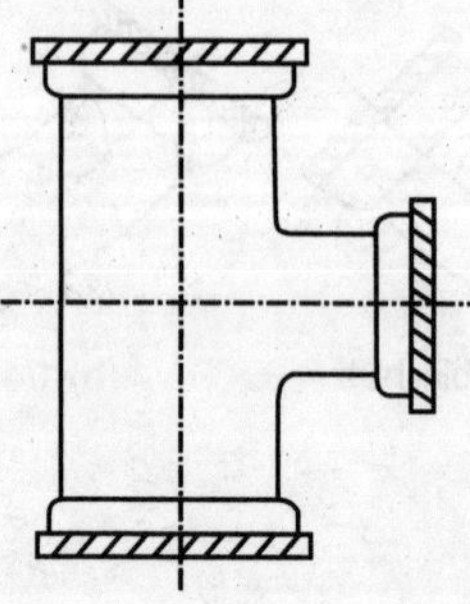

Fig. 9.5. Tee.

(v) **Elbows.** These are used to provide a change in the direction of pipes by 90°, as shown in Fig. 9.6. Elbows are made of galvanised iron.

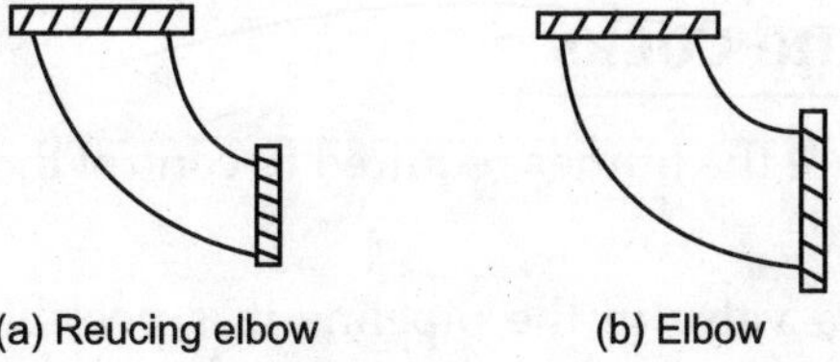

Fig. 9.6. Elbows.

(vi) **Wye.** When the pipes are required to be joined in '*Y*' shape, the wyes are provided, as shown in the Fig. 9.7.

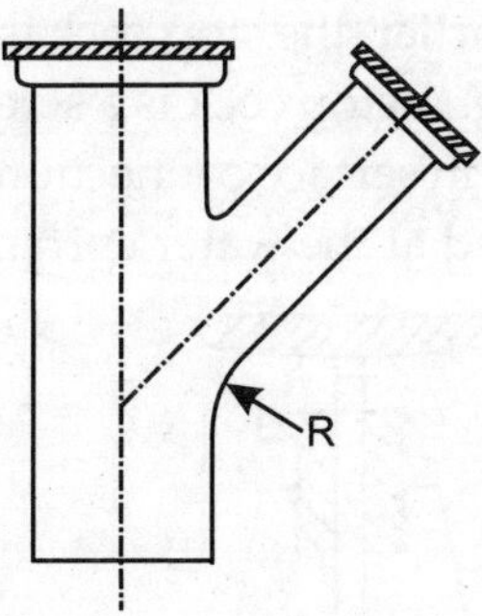

Fig. 9.7. Wye unions.

(vii) **Increaser or Reducer.** When two pipes of different diameters are required to be joined, increaser or reducer are used to join the two pipes, as shown in Fig. 9.8.

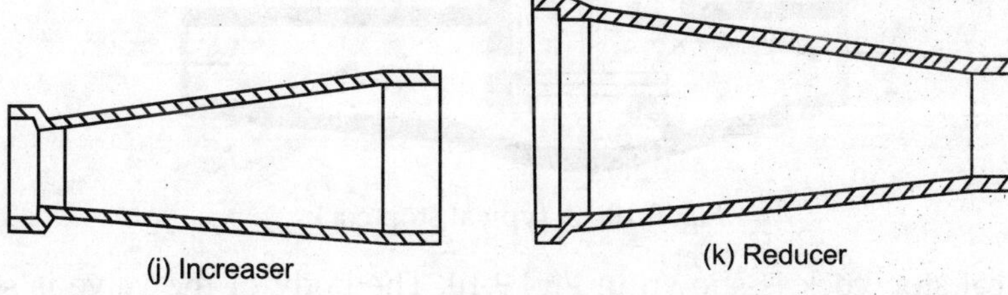

Fig. 9.8.

(viii) **Plugs.** These are used to close the pipe at the end of the flow, shown in Fig. 9.9.

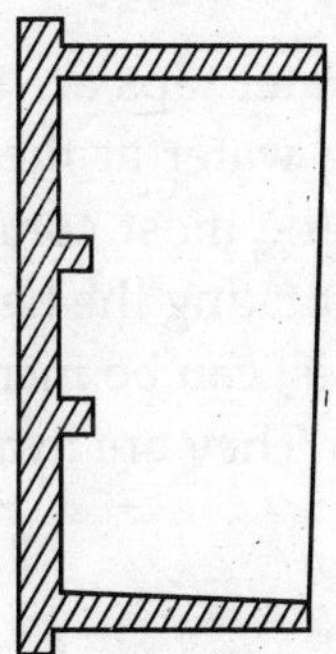

Fig. 9.9. Ordinary plug.

9.3.3. Valves, Taps and Cocks

The valves, taps and cocks are the fittings required to control the flow of water either along or at the end of a pipeline.

1. **Valves.** By fitting a valve in the pipeline it is possible to isolate any sections of a pipeline for the purposes of inspection, repair of a leakage or addition/alteration to the existing functioning. We have already discussed different types of valves, such as sluice valve, air valve, drain valve, pressure relief valve, and check valve in the chapter on "Transmission of Water".
2. **Stop Cock.** As discussed earlier, the stop cock is provided before the water enters the water meter in the building. A stop cock is a screw down type of sluice valve which is used in smaller sized pipes in service connections for stopping or opening the supply. They are generally provided at the water entrance of each building.

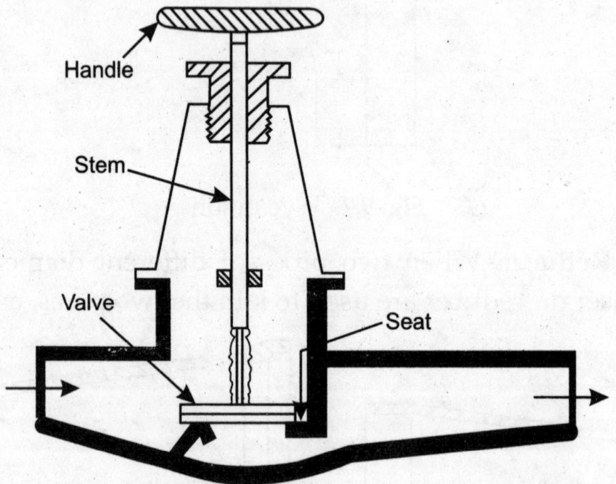

Fig. 9.10. A typical stop cock.

A typical stop cock is shown in Fig. 9.10. The body of the valve is so cost that the water passes through an orifice when the valve stem is raised. When the valve is closed, it rests against the seat, and thereby closing the orifice. They are extensively used in pipes upto 50 mm sizes.

3. **Water Taps and Bib Cocks.** Water taps are the valves provided at the end of house service pipes for withdrawing water at the consumer's house. Several varieties of water taps are available, and the most common of them is bib cock. A typical bib cock is shown in Fig. 9.11. By rotating the handle of the bib cock, the orifice opening, through which the water passes, can be increased or decreased, thereby controlling the outflow through the spout. They are available for different pipe sizes from 10 to 50 mm dia.

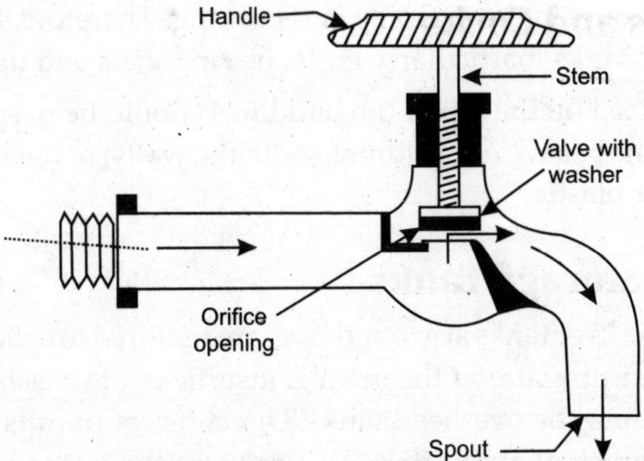

Fig. 9.11. A typical did cock.

The bib cocks may also be of the push type (self closing tape) which opens from a slight push given vertically upward, and closes down automatically due to self-weight when the push is removed. Such valves are used in public taps in order to avoid wastage of water from handle type bib cocks, which may be left open.

The bib cocks should be water-tight and should not leak, as it may lead to considerable wastage of water. The rubber or leather washers used in them may get damaged and lead to leakage, and hence should be replaced as and when needed.

9.4. STORAGE TANKS IN BUILDINGS

Water supply to a building from city mains could be either continuous or intermittent. Generally, due to shortage of water, the local authority plan distribution of water in shifts (*i.e.* morning or/and evening). Therefore provision of storage tank is made in building to ensure availability of water during non-supply hours. The minimum **storage capacity of the tank** is equal to 50 per cent of daily demand if the water supply is in two shifts, and it is equal to 100 per cent if it in one shift per day.

The water storage capacity for a building depends on the following factors:

(*i*) Hours of supply;

(*ii*) Rate of supply, governed by the pressure in the mains;

(*iii*) Demand pattern in the given building; and

(*iv*) Fire storage required.

Water storages required in the buildings are of two types:

(*i*) Overhead storage; and

(*ii*) Underground storage

Both these types of storages are discussed hereunder:

1. Overhead Storage Tanks

When the city water supply is available at sufficient pressure round the clock, as to rise to the highest floor, then no overhead storage is required, and all the water supply fixtures can be connected with the direct supply line. However, in most of the city water supplies are confined only to a few hours in the morning and evening. Therefore, it becomes imperative to store water in overhead tanks, particularly for its use in toilets and urinals.

Overhead tanks placed on the top of the buildings should be properly located as to safely transmit their loads to the beams and columns (or to the walls) in the building. These tanks are made of RCC or steel or plastic.

2. Underground Storage Tanks

Underground storage tanks or tanks at ground level are required to collect water from municipal supply lines, if the water pressure in the main is insufficient to reach the overhead tank. The water is then pumped upto the overhead tanks. Direct use of pumps on the municipal mains is prohibited by the municipal authorities, as it reduces the water pressure in the adjoining houses or buildings.

Such a tank is made of R.C.C. or brick masonry, and has to be structurally designed safe to withstand earth pressure (when the tank is empty). The tank is to be filled from the municipal supply inlet, and is covered from the top to avoid any contamination. The top cover may be provided with a manhole for inspection and cleaning purposes. Suitable pump is installed to lift the water from this tank and upto the overhead tank. Care should be taken to construct an underground tank so that:

(*i*) It is water-tight and does not leak when empty or full;

(*ii*) It is not located in low-lying areas, which may permit entry of surface water from the top;

(*iii*) It should prevent ingress of subsoil water into the tank through leakage;

(*iv*) It should not be located near sewers, septic tanks, soak pits, oil tanks, or underground car parking areas, to avoid seepage of polluted surface waters;

(*v*) The overflow water level in the tank should be above the surrounding ground level, to prevent the surface water from entering the tank through overflows;

(*vi*) The masonry or R.C.C. retaining walls of the tank should be structurally safe to withstand the earth pressure from behind, when the tank is empty.

General Requirements of Domestic Storage Tanks

1. It should be water tight.
2. It should be constructed with non-corrosive and non–toxic materials.
3. These should be provided with vent pipe for ventilation and preventing pressures when water is withdrawn.
4. These should have overflow pipe alongwith ball valve with float.
5. These should have a scour pipe with a plug at the bottom for emptying the tank.
6. These should have manhole with cover on the roof of the tank to prevent entry of dust mosquitoes etc.

9.5. HOME PLUMBING SYSTEMS FOR WATER SUPPLY

Water piping systems must be designed to provide uniform flow and pressure in all floors and places within certain practical limitations. The piping systems to be adopted will differ when the main supply is used or when the overhead tank supply is used, or when the underground-overhead tanks supply is used, as discussed below:

1. Direct Supply Plumbing (Piping) System

This system is also known as upward distribution system, because in this the supply is directly given to the various floors directly from the street main which has enough pressure to feed them directly. The minimum pressure available will limit the number of floors to which the water can be supplied.

This system is very convenient and economical to users, as it avoids the necessity of storage tanks, pumping sets etc. However, in such a system, on overhead tank is provided (of course, small in size) to ensure continuous supply of water to the flushing cisterns of water closet (W.C.) and urinals in case of accidental low pressure or breakdown in the city supplies.

2. Plumbing System Using Overhead Tank

This system is also known as 'Down feed distribution system' where supply from the street, main is drawn into the overhead storage tank directly from where the supply is distributed to various floor by gravity. This system is useful when water supply is limited to the morning and evening shifts. To meet the water requirement during non-supply hours, water is stored in overhead tanks placed on the terrace, which are filled with a direct connection from the mains. Supply to kitchen is taken from the direct main, while supply to taps for bathing and flushing etc., is taken from the overhead tank.

In this system, drinking water is stored by the consumers in their vessels everyday during direct supply hours, or the overhead tank water has to be used for drinking through a R.O filter. This system is common in Indian houses.

3. Plumbing System Using Underground-overhead Tank Supply

This system also follows a 'down feed distribution system'. In this system supply is pumped from ground level storage tank to an overhead tank and then it is distributed by gravity. This system is used when the municipal water supply is at low pressure, whether intermittent or continuous, then it becomes essential to store water in underground tank, and then this water is lifted to the overhead tank.

Generally 3 to 4 storeyed houses install overhead tank system. High rise multi–storeyed buildings, which cannot use low pressure municipal supply have to pump up the entire requirement to the overhead tank by collecting it in the underground tank. Then provide all down take connections from the overhead tank. Use of domestic filters/R.O. filters are essential in such cases to obtain safe drinking water.

9.6. HOT WATER APPLIANCES

Hot water is required in houses for bathing and washing of clothes and utensils, etc. Higher temperatures make the cleaning work easier. Bathing with hot water opens the body pores, washing dirt and sweat easily, and giving a sense of freshness. Evidently, in winter season, the requirement of hot water becomes all the more necessary.

Hot water is used in kitchens and bathrooms in residential areas. The quantum of consumption of hot water widely depends upon the weather conditions of the area, as well as the habits and living standards of the people. In India, natural climatic temperature widely varies from about – 20°C to as high as 40 to 45°C. Hot water requirement therefore varies widely, and are much high in areas of cold climates as compared the hotter regions. Large quantity of hot water may be required for laundries, and industrial kitchens and in buildings like hotels, hospitals, hostels,industrial canteens *et.* Hot water may also sometimes be required as a heating medium for space heating and air conditioning.

Electrical Geysers/Hot Water Heaters

As natural hot water springs are known as **'Geysers'**, modern appliances generating hot water artificially have also acquired the some name *i.e.* geyser. Electrical Geysers generate hot water using **electrical heating elements**. LPG may be used in **gas water heaters**. Electrical geysers are very common appliance of our life. Generally two types of geysers are available in the market, namely;

1. Instantaneous electrical geysers.
2. Storage geysers or storage water heaters.

9.6.1. Instantaneous Electrical Geysers (Water Heaters)

These geysers are designed for direct connection to the water-taps, and the electric heating element is switched on after the flow of water is established. Such an appliance requires a heating element sound round a pipe, which heats the water to a specified temperature matching the rate of flow. The heating elements has to be of a much higher capacity (wattage) as compared to the one required in a storage heater.

An electrical load of 3 kW will given warm water at a rate of about 1.5 litres/min. These heaters not only not require heavy wiring switch board connections but also require very efficient earthing to avoid shocks. Sub-standard instantaneous heaters can lead to fatal electrical shocks. Therefore, it would be advisable to install a storage type of a geyser rather than using an instantaneous geyser.

9.6.2. Storage Geysers

In this type of storage geyser, the inlet is not blocked and its stop cock even if provided for convenience of convections, is kept always open. The outlet is connected to a bib cock (water tap) through a stop cock. The outlet stop cock is usually kept open, and hot water can be drawn by just opening the hot water tap. In such a geyser, therefore,the over-head tank supply is always open, which keeps the geyser always under the pressure of water. This pressure allows hot water to be distributed to a number of point, and permits mixing of hot and cold supply through a mixing fitting (mixer).

A storage geyser consists of a thermally insulated cylindrical vessel, one or more electric heating elements, thermostat to automatically cutoff and cut on the electric current, and pipe connections; and is assembled by the manufacturers ready for installation on site. The rated capacities and power input (kW) requirements of storage heaters of various sized, as per recommendations of IS 2082–1965, are shown in table below:

Rated capacity in lites	6	15	25	35	50	70
Recommended input in kW	1	1	1	1.5	2	2

A section of a closed automatic geyser (pressure type) provided with a pressure release safety valve near the inlet is shown in Fig. 9.12.

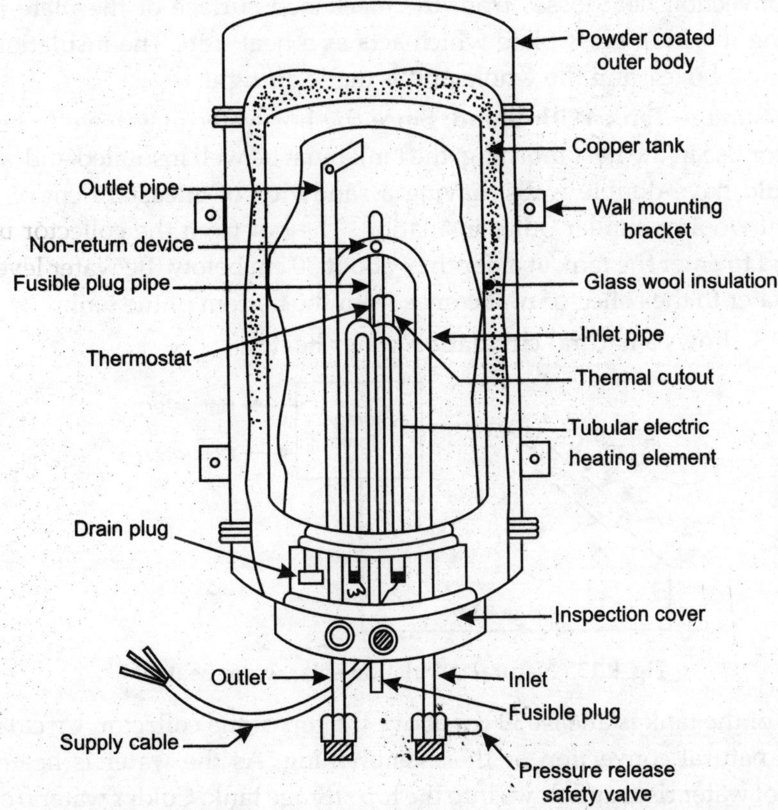

Fig. 9.12. Sectional view of a automatic storage type water heater.

In all storage geysers, the cold water enters at the bottom of the geyser tank, while the hot water from the top is withdrawn. This keeps the geyser always filled up with water, thereby eliminating the danger of the geyser becoming empty and burning of the heating element.

9.7. SOLAR WATER HEATERS

Solar water heating, now-a-days, has become quite cost effective and saves on fossil fuels or electricity. Solar water heaters have therefore, gaining popularity in the country. A solar water heater consists of:

(*a*) A solar energy collector, and

(*b*) A storage tank with a suitable device to minimise heat lost.

1. **The Collector Unit:** The main function of the collector unit is to absorb solar radiation and to transfer this heat energy to the cold water for heating the same. It usually is a blackened plate, constructed in such a way that channels are formed, through which water can circulate. Copper in the from of tubes and sheets proves to be the best material for this unit due to its high thermal conductivity, long life, and high corrosive resistant. Since copper is a costly material, G.I. (galvanised iron) sheets are generally used to make this unit.

   Heat loss from the absorber is reduced by providing insulation at the back and sides. The convection heat losses from the blackened surface of the plate is reduced by covering it with a glass plate which acts as a heat trap. The insulation is protected with cover boxes, and the whole unit is made air tight.

2. **Water Storage Tank With Stand:** Since the hot water may have to be stored over-night for use in the next morning, the tank must be well insulated and weather-proof. It should have double walls, having a sandwich of atleast 10 cm of fibre glass or mineral wool or similar other insulation. The hot from the collector unit is usually allowed to enter the tank at a depth of about 30 cm below the water level in tank. The cold water to the collector unit comes from the bottom of the tank.

   Fig. 9.13. shows a **natural circulation water heater.**

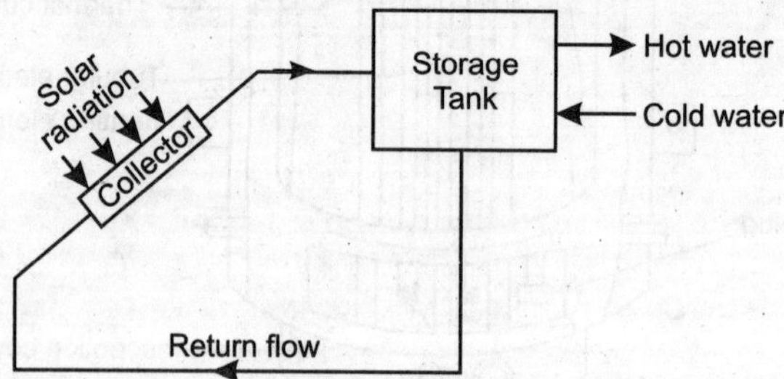

Fig. 9.13. Natural circulation solar water heater.

The bottom of the tank is atleast 30 cm above the top of the collector. Circulation of water occurs through natural convection or thermosiphoning. As the water is heated its density decreases and hot water rises and flows into the top storage tank. Colder water from the bottom of the tank having higher density enters the lower heater of the collector. The difference in densities in hot and cold water provides a driving force for the circulation of water. Hot water is drawn from the top of the tank.

For forced circulation, a circulating pump can be provided with a automatic control unit and a heat exchanger in collector loop. A check valve can also be provided to prevent reverse circulation and thermal losses during night.

9.8. DRAINAGE SYSTEM

As discussed earlier, potable water supply is distributed in a building under pressure to various areas, such as bath, W.C. kitchen etc. through a pipe network. Adequate arrangents are also required for quick collection, conveyance and disposal of used water from the fittings without any risk of health to the users/occupants. Rain or storm water is also essentially required to be collected and discharged from the roof and paved areas of the building and ground surface.

The term drainage includes the system of removal of sullage or waste water (from floor traps, both wash basin and kitchen) soil water (from W.C. and urinals) and storm water from a building and conveying the same upto treatment plant and disposal.

The drainage system can be broadly divided in following two parts:

1. Drainage above the ground.
2. Drainage below the ground.

1. **Drainage Above the Ground:** This consists of a system of vertical stacks, horizontal branches, floor traps etc. provided for conveying sewage (soil water and waste water), storm water etc, to the underground drainage system. This system is also known as building drainage system.
2. **Drainage Below the Ground:** This consists of a system of underground house drain, main drain or sewer, manholes, ventilation shafts etc., provided for conveying. The sewage and storm water for final treatment and disposal. Underground drainage systems are of following three types:
   (i) **Combined system.** In this system, storm water is completely mixed with the sanitary sewage and conveyed through a single drain or sewer.
   (ii) **Separate system.** In this system, storm water is not allowed to be mixed sewage. Two separate drains are provided–one for sanitary sewage and other for drainage of storm water.
   (iii) **Partially combined system.** In this system a part of storm water (generally run-off from roofs, paved yards, surface area and streets etc.) is mixed with sewage and conveyed through sewer and remaining storm water is conveyed through separate surface drains.

Pipes and Fittings

For drainage above the ground, cast iron, asbestos cement or PVC pipes are generally used, while stoneware pipes are used for house drain. These are jointed either with cement mortar or lead. For sewers, stoneware pipes, cement concrete pipe or CI pipes are generally used.

Sanitary Fittings and Appliances

Various sanitary fittings and appliances commonly used are traps, water closets, washbasins, sinks urinals, flushing cistern etc. These have been discussed later on in the separate chapter on "sewage collection from houses and buildings".

Types of Home Plumbing

Common types of plumbing adopted in drainage system of building are:

1. Two pipe system,
2. One pipe system,
3. Single stack system, and
4. Partially ventilated single stack system,

These methods have been discussed in the chapter on "Sewage Collection from Houses and Buildings".

9.9. WATER SUPPLY IN HIGH-RISE BUILDINGS

In the case of high - rise or multi - storied buildings, **Down-take supply system** (as discussed in section 9.5 above) is used. Down-take or down-feed system are of following types, and is used in one or combination of these:

1. **Over-head Storage System:** In this system, mild steel or RCC tanks are provided on the terrace floor each zone of four storey's one common **down-take** pipe, which is linked horizontally in a loop formed on terrace. Each down-take and this loop are designed for peak load demand. A **pressure reducing valve** or throttling disc are to be provided in the down-takes of lower zone for limiting the head to a maximum of 25 m.
2. **Break Pressure System:** In this system entire building is sectionalized in suitable zones consisting of 4 to 8 stories each. For each zone a break pressure tank is provided on the topmost story of that particular zone. The capacity of these tanks is equal to 10 to 12 minutes of water demand during peak hours of that particular zone and shall not be less than 2000 litres each for flushing and other domestic purposes separately. The submain downtakes from master overhead tank feeds into break pressure tank.

In case of multi-storeyed buildings, there should invariably be provision of underground reservoir, overhead tanks from municipal mains plus 15 to 30 minutes of normal consumption as residual storage plus amount of water required for fire-fighting purposes.

QUESTIONS

1. Explain with a neat sketch as to how municipal water mains are connected to buildings and houses for giving water supply connections.
2. Discuss the water storage requirement of building. Also discuss the underground and overhead storage tanks.
3. Discuss various types of water piping systems employed in buildings for fulfilling the water demands of its residents, giving merits and demerits of each system.
4. How would you assess the water demand of the buildings to be used for different purposes?
5. Discuss the 'flow rates' and fixture units' in relation to computation of gross water demand of a building. How is probable simultaneous demand computed?
6. Briefly discuss the principles involved in the design of water supply network to be laid in a multi-storeyed residential building.

7. Discuss the following in brief:
   (*a*) Stop cock.
   (*b*) Water taps and Bib cocks.
   (*c*) Pipe fittings.
8. Explain the construction and working of:
   (*a*) Domestic under-ground storage tank, and
   (*b*) Domestic overhead storage tank.
9. Discuss various requirement of domestic water storage tank.
10. Discuss various types of domestic appliances used for hot water supply in houses.
11. Explain and differentiate between the pressure type and non-pressure type of electric storage water geysers with neat sketches.
12. Discuss (*a*) pressure release valve, and check valve in a domestic water geyser, being used in modern bathrooms for giving hot and cold (mixed) supplies.
13. Discuss the storage geysers for individual bath rooms.
14. Write short notes on:
   (*a*) Instantaneous geysers.
   (*b*) Solar water heaters
   (*c*) Non-pressure v/s pressure type of storage geysers.
15. Discuss types of drainage systems, in detail.
16. Discuss water supply system in high-rise buildings. Also explain break pressure system.

Rural Water Supply

10.1. INTRODUCTION

Majority of the Indian population live in rural areas but only about 60% of them have some from of potable water supply, with only about 20% having piped water supply. It is not feasible to cover all these villages with potable supplies because of various constraints, such as :

1. Scattered and inaccessible nature of villages.
2. Non-availability of nearby water sources.
3. Non-availability of adequate funds to take up those projects where only a few people are living and hence may not be financially viable.

Due to these constraints, it becomes very difficult to supply piped potable water to the villages through a proper water-treatment plant; and hence, simple methods are devised and adopted to treat the available water supplies in such areas.

Sources of rural water supply can be categorised in following forms :

1. Water **requiring no treatment** except disinfection. These can be either:
   (*a*) Supplied by gravity from springs and streams in protected drainage areas as in hilly terrains; or
   (*b*) Supplied by pumping from wells, tube-wells, infiltration gallaries etc.,
2. Water **requiring simple treatment** like plain sedimentation, slow sand filtration or a combination of these two, and supplied either;
   (*a*) By gravity; or
   (*b*) By pumping.
3. Water **requiring special treatments** like sedimentation coagulation, rapid sand filtration and supplied either;
   (*a*) By gravity; or
   (*b*) By pumping.

10.2. SOURCES OF WATER FOR SMALL COMMUNITIES IN RURAL AREAS

Since most of the Indian rivers are non-perennial with meagre storages, and rural villages are generally located at far off distances from these rivers, it has generally not been feasible to supply surface waters to the scattered and isolated rural areas. Most of our villages depend upon ground water supplies, to be extracted through *wells* and *tubewells.* In hilly regions, sometimes, natural *springs* are available in exceptional cases.

The degree and type of treatment to be given depend upon the quality of water source available. Various water sources available in rural areas are:

1. **Rivers and Streams:** Generally, the waters of rivers and streams are soft, containing relatively low concentrations of dissolved salts. They are often affected by pollution in small streams near settlements. The quality of the water can vary considerably with rainfall, but turbidity can be expected at most times. Slow flowing rivers can contain considerable amounts of organic matter.
2. **Lakes:** Large lakes, in general, have the water quality as fairly good and consistent. Near the areas of industrial and agricultural development, and sewage and industrial effluent discharges may affect the water to such an extent that it becomes impracticable to consider it as a source for public water supply.
3. **Springs and Boreholes:** Springs are natural sources of ground water, while the boreholes are dug or drilled to lift ground water. These sources can provide reliable water supply. Moreover, if properly maintained and kept sufficiently away from pit privies and soakpits, they can provide us water almost free from pollution. Proper protection of the site of spring from defecation and drainage is very important to avoid its contamination. Such a source can be utilised to provide supply to a community without pumping, at minimum cost and maintenance.

   A well or a bore hole is dugged or drilled into the permeable strata, and the water lifted manually or by mechanical pumps. It is usually advisable to prevent surface water entering the well by sealing the top section with brick or concrete, as shown in Figs. 10.1 and 10.2 for dug well and driven tube well respectively.

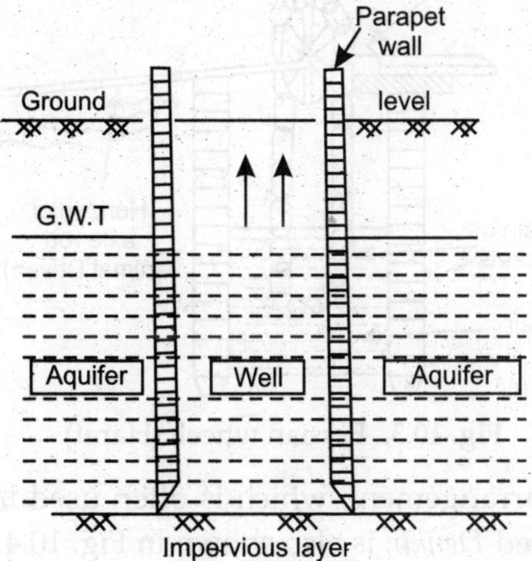

Fig. 10.1. A dug well to obstain water supply in rural areas.

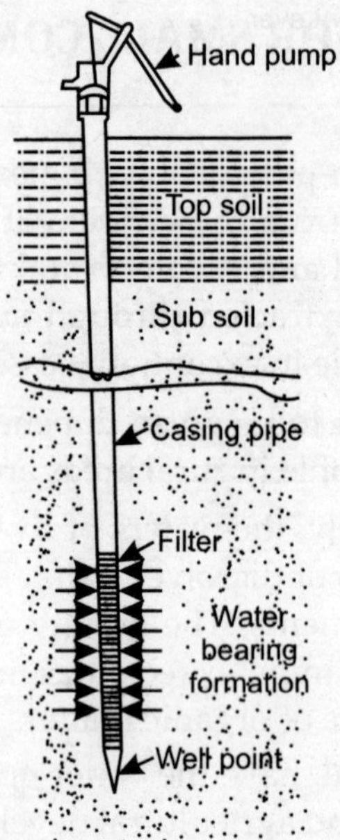

Fig. 10.2. A shallow driven tube well provided with a hand pump.

In order to obtain larger centralised discharges for distribution among the villagers through piped supply after treatment, deep tube wells may be drilled. Water from such tube wells shall be lifted by means of pumps, which may be run on electrical diesel or wind power. The chances of contamination of such deep supplies are very meagre, although they may contain large scale minerals and dissolved salts.

Animal power has also been used traditionally in our villages to lift water from open wells through the use of a persion wheel, also known as 'Harat' in villages as shown in Fig. 10.3.

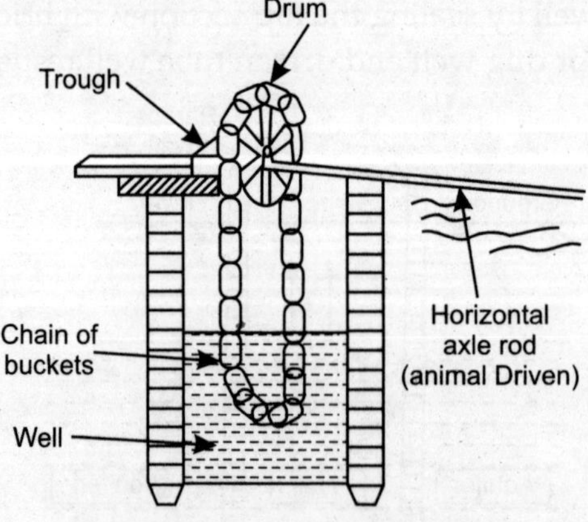

Fig. 10.3. Persian wheel. (Harat)

The simple manual arrangement, which is often used in villages to lift water from often open wells, called *Denkli*, is also shown in Fig. 10.4.

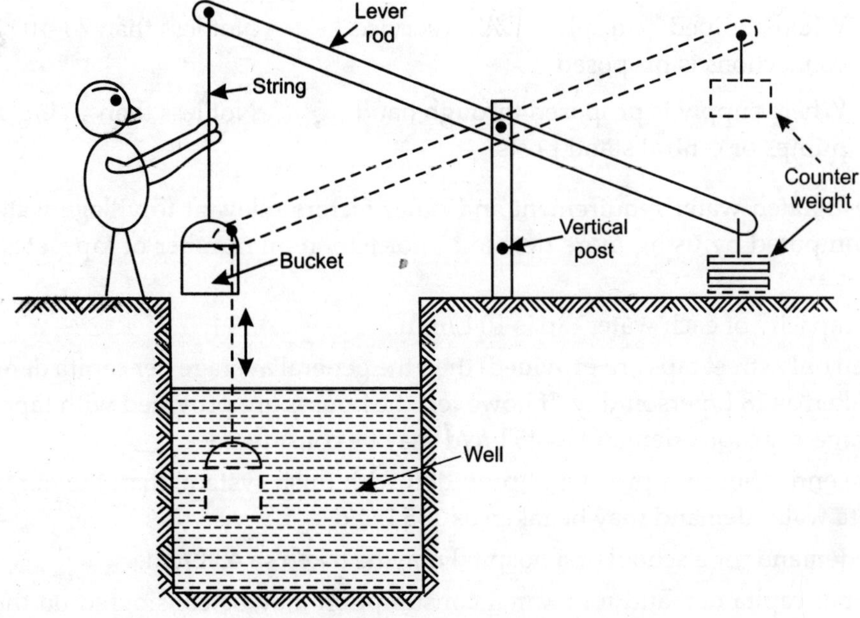

Fig. 10.4. Denkli.

10.3. SELECTION OF SUITABLE SOURCE OF WATER

Where more than one possible source is available the choice should be made in order of the following priorities :

(*i*) The reliability of the source;

(*ii*) The purity of water at the source; and

(*iii*) The ease with which the water from the source can be supplied to the consumers

Samples of water should be collected in sufficient number from time to time for testing. Quality checks on the water source can also be made with the help of some mobile *analytical water testing kits,* which are easily available these days.

The parameter to be tested depends upon the type of the water source. Say for example in the case of surface sources and shallow wells, fecal pollution may be widely present, thus necessitating testing of parameters like *Chlorine demand* or *BOD*, etc., Testing the presence of *Ammonia* by using Nesslers solution can help in detecting recent fecal pollution. In case of deep bore waters, however, parameters like *hardness, pH value,* dissolved metals like lead, *iron fluorides, arsenic, nitrates,* etc., will become more important.

10.4. QUANTITATIVE REQUIREMENT OF WATER

Bureau of Indian Standards, has laid down the following minimum per capita domestic consumption for rural communities with population upto 20,000 :

(*i*) Where piped supply with service connections is proposed — Not less than 70–100 L/c/d

(*ii*) Where supply is proposed through hand pumps or central stand posts — Not less than 40 L/c/d

The estimated water requirement and other factors relevant to village water supply can also be computed by using more detailed information on number of taps, etc., as given below :

1. The capacity of each water tap is 10 L/min.
2. When only street taps are provided, then the general average per capita demand may be taken as 18 L/person/day. If however, the houses are provided with taps, then the average per capita demand of 45 L/c/d may be considered.
3. If the entire house is piped and provided with flush system, than the minimum per capita water demand may be taken as 135 L/c/d.
4. The demand for a school or a hospital may be taken sa 45 L/c/d.
5. The per capita demand for animal consumption may be considered on the basis of factors indicated below.

Livestock	
Horse	50 L/animal/d
Dairy cow	70 L/animal/d
Sheep	10 L/animal/d
Chicken (100 Nos)	20 L/100 No./d

6. For every 200 people, atleast one tap must be provided.
7. The design population for computing water demand should be taken as 1.5 times the present population.

10.5. WATER TREATMENT METHODS FOR RURAL SUPPLIES

As discussed earlier, the treatment of water for rural supplies will largely depend upon the quality of available raw water, and also upon the availability of funds to meet the capital and Operation and Maintenance cost of the treatment plant. Most of the Indian rural supplies are derived from wells and tube wells (waters of which usually do not require much treatment) and are usually supplied after simple **chlorination**, which ensures elimination of bacterial contamination present in raw water, or of enroute recontamination in piped supply to save from waterborne diseases.

When deep bore waters are to be used for drinking purposes, testing for the dissolved metals like *fluorides, iron, manganese, nitrates, arsenic*, etc., becomes a virtual necessity. Many of the Indian deep well waters are found to contain excess concentrations of *fluorides* and *iron*.

Excess concentration of *nitrates*, if present in the raw water, cannot be easily removed. In such a case, we will have to look out for an alternate source. Tests for the nitrate concentrations can be made, and if no alternate source is available, then even higher nitrate concentrations can be allowed.

Acidic and *alkaline* waters can be easily tested with pH papers. pH values below 6.5 indicate a *corrosive water* (acidic water) requiring **treatment with limestone.** pH values above 9.0 or so, may indicate an alkaline water, and should be rejected, if possible.

For *suspended matter*, if present in raw water, the usual methods of simple treatment are **settlement** and **filtration.**

10.6. TREATMENT FOR HIGH FLUORIDE WATERS

Ground waters in many parts of India are found to contain very high concentrations of fluorides, which cause mottling of teeth initially, and *fluorosis disease* with continued use of fluoride rich waters. Removal of excess fluorides from water, called **defluoridation,** therefore becomes very important of in rural areas, since most of our rural supplies are made from the ground waters.

A simple method, called the Nalgonda method, not only helps in removal of fluorides, but also helps in removing colour, odour, turbidity, bacteria and organic contaminations from raw waters.

The Nalgonda technique uses aluminium salt (alum) for removing fluoride. The raw water is firstly mixed with adequate amount of lime (CaO) or Sodium carbonate (Na_2CO_3) and thoroughly mixed. Alum solution is then added, and water is stirred slowly about 10 minutes, and allowed to settle for nearly one hour. The precipitated sludge is discarded and supernatant containing permissible amount of fluoride is withdrawn. Bleaching powder is also generally added with lime prior to addition of lime, to achieve simultaneous disinfection of treated water and also to keep the system free from undesirable biological growth.

The Nalgonda method, thus involves *precipitation, settling* and *filtration*.

10.7. TREATMENT FOR REMOVING IRON

Many deep borehole waters are found to contain excess concentration of iron than the permissible value of 0.3 mg/lit. The excess concentration of iron does impart bitter characteristics metallic taste to the water. The oxidised precipitation of iron, if present, can cause colouration of water, which may be yellowish brown to black, thereby rendering the water unsuitable and objectionable for domestic purposes. The iron rich waters also cause staining of the plumbing fixtures and clothes washing in laundries. Deposition of iron oxide in the pipelines may lead to reduced carrying capacity.

Iron in ground waters is usually contributed by the dissolution of rocks and minerals, mainly oxides, sulphides, carbonates and silicates of iron.

A small iron removal plant contains two tray aerators, operating at a rate of about 1260 L/ m^2/h to aerate the water. The water is then settled in a sedimentation basin having a detention period of about 3 hours, and the clarified water finally passes through a rapid sand filter. Provision could be made for adding sodium carbonate to the raw water wherever essential.

Based upon the above principles, NEERI has designed package iron removal plants having different capacities, such as 500 L/hr, 1000 L/hr, 1500 L/hr, 2000 L/hr depending upon the requirement of treated water and discharge of hand pump. Such a plant is designed in a circular (or a rectangular) shape, having *aeration chamber, collection chamber, settling chamber,* and *filter*. The *settling chamber* is provided with a *plate settling: device* to enhance settling and reduce detention time, thereby reducing the dimensions of the settling chamber.

The aeration chamber contains 2.0–5.0 cm gravel or stone chips to increase the surface are of air-water interface. The iron contaminated hand pump water trickles over the aeration media through spraying device. The aerated water flows through pores over baffle plate to collection chamber, to settling chamber, and to filter.

The sedimentation chamber, having a detention period of one hour, is provided with two plate *settlers* at 45° angles.

The filter bed of 20 cm depth contains sand media of size 0.8–1.4 mm, supported by 5 cm depth of gravel of size 0.8–1.0 cm. The treated water is taken out from the water tap attached to it.

The filter is cleaned by making back-wash connection with hand pump, scrapping sand bed manually and opening the *sludge scouring valve*. The hand pump is operated with possible high speed till clean water gets accumulated over the sand bed.

Continuous iron removal plants in capacitance of 5000 L/hr, 1000 L/hr and 20000 L/hr, which can be used for larger communities, have also been designed by NEERI. Such a continuous community plant not only removes iron from the raw water, but also helps in removing the free carbon dioxide and toxic substances, which are usually present in ferruginous waters. Moreover, since simple aeration may not remove the iron which is associated with organic matter, lime addition to raise the pH value to about 8.5–9, may also become necessary to treat community water supplies. Such a plant comprises of : *raw water overhead storage tank, cascade tray aerators, chemical dosing cum sedimentation tank, sand filters, and clear water tank* with provision of *disinfection* (chlorination).

Tray aerators are commonly used in such a plant for aerating. The trays are designed for aeration rate of 1260 L/m^2/hr and spaced at intervals of 1 m. The aerated water is *settled in sedimentation* tank having a detention period of 2.5 hours. The clarified water is filtered through a *rapid sand filter* having sand of effective size 0.6–0.8 mm and uniformly coefficient 1.3, with an effective depth of 1.2 m.

10.8. CHLORINATING SMALL COMMUNITY SUPPLIES

It is necessary that public water supplies are bacteriologically safe. Some deep well waters are naturally safe, while most surface waters are polluted, and must be sterilized. This sterilization is usually carried out by chlorination. The chlorination is usually carried out by using solutions of chemicals containing free chlorine, such as bleaching powder, chlorinated lime, sodium hypochlorite etc.

Chlorinating equipment have been discussed in the chapter on "Purification of Water".

10.9. RAIN WATER HARVESTING

After the electricity became available on a large scale, a large number of wells and tube wells installed to lift large quantities of ground water for using it for irrigation as well as for domestic needs, has resulted in lowering the ground water and reduced has lowered the water-tables almost throughout the country. The fast wihdrawal of ground water and reduced recharge has lowered the water-tables drastically, we have exhausted the good quality water, leaving salty deeper water. This water at various places is found to contain excessive quantities of flouride and arsenic, which are extremely harmful to human health.

Under the rain water harvesting, the primary and, therefore is to let the rain water infiltrate into the underground reservoir (acquifer) at a rate exceeding that under natural conditions of replenishment. Rain water harvesting, thus, primarily aims at artificial recharges of groundwater.

Rain water harvesting is an activity of collection and storage of rain water either for direct use or recharging the ground water. The term 'water harvesting' also includes prevention of losses through evaporation and surface runoff and other engineering activities aimes at conservation and efficient utilization of limited water endowment of an area.

Measures for Rain Water Harvesting

Following **measures** are generally adopted for rain water harvesting to recharge the underground reservoir (aquifer):

1. On channel storage of storm water in drains and streams by using the gated regulator structures. This is done to promote infiltration of the impounded water to the underground reservoir.
2. Development, deepening and restoration of old village ponds removing encroachments and stopping the inflow of domestic sewage.
3. Creation of new water bodies by using old abondoned water courses of drains or escape channels.
4. Storage of storm water in streams by constucting check dams.
5. Construction of soil conservation bunds to create percolation tanks. These are constructed across streams, or lakes, duly provided with masonry spillway.
6. Construction of recharge shafts of 0.5 to 3 m dia and 10 m to 15 m deep for easily and speedly recharging to the ground water in village ponds or in storm water drains.

Roof-top Rain Water Harvesting

This is a technique through which rain water is captured from the roof catchments and stored in reservoirs to meet the household needs or recharge the subsurface ground water aquifer. This enchances the availability, reliability and quantity of ground water. Roof catchment includes roof top, compounds and rocky areas.

This collected water serves as a good source of water in rural and water scare urban areas.

This technique is also used where ground water levels have gone down, such as in Rajasthan, Gujarat, Haryana, Delhi, Madhya Pradesh etc.,

Realising the importance of rain water harvesting, Government through bye-laws from August, 2001 has made water harvesting in all new buildings on plots of 100 sq.m and above mandatory in the National Capital Territory of Delhi.

Advantages of Roof-top Rain Water Harvesting.

1. Reduce the surface run-off that chokes the storm water drains.
2. Avoid flooding of roads.
3. Reduce soil erosion.
4. Reduce the cost of pumping of ground water.
5. Rise in ground water levels in wells.
6. Increase availability of water wells.
7. Reduction in the use of energy for pumping water.
8. Effective use of defunct wells and tube wells as recharge structures.

In urban areas where adequate space for surface storage is not available and ground water levels are deep enough to accommodate additional rain water to recharge the aquifers, roof-top rain water harvesting is ideal solution to solve the water supply problems.

Procedure for Rain-Water Harvesting

Roof-top water harvesting involves taking down a PVC or M.S. pipe of 90-120 mm dia from the roof's outlet to the ground floor, which can be connected to a water tank (placed either above or below the ground level) or to the underground water-table. The rain water before collection should be passed through simple sand or charcoal filters [Fig. 10.5] for the removal of suspended particles and micro-organisms from the rain-off being collected. The roofs or court yards should also be kept as clean as possible at the time of rains. These filters can be vertical or horizontal and needs to be periodically cleaned to prevent clogging of its pores.

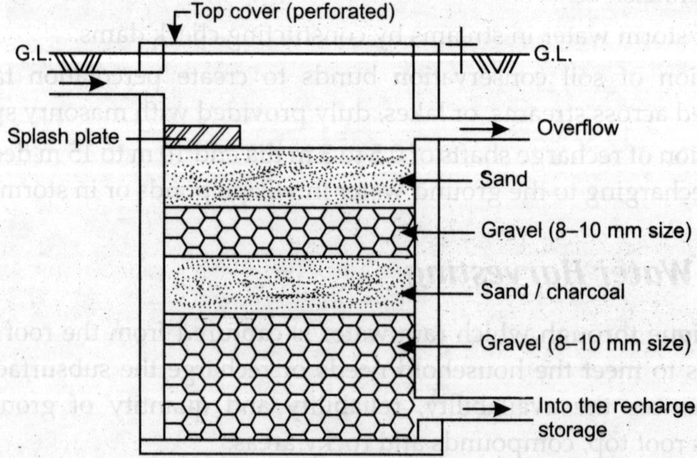

Fig. 10.5. Sand filter.

Water Collection Techniques

Various techniques for collection of water for recharging the ground water are :

1. Rain Water Collection for Ground Water Recharge

This option can be easily practised to supplement the falling ground water table of the area. This can be done by detecting the rain water to infiltrate into the ground to join the water table through:

(*a*) *Recharge pit.* (Refer Fig. 10.6)

(*b*) *Recharge well* or Borewell (Refer Fig. 10.7 and 10.8)

(*A*) Rain Water Harvesting Through Recharge (Percolation) Pit.

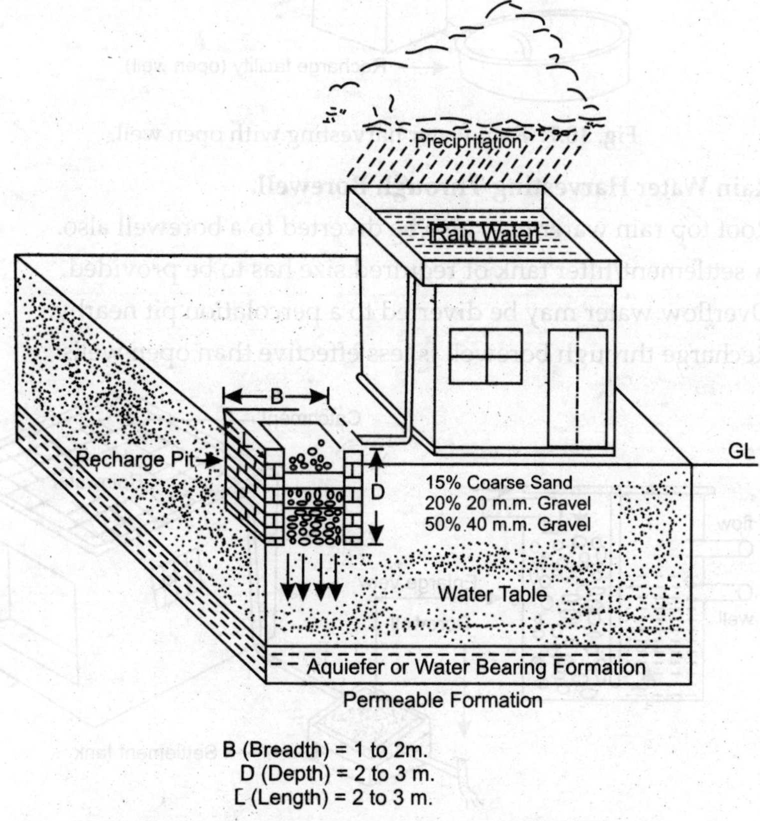

Fig. 10.6. Rain water harvesting through recharge pit.

(*B*) Rain Water Harvesting by Open Wells

— Rain Water collected from the terrace is diverted to the existing open well using PVC pipes through a fitter chamber/sump.

— The size of the fitter chamber may be 0.6 m × 0.6 m × 0.9 m filled with broken bricks in the bottom and sand on the top.

— The chamber may be covered with RCC slab.

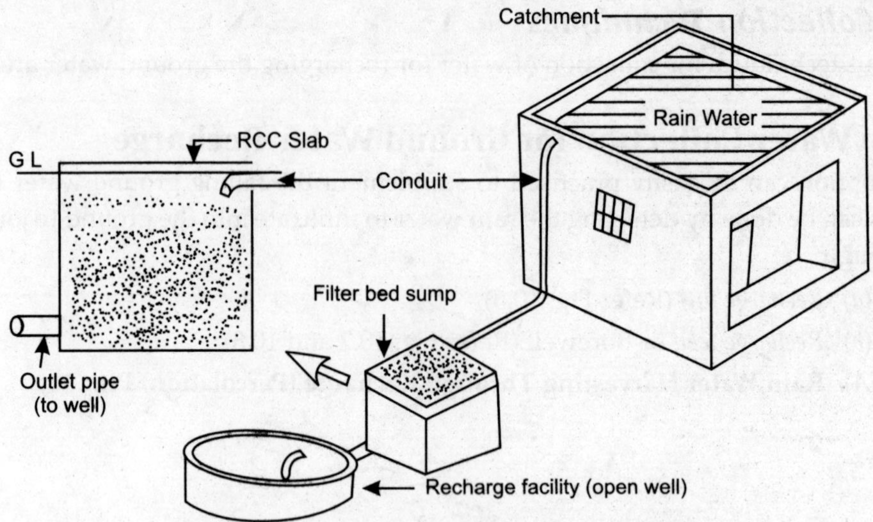

Fig. 10.7. Rain water harvesting with open well.

(*C*) Rain Water Harvesting Through Borewell.

— Roof top rain water may also be diverted to a borewell also.

— A settlement/filter tank of required size has to be provided.

— Overflow water may be diverted to a percolation pit nearby.

— Recharge through borewell is less effective than open wells

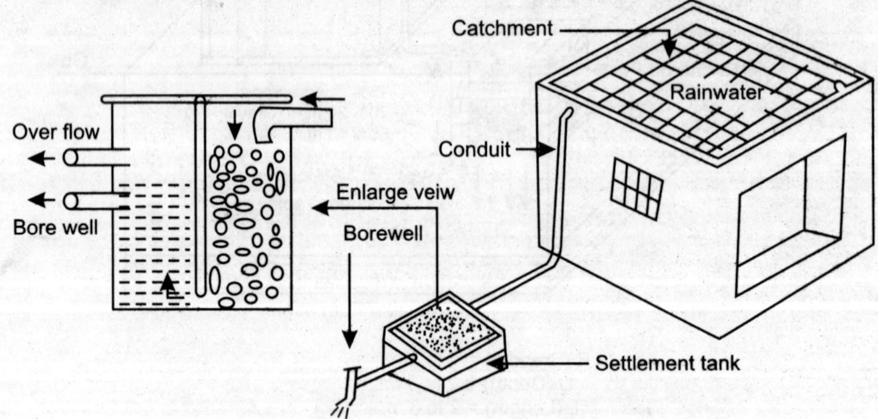

Fig. 10.8. Rain water harvesting through borewell.

2. Roof-Top Water Collection for Direct Use

Rain water for direct use can be collected from the roof into a storage tank of plastic, RCC or masonry. The tank can be placed either above the ground (Fig. 10.9) or below the ground depending on the availability of space are cost considerations. Cleaning and disinfecting the tank at regular intervals is necessary.

Fig. 10.9. Roof-top water collection for direct use.

QUESTIONS

1. Enumerate and discuss various sources of water for rural water supply schemes.
2. Discuss various methods employed to lift water from wells and tube wells in rural areas in India.
3. Discuss the method employed for rain water harvesting for drinking purposes.
4. Discuss the factors considered for selection of a source of water for a rural water supply scheme. What quality checks you will like to perform to assess their usefulness before making a final choice?
5. How will you estimate the water requirement of a village, while planning a water supply scheme for rural areas? Also discuss as to how you will assess the necessary treatment which should be given to the available water source?
6. Discuss the suggested treatment method that is generally used in India for water supply containing high concentration of fluoride.
7. How does excess concentration of iron in water affect the consumers?
8. Discuss the necessity of iron removal from bore-hole waters and the techniques to be used for doing so.
9. Explain as to how you will chlorinate rural water supplies by using bleaching powder and a doser.
10. Write short notes on :

(*a*) Categories of sources of rural water supply.

(*b*) Protection of spring supply to prevent pollution from surface run-off.

(*c*) Rain water harvesting and its advantages

(*d*) Discuss various water collection techniques for ground water recharging.

11. What is a dug well and what are their types? How are dug wells constructed?
12. What is meant by tube wells? What are their types. Describe the most widely used type of tube well with a neat sketch.

PART – 2

Sewage Disposal and Solid Waste Management

Waste Water (Sanitation) System

11.1. INTRODUCTION

After use the wholesome water gets polluted and has to be drained off from the building, for which an elaborate pipe system, and storage have to be provided in the building to serve sanitary fittings and after use elaborate underground drains have to be provided to drain off contaminated water. The system of sanitary drainage include fittings and fixtures like wash basins, sinks, water closets urinals, traps, waste pipes, sewers, septic tank etc.,

Efficient and quick disposal of sewage and waste water from an occupied premise is a primary requirement and ensure the proper maintenance of hygienic and healthy environmental conditions in the building.

The sewage contains human excreta, and waste waters from bathroom, cloth washings etc., The sewage contains about 99% water and 1% solids. It contains large quantity of unstable organic matter which decays and emit offensive smell. Sewage also contains harmful bacteria and in case it is left in raw state, it can cause real danger to human life. In eases where municipal sewage system does not exist or where the buildings are located in isolated places, sewage from individual house or cluster of buildings can be disposed off by providing *septic tanks*. In urban areas, the house drains are connected to municipal sewers.

Water, used for domestic purposes, collects, impurities and becomes **waste water**. Thus domestic waste water (also known as **sewage**) consists of faecal matter and used water from baths, kitchen etc., (known as **sullage**). This waste water (consists of about 99% of water and 1% solids) is transported through a network of sewers, away from residential units. Collection, conveyance, treatment and disposal of sewage (and storm water) from a domestic environment is of primary importance. Municipal or government organisations must take care of this. Various process industries also discharge their untreated liquid effluents into the land and water bodies.

This old method of collecting and disposing of the society's wastes, has now been modernised and replaced by a system, in which these wastes are mixed with sufficient quantity of water and mixture of water and waste products, popularly called **sewage,** is disposed of after giving it suitable treatments; thus avoiding the carriage of washes on heads or carts. The treated sewage effluents may be disposed of either in a running body of water, such as a stream, or may be used for irrigation purposes.

India being a developing country, still uses the old conservancy system at various places, particularly in villages. The metropolitan cities and bigger towns of our country have been equipped with the facilities of this modern water carriage sewerage system; and attempts are being made to equip the remaining towns with this system.

11.2. DEFINITIONS

1. **Sewage:** Sewage is the liquid waste from a community and generally conveyed by a sewer. Sewage is differently classified depending on its nature and sources, such as domestic and industrial sewage.
2. **Domestic Sewage:** It consists of liquid wastes originating from urinals, latrines, bathrooms, kitchen sinks, wash basins, etc., of the residential, commercial or institutional buildings. This sewage is generally extremely foul, because of the presence of human excreta in it.
3. **Industrial Sewage**: It consists of liquid wastes originating from the industrial processes of various industries, such as Dyeing, Paper making, Brewing, etc., The quality of the industrial sewage depends largely upon the type of industry and the chemicals used in their process waters. Sometimes, they may be very foul and may require extensive treatment before using disposed of in public sewers.
4. **Sanitary Sewage:** The sum total of domestic and industrial sewage, may be termed as **sanitary sewage** or simply **sewage.**
5. **Storm Drainage:** The run-off resulting from the rain storms was used to be called **storm sewage,** but the modern approach is to call it **storm drainage** or simply **drainage,** so as to differentiate if from *sewage,* which is much more foul as compared to the *drainage,* and requires treatment before disposal.

   Since heavy rain storms concentrated for a period of 3 months or so, and there are poor water supplies, the ratio of the 'drainage' to 'sewage' works out to be as high as 20 to 20. Thus, during non-monsoon periods, only 1/20th to 1/30th of the designed discharge will be passing through the sewers, if the combined system has been adopted.
6. **Sewage:** The term sewerage is applied to the art of collecting, treating and finally disposing of the sewage.
7. **Sewer:** A sewer is a pipe or conduit generally closed but normally not flowing full, used for carrying sewage. A sanitary sewer is one which carries sanitary sewage and excludes as far as possible storm sewage. A storm sewer carries storm water only, including surface runoff and street wash. A combined sewer is designed to carry all types of sewage, such as domestic sewage, industrial wastes, and storm water.
8. **Sewerage System:** This is the system of sewers including all appurtenances required for sewage disposal. A sewerage system will consist of namely:

   (*i*) Collection works provided for collecting sewage from different points.

   (*ii*) Treatment works to carry out the process of removing the objectionable constituents so as to render it less dangerous or offensive.

(*iii*) Sewage disposal works.

9. **Sullage:** It is the spent up water from kitchen, wash basin etc. It does not include water from urinals, water closets etc.
10. **Drain:** It includes surface drains of different shapes and sections, and are used for carrying storm water, sullage etc.
11. **Refuse:** It includes all kinds of dry wastes produced from a community. It may consist of house sweeping, garbage etc.
12. **Garbage:** It includes all sorts of semi solid and solid wastes out of food products such as vegetables, peeling of fruits etc.
13. **Rubbish:** It means all sundry solid wastes as paper sags, broken pottery, cans and other material comparatively inert in chemical action

11.3. TYPES OF SEWERAGE SYSTEMS

Nature of sewers provided for carrying different types of liquid wastes indicate sewerage systems. There are three systems of sewerage adopted in practice. They are of following types :

1. Combined system.
2. Separate system.
3. Partially separate system.

1. **Combined System:** Modern water carriage sewerage system not only helps in removing the domestic and industrial wastewaters, but also helps in removing storm water drainage. The run off resulting from the storms is also sometimes carried through the sewers of the sewerage, or more generally is carried through separate set of drains (open or closed) discharging their drainage waters into a body of water, such as a lake or a river. Since the rain run-off is not so foul as the sewage is, no treatment is generally required to be given to the drainage discharge. When the drainage is taken along with sewage, it is called a **combined system.**
2. **Separate System:** When the drainage and sewage are taken independently of each other through two different sets of conduits, it is called a separate system.
3. **Partially Separate System:** Sometimes a part of drainage water, especially that originating from the roofs or paved courtyards of buildings, is allowed to be admitted into the sewers; and similarly sometimes, the domestic sewage coming out from the residences or institutions, etc., is allowed to be admitted into the drains, the resulting system is called a **partially separate system.**

It is generally advantageous and economical to construct a 'separate system' atleast in the bigger cities and towns. But in practice, some rain water may always find its way into the sewers either through wrong house sewer connections or through open manhole covers.

Selection of a Sewerage System

A 'separate system' is generally preferred to a 'combined system', although each individual case should be decided on merits, considering the following points:

(*i*) A separate system will require laying two sets of conduits, whereas, a combined system requires laying only one set of bigger sized conduits, thus making the former system costlier. Moreover, the separate conduits cannot be laid in congested streets and localities, making it physically unfeasible.

(*ii*) The sewer pipes in the combined system are liable to frequent silting during the non-monsoon season (when the flows in them are quite less), thereby making them costlier.

(*iii*) In a combined system, the less-foul drainage water gets mixed with the highly foul sewage water, thus necessitating the treatment of the entire flow, needing more capacity for the treatment plant, thereby making it costlier. Whereas, in the separate system, only the sewage discharge is treated and the drainage discharge is disposed of without any treatment.

(*iv*) In case of excessive rains, more foul and insanitary conditions will prevent in the case of combined sewage than in the case of storm drainage alone.

(*v*) Since the sewer lines are generally laid deep and at stepper slopes, as compared to the storm water surface drains, pumping of sewage and often no pumping of drainage is required in a separate system. Whereas, the entire discharge will have to be pumped, if the sewage and drainage discharges are mixed together; thereby making the combined system more costly.

The economy of the two systems must be worked out for each individual project, and the economical system be adopted.

11.4. COMPONENTS OF A SEWERAGE SYSTEM

A sewerage system consists of :

A network of sewer pipes laid in order to carry the sewage from individual homes to the sewage treatment plant. This network of sewers may consist of *house sewers* (or individual house connections); *lateral sewers; branch sewers* (or submains); *main sewers* (generally called *trunk sewers*); *outfall sewer* (*i.e.* the sewer which transports sewage to the point of treatment); etc.,

Manholes are provided in every sewer pipe at suitable intervals, so as to facilitate their cleaning and inspection.

Inlets called *catch basins* are provided to permit entrance of storm water from street gutters.

A complete treatment including screening, *sedimentation, biological filtration* (or *activated sludge treatment*), *sludge digestion*, etc., is given to extremely foul sewage, so as to bring down its BOD and concentrations of other constituents of safer values, before discharging it into a national river resource.

A recent use of sewage is being made for irrigation purposes. For which the sewage has to be treated so as to bring down its constituents to permissible values, as per the requirements of I.S. 3307–1965.

All these aspects are explained in details in subsequent chapters.

11.5. DESIGN AND PLANNING OF A SEWERAGE SYSTEM

The sewerage system must be planned and designed, so as to remove the entire sewage effectively and efficiently from the house, and upto the point of disposal. The sewers must be of adequate size, so as to avoid their overflow and subsequent health hazards. In order to provide economically adequate sized sewers, it is necessary that the likely sewage discharge be estimated as correctly as possible. The sewer pipes should be designed to be laid on a slope that will permit reasonable velocity of flow.

The sewers are generally designed to flow under gravity with $\frac{1}{2}$ to $\frac{3}{4}$ th full. For seeking gravity flow, the sewage treatment plant should generally be located in a low lying area. For adequate and economical treatment, it is necessary to thoroughly study the constituents of the sewage produced in a particular project, and also the quality and other characteristics of the body of water that will receive the sewage. The permissible standards for the effluents, and the possible uses of water downstream, should also be studied. The legal bindings, if any, will also have to be taken into consideration, while deciding the quantum of treatment required to be given.

Since the treatment plant have to be located at low level, the flood protection devices should also be taken care of.

In view of the above, following data need to be collected in **planning a sewage scheme:**

1. **Present population and future population** at the end of design period.
2. **Quantity of water supply in a city** will help in determining the quantity of sewage flow.
4. The **nature and number of various industries** to assess the quality and quantity of industrial waste at present and that expected at the end of the design period.
4. Estimating the quantity of flow by knowing the **rainfall** data.
5. Determine the contour map, location of pumping station and treatment plant etc., and location of disposal point.

11.6. CHARACTERISTICS OF SEWAGE (WASTE WATER)

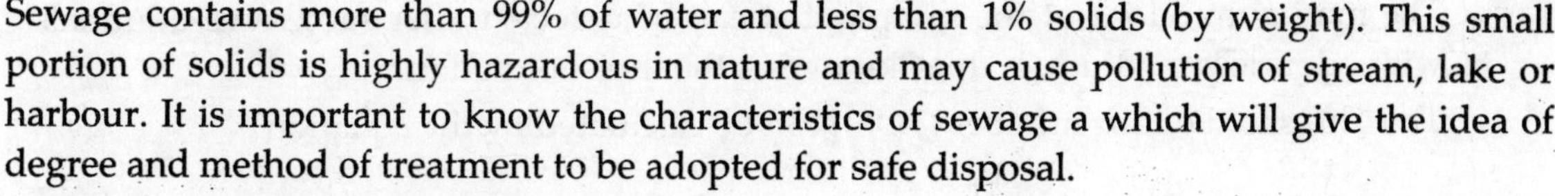

Sewage contains more than 99% of water and less than 1% solids (by weight). This small portion of solids is highly hazardous in nature and may cause pollution of stream, lake or harbour. It is important to know the characteristics of sewage a which will give the idea of degree and method of treatment to be adopted for safe disposal.

Contaminants in waste water are :

1. Suspended solids and sludge disposits.

2. Organic matter (measured by BOD, COD and TOC). which deplete oxygen from water sources.
3. Pathogenes which are responsible for communicable diseases.
4. Nutrients (N and P) that cause undesirable aquatic growth and contaminate water sources.
5. Surfactants, phenols, pesticides etc., resisting conventional treatment.
6. Heavy metals.
7. Dissolved inorganic solids.

The quality of sewage can be checked and analysed by studying and testing its physical, chemical and bacteriological (biological) characteristics, as explained in following paragraphs.

11.7. PHYSICAL CHARACTERISTICS OF SEWAGE

Physical examination of sewage is carried out so as to determine its physical characteristics. This include : tests for determining *(i)* turbidity; *(ii)* colour; *(iii)* odour; and *(iv)* temperature, as explained below :

1. Turbidity

Sewage is normally turbid, resembling dirty dish water or wastewater from baths having other floating matter like feel matter, pieces of paper, greases, vegetable debris, fruit skins, soaps, etc., The turbidity increases as sewage becomes stronger.

The degree of turbidity can be measured and tested by **turbidity rods** or by **turbidimetries,** as is done for testing raw water supplies.

2. Colour

The colour of sewage can normally be detected by the naked eye, and it indicates the freshness of sewage. If colour is yellowish, grey, or light brown, it indicate fresh sewage. However, if the colour is black or dark brown, it indicates stale and septic sewage. Other colours, may also be formed due to the presence of some specific industrial wastes.

3. Odours

Fresh sewage is practically odourless. But, however, in 3 to 4 hours, it becomes stale with all oxygen present in sewage being practically exhausted. It then starts omitting offensive odours, especially that of hydrogen sulphide gas, which is formed due to decomposition of sewage.

The odour of water or wastewater can be measured by a term called the **Threshold odour number (TON),** which represents the extent of dilution required to just make the sample free of odour.

The minimum odour of the sample that can be detected after successive dilutions with odourless medium, is thus, known as the threshold odour.

The Threshold odour number (TON) can be calculated by the equation :

$$\text{TON} = \frac{V_s \circledcirc V_d}{V_s} \quad \text{...(7.8)}$$

where TON = Threshold Odour Number

V_s = vol. of the sewage

V_d = Vol. of distilled or odourless water added to just make the sewage sample loss its odour.

4. Temperature

The temperature has an effect on the biological activity of bacteria present in sewage, and it also affects the solubility of gases in sewage. In addition, temperature also affects the viscosity of sewage, which, in turn, affects the sedimentation process in its treatment.

The normal temperature of sewage is generally slightly higher than the temperature of water, because of additional heat added during the utilisation of water. The average temperature of sewage in India is 20°C, which is near about the idea temperature for the biological activities. However, when the temperature is more, the dissolved oxygen content (D.O.) of sewage gets reduced.

11.8. CHEMICAL CHARACTERISTICS OF SEWAGE

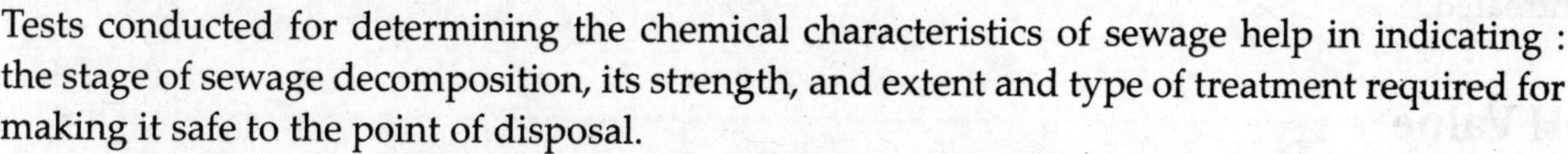

Tests conducted for determining the chemical characteristics of sewage help in indicating : the stage of sewage decomposition, its strength, and extent and type of treatment required for making it safe to the point of disposal.

Chemical analysis carried out on sewage includes *tests for determining* :

(*i*) Total solids, suspended solids, and settleable solids;

(*ii*) pH value;

(*iii*) Chloride content;

(*iv*) Nitrogen content;

(*v*) Presence of fats, greases, and oils;

(*vi*) Sulphides, sulphates and H_2S gas;

(*vii*) Dissolved oxygen;

(*viii*) Chemical oxygen demand (C.O.D.);

(*ix*) Biochemical oxygen demand (B.O.D.).

These tests are briefly discussed below :

Total Solids, Suspended Solids and Settleable Solids

Sewage normally contains very small amount of solids in relation to the huge quantity of water (99.9 %). It only contains about 0.05 to 0.1 per cent (*i.e.* 500 to 1000 mg/L) of total solids. Solids present in sewage may be in any of the four forms : suspended solids, dissolved solids, colloidal solids, and settleable solids.

Suspended solids are those which remain floating in sewage. *Dissolved solids* are those which remain dissolved in sewage just as salt in water. ***Settleable solids*** are that portion of solid matter which settles out, if sewage is allowed to remain undisturbed for a period of 2 hours.

Particles more than 1 μm (micron) in size are classified as *suspended solids,* particle less than 1 μm and upto 10^{-3} μm are classified as ***colloidals;*** and practice smaller than 10^{-3} μm are considered as ***dissolved solids.*** Particles than 10 μm in size without coagulation, and hence are called *settleable solids.* Particles smaller than 10 μm are removed by coagulation.

Further, the solids in sewage comprise of both : the organic as well as inorganic solids. The organic matter works out to be about 45 per cent of the total solids, and the remaining about 55 per cent is the inorganic matter.

Inorganic matter consists of minerals and salts. like : sand, gravel, debris dissolved salts, chlorides, sulphates, etc.,

Organic matter consists of : (*i*) carbohydrates such as cellulose, cotton, fibre, starch, sugar, etc., (*ii*) fats and oils received from kitchens, laundries, garages, shops, etc., (*iii*) nitrogenous like proteins and their decomposed products, including wastes from animals, urea, fatty acids, hydrocarbons, etc.,

As a general rule, the presence of inorganic solids in sewage is not harmful. They require only mechanical appliances for their removal in the treatment plant. On the other hand, suspended and dissolved organic solids are responsible for creating nuisance, if disposed of, untreated.

pH Value

The pH value of sewage indicates the alkalinity of sewage. *If the pH value is less than 7, the sewage is acidic, and if the pH value is more than 7, the sewage is alkaline.* The lesser is the pH value, the lesser is the alkalinity.

The determination of pH value of sewage is important, because of the fact that efficiency of certain treatment methods depends upon the availability of a suitable pH value. The pH value can be measured quickly and automatically with the help of a potentiometer.

The fresh sewage is generally alkaline in nature (with pH more than 7); but as time passes, its pH tends to fall due to production of acids by bacterial action. The pH, however, rises upon treatment of sewage.

Chloride Contents

Chlorides are generally found present in municipal sewage, and are derived from the kitchen wastes, human feces, and urinary discharges, etc., The normal chloride content of domestic sewage is 120 mg/l, whereas, the permissible chloride content for water supplies is 250 mg/L. However, large amounts of chloride may enter from industries like ice cream plants, meat salting, etc., Hence, when the chloride content of a given sewage is found to be high, it indicates the presence of industrial wastes or infiltration of sea water.

Nitrogen Contents

The nitrogen in sewage indicates the presence of organic matter, and may occur in one of the following forms :

(*a*) Free ammonia, called Ammonia nitrate.
(*b*) Albuminoid nitrogen, called organic nitrogen.
(*c*) Nitrites, and
(*d*) Nitrates.

Presence of Fats, Oils and Greases

These are derived in sewage from the discharges of animals and vegetables matter, or from the industries like garbage from kitchens of hotels and restaurants etc.

Such matter form scum on the top of the sedimentation tanks and clog the voids of the filtering media. They thus interfere with the normal treatment methods, and hence need proper detection and removal.

Sulphides, Sulphates and Hydrogen Sulphide Gas

Sulphides and sulphates are formed due to decomposition of various sulphur containing substances present in sewage. The decomposition also leads to evolution of hydrogen sulphide gas, causing bad smells and odours, besides causing corrosion of concrete sewer pipes.

Dissolved Oxygen (D.O.)

While discharging the treated sewage into some river stream, it is necessary to ensure atleast 4 ppm of D.O in it, as otherwise, fishes are likely to be killed, creating nuisance near the vicinity of disposal. The D.O. test performed on sewage before treatment, helps in indicating the condition of sewage.

It is well known that only very fresh sewage contains some dissolved oxygen, which is soon depleted by aerobic decomposition. Also, the dissolved oxygen in fresh sewage depends upon temperature. If the temperature of the sewage more, the D.O. content will be less.

Chemical Oxygen Demand (C.O.D)

Organic matter is most often assessed in terms of oxygen required to completely oxidise the organic matter to CO_2, H_2O etc., If the chemical formulae and the concentrations of the chemical compounds present in water are known to us, we can easily calculate the theoretical oxygen demand of each of these compounds by writing the balanced reaction for the compound with oxygen to produce CO_2, H_2O and oxidised inorganic components.

Biochemical Oxygen Demand (B.O.D.)

Organic matters are of two types, *i.e.* :

(*i*) That which is biologically oxidised (*i.e.* oxidised by bacteria) and is called **biological active** or **biological degradable**; and

(*ii*) That which cannot be oxidised biologically, and is called **biological inactive.**

While testing a waste water, we are mainly interested in finding out the amount of biologically active organic matter present in it, whereas, the above COD test gives us the total of biological active as well as biological inactive organic matter. Therefore, further testing is carried out to determine the biochemical oxygen demand (B.O.D.) of sewage, which directly gives us the amount of biologically active organic matter present in sewage.

If sufficient oxygen is available in waste water, the useful aerobic bacteria will flourish and cause the aerobic biological decomposition of waste water, which will continue until oxidation is completed. The amount of oxygen consumed in this process is BOD.

Therefore the BOD (Biochemical Oxygen Demand) can be defined as the oxygen requirement of bacteria to decompose the organic matter (present in a given sample of sewage) under aerobic conditions.

11.9. ESTIMATION OF SEWAGE DISCHARGE

The sewage discharge through a sewer must be estimated correctly, otherwise the sewers may either prove to be inadequate, or of too much of size, resulting in unnecessary wasteful investments. The quantity of sewage (*i.e.*, domestic sewage + industrial sewage) that is likely to enter the municipal sewers under design, in actual practice, is not the precise quantity which appears as sewage, but certain additions and subtractions do take place from it, as explained below :

1. **Additions due to Unaccounted Private Water Supplies.**

   Some private wells and tubewells may sometimes be used by the public for their domestic needs; and similarly, certain industries may utilise their own sources of water. This quantity can be estimated by actual field observations.

2. **Additions due to Infiltration.**

   Whenever, the sewer pipes are laid below the ground water-table, certain amount of ground water generally seeps into them, through their faulty leaky joints or cracks formed in the pipes. The quantity of the ground water entering these sewer pipes depends mainly upon the height of the water-table above the sewer invert level, and the nature and extent of faults and fissures present in the sewer pipes.

3. **Subtractions due to Water Losses.**

   The water lost, due to leakage in the distribution system and house connections of the water supply scheme, does not reach the consumers, and hence, never appears as sewage.

4. **Subtractions due to Water Not Entering the Sewage System**

   Certain amount of water may be used by the public and industries for such uses which may not produce any sewage at all. For example, the water used in boilers for steam generation; the water sprinkled over the roads, streets, lawns and gardens; the water used for automobile washings; the water consumed in industrial products, such as beverages, etc.,, the water used in air cooling etc.,, does not normally produce any sewage.

Net Quantity of Sewage Produced.

The net quantity of sewage produced will be equal to the accounted quantity of water supplied from the water-*works plus the additions due to factors* (*i*) and (*ii*) *minus the subtractions* due to factors (*iii*) and (*iv*), described above. *This net value may be taken as 70 to 80 per cent of the accounted water suppliea from the water works.*

11.10. DESIGN PERIOD OF A SEWAGE SCHEME

A sewerage scheme involves the laying of underground sewer pipes and construction of costly treatment units, which cannot be replaced or increased in their capacities easily or conveniently at a later date. To take care of the future expansions of the city and consequent increase in the quantity of sewage produced, it is necessary to design the scheme larger than their present day requirements and of such sizes, as to serve the community, satisfactorily, for a reasonable number of years to come. *This future period which the provision is made in designing the capacities of the sewerage scheme is known as the design period.*

The design period is generally guided by the following considerations :

(*i*) Useful life of component structures, and the chances of their becoming old and obsolete.

(*ii*) Ease and difficulty, that is likely to be faced in expansion, if undertaken at future dates. For example, more difficult expansions mean choosing a higher value of the design period.

(*iii*) Amount and availability of additional investment, likely to be incurred for additional provisions.

(*iv*) The rate of interest on the borrowings and the additional money invested.

(*v*) Anticipated rate of population growth, including possible shifts in communities, industries and commercial investments. For example, if the rate of increase of population is less, a higher figure for the design period may be choosen.

The following design periods are generally used in designing the different components of a sewage scheme.

S. No.	*Component structure*	*Reasons for selected design period*	*Design period in years*
1.	Lateral sewers, less than 15 cm in dia.	Requirements may change faster in limited area.	Full development
2.	Branches, mains and Trunk sewers	Difficult and costly to enlarge.	40—50
3.	Treatment units	Growth and interest rates being high to moderate.	15—20
4.	Pumping plant	The additional pumps can be installed in future, very easily.	5—10

11.11. FUTURE FORECASTS FOR SEWAGE DISCHARGE

The quantity of sewage that is likely to pass through a sewer at the end of the design period, can be easily computed by multiplying the *per capita production of sewage by the expected population at the end of the design period.*

The **per capita sewage** which is produced in a community can be easily estimated by assuming it as 75 to 80 per cent of the per capita water supplied to the public. However, is should also be kept in mind that the future increase in population may also increase the per capita water demand, and consequently increasing the per capita production of sewage. The increase in per capita water supply or sewage production with the increase in population, obviously occurs due to improved economical conditions in the city, implying higher standards of living and greater consumption of water. For normal Indian conditions, the following norms may be adopted :

Per Capita Water Demand and Sewage Production with Population

S. No. (1)	*Population* (2)	*Per capita water demand in litres/ day/ person* (3)	*Per capita sewage production in litres/ day/person ~ 80% of col. (3)* (4)
1.	Less than 20,000	110	90
2.	20,000—50,000	110—150	90—120
3.	50,000—2 lakhs	150—180	120—150
4.	2 lakhs—5 lakhs	180—210	150—170
5.	5 lakhs—10 lakhs	210—240	170—190
6.	Over 10 lakhs	240—270	190—200

The expected population at the end of the design period can be estimated by collecting the data of the past populations from the Census Department, and then by extrapolating the future population.

After determining the excepted population as well as the per capita sewage contribution of the town, by the end of the design period, the average quantity of sewage produced can be easily determined by multiplying both these figures.

11.12. DESIGN OF SEWERS AND STORM WATER DRAINS

In a separate sewerage system, which is common in modern days, the circular sewer pipes are laid below the ground level, sloping continuously at sufficiently steeper gradients towards the outfall point. The storm water drains (S.W. drains) are the separately constructed rectangular or trapezoidal surface drains at suitable gradients, and may be covered or kept open. The sewer pipes are designed to carry the maximum quantity of sanitary sewage likely to be produced from the area contributing to the particular sewer, while the S.W. drains are designed to carry the maximum drainage discharge.

Provision of Free Board

The *sanitary sewers,* as pointed out earlier, are designed large enough to carry the maximum sewage discharge while flowing half to three-fourth full. Generally, the sewer pipes of sizes less than 0.4 m dia are designed as running half full at maximum discharge, and the sewer pipes greater than 0.4 m in dia are designed as running $\frac{2}{3}$ *rd* or $\frac{3}{4}$ *th* full at maximum discharge.

The extra space, thus left, above the designed full supply line, will leave an ample margin, as to act as a factor of safety to counter-act against the factors, such as :

(*i*) Low estimates of the average and maximum flows, made due to some wrong data obtained regarding the private water consumption by industries, or public, or about the quantity to be supplied from the water-works at the end of the design period.

(*ii*) Large scale infiltration of storm water due to the wrong or illegal connections, and that of underground water through cracks or open joints in the sewers.

(*iii*) Unforeseen increase in population or water consumption and the consequent increase in sewage production.

The *storm water drains,* on the other hand, are generally not provided with so much margins, because the over-flowing of drains in not so much harmful, as is the overflowing of sewers, mainly because sewage is highly polluted as compared to the storm water. The storm water drains are, therefore, provided with nominal provisions of freeboard above their designed full supply lines.

Design of Water Supply Pipes V/S Sewer Pipes

Design of sewers and drains, means finding out their sections and gradients, and is generally carried out on the same lines as that of the water supply pipes. However, there are two major differences between the characteristics of flows in sewers and water supply pipes. These differences are :

1. The water supply pipes carry pure water without containing any kind of solid particles, either organic or inorganic in nature. The sewage, on the other hand, does contain such particles in suspension; and the heavier of these particles may settle down at the bottom of the sewers, as and when the flow velocity reduces, thus ultimately resulting in the clogging of the sewers.

In order to avoid such clogging or silting of sewers, it is necessary that the sewer pipes be of such a size and laid at such a gradient, as to generate self-cleansing velocities. The sewer materials must also be capable of resisting the wear and tear caused due to abrasion of the solid particles present in the sewage.

2. The water supply pipes carry water under pressure, and hence, within certain limits, they may be carried up and down the hills and the valleys; whereas, the sewer pipes carry sewage as gravity conduits (or open channels), and they must, therefore, be laid at a continuous

gradient in the downward direction upto the outfall point, from where the sewage will be lifted up, treated and disposed of.

QUESTIONS

1. Discuss the relative merits of the separate and combined systems of sewage. Give the conditions favourable for the adoption of each one of them.
2. Differentiate between domestic sewage, industrial sewage, and sanitary sewage.
3. Differentiate between combined and separate systems of sewage.
4. Write short notes on :
   (*a*) Types of sewages
   (*b*) Systems of sewages
   (*c*) Estimate the design sewage discharge.
5. Explain the following in brief :
   (*i*) Design periods for different components of a sewage scheme.
   (*ii*) Per capita sewage
   (*iii*) Variations in sewage flow and their effects on the design of sewer capacity.
6. What do you mean by variation in flow of sewage? Explain average flow, dry weather flow, and maximum flow.
7. Differentiate between (*i*) 'Sewage' and 'Drainage'. (*ii*) 'Sanitary sewage' and 'Storm sewage'.
8. Mention various aspects you would keep in view when designing a sewer.
9. Discuss various factors that affect the hydraulic of sewer lines.

Sewer Construction, Maintenance and Appurtenances

12.1. INTRODUCTION

The physical structure required for the collection and disposal of both sanitary sewage and storm water sewage should be constructed according to various parameters, such as the shape of the sewers, material used for the pipes, and the method of saying the sewers.

Sewer pipes, as pointed out earlier, are generally circular pipes laid below the ground level, and generally sloping continuously towards the outfall. They are designed to flow under gravity, except for the outfall sewer which carries the treated and pumped sewage into the discharging source, and hence flows full under pressure.

12.2. SHAPES OF SEWER PIPES

The sewer pipes, as mentioned above, are normally circular in section, although some other sections, such as egg shape, horse shoe shape, parabolic shape, semicircular shape, semi elliptical shape, rectangular shape, etc., may also be used under special circumstances. Out of the remaining shapes also, egg shape sections may be preferred for combined sewers, and rectangular shaped sewers are preferably constructed at site and normally, used as independent covered storm water surface drains, and not as sewers. All other forms of sewers, are almost out-dated and rarely used these days. The various forms of sewers, which can possibly be constructed and the circumstances in which each one is preferred, are shown in Fig. 12.1.

Circular sewers are economical and are suitable for small and medium diameters (both for laterals and mains).

The shapes and design of the sewer pipes should be such as to enable them to withstand the various forces likely to come on them. The following forces generally come into play in the sewer pipes :

(1) Internal pressure of sewage.

(2) Pressure due to external loads.

(3) Temperature stresses.

(4) Flexural stresses.

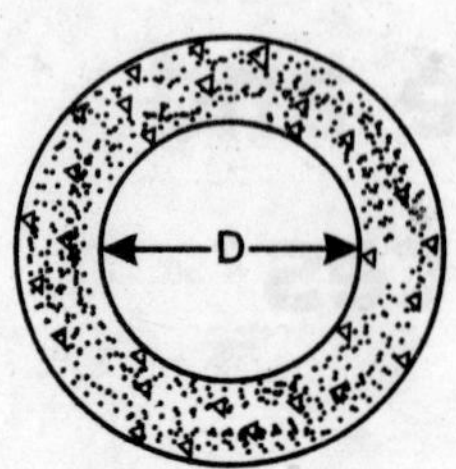

(a)
Circular shaped sewer (most widely used for all types of sewers).

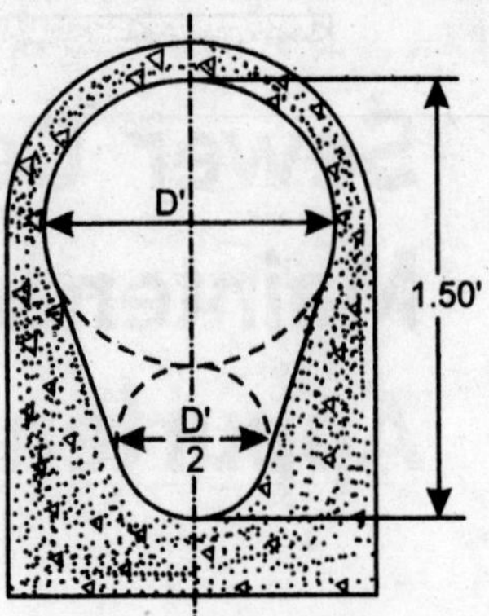

(b)
Standard Egg-shaped sewer (may be preferred for combined sewers).

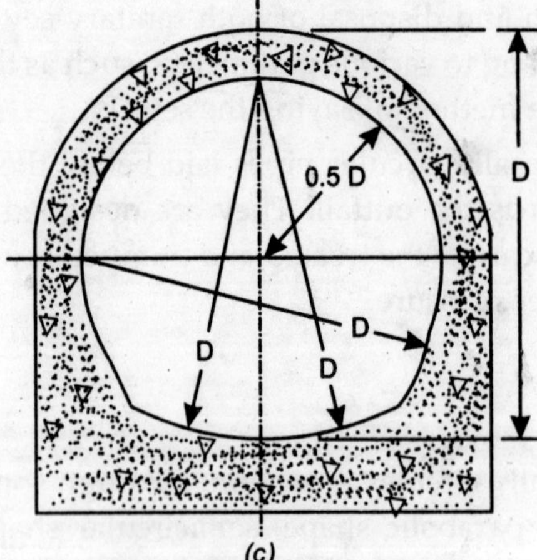

(c)
Horse shoe shaped sewer (may be used for large sewers with heavy discharge).

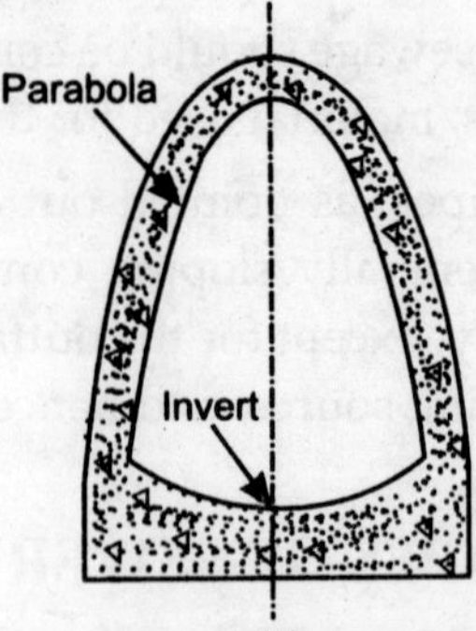

(d)
Parabolic shaped Sewer (may be used for carrying smaller quantities of sewage).

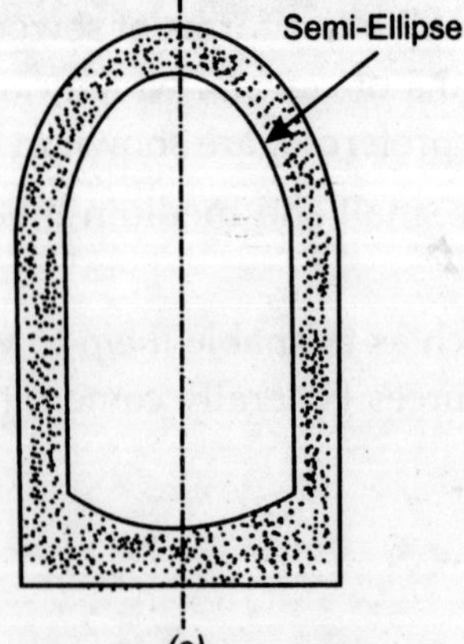

(e)
Semi-Elliptical Section useful for carrying large amounts of sewage, in sewers greater than about 1.8 m in a diameter.

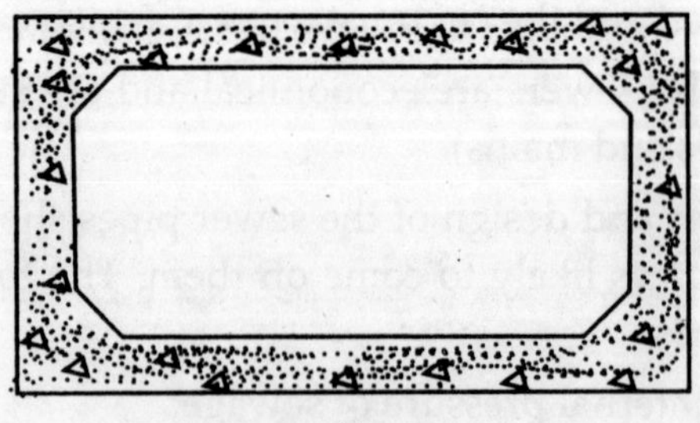

(f)
Rectangular shaped section (generally used to covered storm water drains).

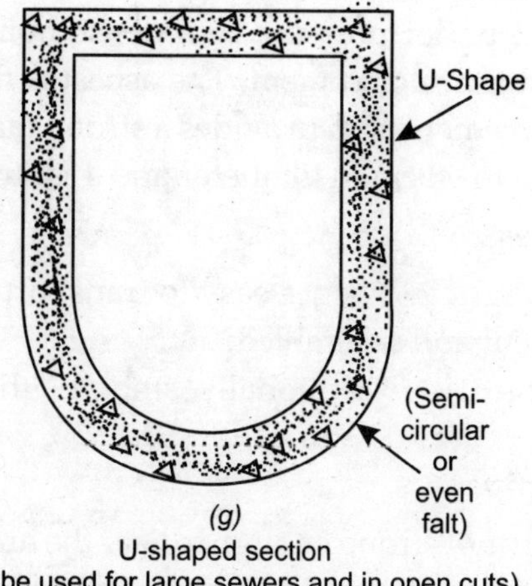

(g)
U-shaped section
(may be used for large sewers and in open cuts).

Fig. 12.1. Shapes of Sewer Section.

12.3. MATERIALS FOR SEWER CONSTRUCTION

Cement concrete, asbestos, cast iron, vitrified clay (or stoneware) are the most common materials used for constructing sewer pipes.

Factors considered while **selecting a material for sewer pipes :**

(i) ***Resistance to Corrosion***. The sewer pipes are likely to get corroded due to the presence of acid and other impurities in sewage. The sewer material should therefore, be resistant to corrosion so as to last for a longer life.

(ii) ***Resistance to Abrasion***. When the sewage contains a lot of grit and sand particles, moving at a high velocity at the sewer invert, a lot of wear and tear of the sewer material may be caused due to abrasion. To avoid this erosion or wear and tear of the sewer pipe, the sewer material must be strong enough to withstand abrasions.

(iii) ***Strength and Durability***. The sewer pipes should be strong enough to withstand all the forces that are likely to come on them. Since they are laid well below the ground level, they are subjected to considerable external loads. However, they are generally not subjected to the internal pressure of water (as the water pipes are subjected to).

(iv) ***Light Weight***. The material used for sewers should be light, so that the sewers can be easily handled and transported.

(v) ***Economical***. The sewer material must be overall economical in their construction.

(vi) ***Hydraulically efficient***. The sewer material should be such as to provide a smooth interior surface so as to provide an hydraulically efficient material.

Sewer Materials

12.3.1. Asbestos Cement Sewers

Asbestos cement pipes (A.C. pipes) are manufactured from a mixture of asbestos fibre, silica and cement, converted under pressure to a dense homogeneous material, possessing considerable strength, called asbestos cement. The asbestos fibre which is thoroughly mixed with cement serves as reinforcement, and provides a strong material. These pipes are normally available in sizes say from 10 to 90 cm in diameter and 4 metres in length.

Advantages of A.C. pipes

(*i*) They are light in weight and hence easy to transport.

(*ii*) They can be easily cut and assembled.

(*iii*) Their interior surface is exceptionally smooth, thus providing an hydraulically efficient sewer.

Disadvantages of A.C. pipes

(*i*) They are structurally not strong enough to bear the huge compressive stresses induced by the heavy external loads to which the deeply buried sewers may be subjected to.

(*ii*) They are susceptible to corrosion by sulphuric acid from hydrogen sulphide gas generated in sanitary waste water or by some industrial chemicals. Hence in all such cases, *vitrified clay (popularly called stone ware) pipes* should be used for sewers of less than 1 m in diameter, and cement concrete pipes with cast *insitu* plastic linings may be used for larger diameter sewers.

In view of their disadvantages, asbestos cement pipes are best suited to be used as verticals for bringing down either the rain water from the roofs, or the comparatively less foul sullage from kitchens and bath rooms situated at the upper floors of the buildings. The use of A.C. pipes for these purposes in place of cast iron pipes are economical.

12.3.2. Plain Cement Concrete and Reinforced Cement Concrete Sewers

Plain cement concrete pipes are manufactured in small sizes (*i.e.*, upto say 0.45 m diameter). Plain cement concrete pipes are non-pressure pipes, and classified as NP1 pipes according to IS. 458-1998; their thickness varying from 2.5 to 3.5 cm.

These are reinforced with steel reinforcement for larger diameter pipes. RCC pipes are easily available in sizes upto diameters say 1.8 metres, and may be got manufactured for larger diameter say upto about 4.5 metres, on special orders. These pipes may either be prepared at site by transporting ingredients (*i.e.*, cement, steel, aggregates, water, etc.,) or can be manufactured in factories, and then transported to site. They are known as *cast in situ* pipes in the former case, and *precast pipes* in the latter case. The precast pipes are cast in the factories and thus subjected to greater quality control and supervision.

The **non-pressure R.C.C. pipes** normally used for drainage and irrigation are classified according to IS : 458—1988 into the following three categories :

(*i*) **NP2 pipes.** They are light duty R.C.C. non-pressure pipes, normally used for culverts carrying light traffic. The thickness of NP2 pipes vary from 25 mm for 8 cm diameter pipe to 110 mm for 2.2 m diameter pipe.

(*ii*) **NP3 pipes.** They are medium duty non-pressure pipes, normally used for culverts medium traffic. The thickness of NP3 pipes vary from 25 mm for 8 cm dia pipe to 215 mm for 2.6 m dia pipe.

(*iii*) **NP4 pipes.** They are heavy duty non-pressure pipes, normally used for culverts carrying heavy traffic, such as railway loadings.

The unreinforced as well as reinforced concrete pipes shall be capable of withstanding a test pressure of 7 m head of water.

R.C.C. pressure pipes, classified as P1, P2 and P3 pipes, are generally used for carrying water supplies under pressure and are usually not used as sewers which are designed gravity conduits.

Hume steel pipes are also the R.C.C. pipes patented under this name, and consist of thin steel shells coated from inside with *cement mortar* by centrifugal process. The thickness of the inside coating varies from 12 mm to 30 mm depending upon the size of the pipe. They are also coated from outside, so as to protect the steel shell from external weather or soil action. The thickness of external coating is 25 mm for pipes upto 1 metre in diameter, and is 37.5 mm for pipes of larger diameters. The thickness of steel shell depends upon the size of the pipe, and also upon the pressure to be borne by the pipe. They are available in sizes ranging between 10 cm to 2.4 m in diameter, and in lengths varying from 0.9 to 2.4 m.

12.3.3. Vitrified Clay or Stoneware Sewers

These sewer pipes are widely used for carrying sewage and drainage, as *house connections* as well as *lateral sewers*. They are available in size of 5 cm increments from 10 to 30 cm, and in 7.5 cm increments from 30 cm to 90 cm. They are, however, rarely made in sizes bigger than 90 cm diameter.

These pipes are manufactured from clays and shales of special qualities, which are, first of all, pulverised and mixed thoroughly with water. This mixture is then used for casting standard pipe sections in a pipe press. These moulded pipe sections are dried in warm air and then brunt in hot kilns under controlled temperatures. The temperature of the kiln is maintained at about 150°C, in the beginning for several hours, and then raised to about 700°C, and finally to about 1200°C, when fusion or *vitrification* of clay takes place. This makes it very dense and hard. Near the end of the burning period, sodium chloride (*i.e.* common salt) is placed in the kiln. The intensive heat causes the salt to vaporise, which reacts with the clay, forming a thin smooth, hard, and a waterproof glazed layer on the pipe surfaces.

The advantages of pipes are :

(*i*) The stone-ware pipes are highly resistant to sulphide corrosion.

(*ii*) Their interiors are very smooth and they are hydraulically very efficient.

(*iii*) They are, through weak in tension, yet quite strong in compression, and hence they are quite suitable for withstanding compressive stresses caused by traffic and back-fills. They, however, can withstand only very small tensile stresses caused by internal pressures, hence not suitable for sewers flowing under pressure.

(*iv*) These pipes are quite cheap, durable, easily available, and can be easily land and jointed.

The disadvantages of these pipes are :

(*i*) They are heavy, bulky, and brittle, and, therefore, difficult to transport. Due to this reason, they are cast only in smaller sizes and smaller lengths. Due to their shorter lengths, they cannot be utilised as branch or main sewers.

(*ii*) They cannot be used as pressure pipes, because they are weak in tension.

12.3.4. Cast Iron Sewers

Cast iron pipes are structurally stronger and capable of withstanding greater tensile, compressive, as well as bending stresses, but are costlier, compared to cement concrete or stone ware pipes. They are, therefore, used as sewers, only under special circumstances, such as :

(*i*) For outfall sewers, for rising mains of pumping stations, or for inverted siphon, all running under pressure.

(*ii*) For sewers to be laid below heavy traffic loads, such as those laid below highways or railways.

(*iii*) For sewers carried over piers or trestle while crossing low lying areas.

(*iv*) For sewer which are to be 100% leak proof, so as to avoid possible contamination of under ground water supplies.

Cast iron pipes are though structurally quite stronger and durable, yet cannot withstand the corrosive action of gases and other acids present in sewage; and hence generally lined from inside with cement concrete, or painted with coal tar, etc., The sewer pipes are, therefore heavier than the water pipes, so as to resist the corrosive action of sewage.

Cast iron pipes can be manufactured by two methods. One is the ordinary **sand moulding method,** and the other is the **centrifugal process.**

12.4. LAYING AND TESTING OF SEWER PIPES

During planning a sewerage system for a city or town, the alignment of various laterals, branches and main sewers are, first of all, decided on a contour map of the area. The populations or catchment areas, to be drained by each pipe, are then marked. The required sizes of the pipe sections and their gradients are then computed, so as enable each pipe to pass the maximum sewage that may enter into it. Separate `L` sections are then drawn for each sewer line, showing the invert levels at different reduced distances and the typical cross-sections between them. The positions of manholes at suitable intervals along the route of the sewer line are also decided and marked. The positions of other appurtenances are also decided and marked on the maps.

The sewer pipes are then paid at their designed levels, the different pipe lengths to be suitably jointed, and finally tested for their corrected alignments and water tightness.

From the above we see that **detailed working drawing indicate the following;**

1. The alignment of sewers.
2. The size of sewers.
3. The position of manholes and other appurtenances.
4. The invert levels and grades at different positions.
5. The width and depth of trenches, and
6. The nature and type of bedding.

For laying of sewers following steps are involved.

1. Maintain proper slopes, as the sewage flows under gravity.
2. Sewers are laid on the foundation beds to avoid setting.
3. In water logged areas or when the trench goes below the groundwater table, the dewatering has to be made. The sides of the trenches should be provided with timber planks and streets.
4. The invert level of the sewer has to be properly verified.
5. Sewers are laid, jointed and tested for leakage before the trench is filled back with excavated earth.
6. To avoid deeper excavations, sewage may be lifted (pumped) to higher levels and again allowed to flow by gravity.
7. Flushing tanks are provided to maintain the self cleaning velocities, if slope is not sufficient.
8. Manholes are provided at regular intervals; bends, junctions etc., are also be provided at appropriate places, so that inspection, cleaning and general maintenance can be done with ease.
9. Sewers are likely to contain toxie gases like, H_2S, CO_2, CH_4, SO_2, NH_3, etc., which may cause explosions. Sufficient care must be taken to provide ventilation during inspection.

Sewer pipe lengths are usually laid from the lowest point with their socket ends facing upstream. In this way, the spigot of each pipe can be easily inserted in the socket of the already laid pipe. After fitting the socket and spigot into each other, the proper jointing is done with lead caulking or cement mortar as required. After the sewer has been laid in this way, it is tested for leakage and correct alignment.

Testing of the Sewer Pipes

The sewers after being laid and jointed, are tested for (*i*) watertight joints, and also (*ii*) correct straight alignment.

1. **Test for Leakage (Water Test):** Sewer pipe sections are tested between manhole to manhole under a test pressure of about 1.5 m of water head. The sewer line, under this pressure, is watched by moving along the trench, and the joints which leak or sweat, are repaired, the leaking pipe, if any, will also be replaced.

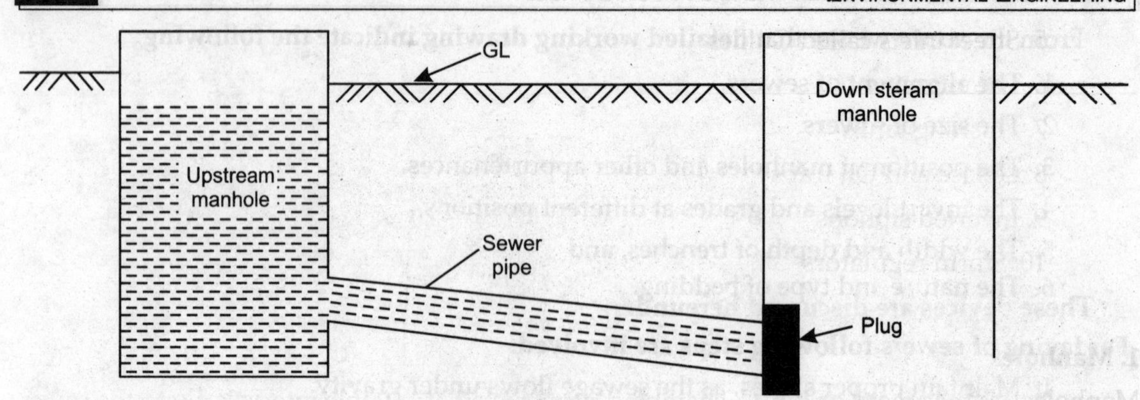

Fig. 12. 2. Sewer Water Test

2. **Test for Straightness of Alignment:** The straightness of the sewer pipe can be tested by placing a mirror at one end of the sewer line and a lamp at the other end. If the pipe line is straight, the full circle of light will be observed. However, if the pipe line is not straight, this would be apparent and the mirror will also indicate any obstruction in the pipe barrel.

Any obstruction present in the pipe can be tested by inserting at the upper end of the sewer, a smooth ball of diameter 13 mm less than the internal diameter of the sewer pipe. In the absence of any obstruction, the ball shall roll down the inverts of the sewer pipe and emerge at the lower end.

Back-filling of the Trenches

After the sewer line has been laid and tested, the trenches are back filled. While back filling the excavated earth in the trench, the earth should be laid equally on either side of the sewer, and the earth should be filled up in layers of about 15 cm thickness. Each layer should be properly watered, tamped and rammed.

However, the earth filling above the crown of sewer pipe should be carefully carried out by hand shovelling in layers and using selected soils (*i.e.* soft earth without any stone rubbish or old lumber). After a few months of exposure, when this top layer gets fully settled, the road pavement may be constructed. This will prevent their subsidence and cracking.

12.5. SEWER APPURTENANCES

Sewer appurtenances are those structures which are constructed at suitable intervals along a sewerage system, and help in its efficient operation and maintenance. The common sewer appurtenances are.

1. Manholes
2. Drop manholes
3. Lampholes
4. Clean-outs

5. Street inlets called Gullies
6. Catch basins
7. Flushing tanks
8. Grease and Oil traps
9. Inverted siphons
10. Storm regulators

These devices are discussed hereunder:

1. Manholes

Manholes are masonary or R.C.C. chambers, constructed at suitable intervals along the sewer lines, for providing access into them.

The manholes, thus, help in joining sewer lengths, and also help in their inspection, cleaning and maintenance. If the manhole covers are perforated, they may also assist in ventilating the sewers, but will cause release of pungent gases.

The manholes are generally provided at regular intervals in a straight sewer line, and also at points of every bend, junction, change of gradient, or change of sewer dia. The spacing between the manholes, in such a case, however, depends mainly upon the size of sewer line. The larger is the diameter of sewer, the greater will be the spacing between the manholes.

Manholes spacings as per IS 1742-1960

Size of the Sewer	*Recommended Spacing of Manholes on straight reaches of Sewer lines*
Dia upto 0.3 m	45 m
Dia upto 0.6 m	75 m
Dia upto 0.9 m	90 m
Dia upto 1.2 m	120 m
Dia upto 1.5 m	250 m
Dia greater than 1.5 m	300 m

Classification of Manholes. Depending upon their depth, the manholes may be classified as :

1. Shallow manholes – about 0.8 m in depth.
2. Normal or medium manholes – about 1.5 m in depth.
3. Deep manhole – having more than 1.5 m depth.

2. Drop Manholes

When a branch sewer enters a manhole by more than 0.5 m above the main sewer, the sewage is generally not allowed to fall directly into the manhole, but is brought into it through a down pipe taken from the branch sewer to the bottom of the manhole. The manhole, in which a vertical pipe is used, is called a drop manhole, whereas the one using an inclined pipe is called a ramp.

3. Lamp Holes

Lamp holes are the small openings on sewers to permit the insertion of a lamp into the sewer. The lamp light is then viewed from the upstream as well as the downstream manholes. The obstructed light confirms the obstructions in the sewers.

4. Clean-outs

A clean-out is an inclined pipe extending from the ground and connected to the underground sewer. A clean hyping out is used for cleaning sewer pipes. A clean out is generally provided at the upper ends of lateral sewers in place of manholes.

The functioning of a clean out is very simple, and consists in removing the top cover and forcing water through the clean out pipe in lateral sewers to remove obstacles in the sewer line. If obstructions are large enough, a flexible rod may be inserted through the clean out pipe and pushed forward and backward to remove such obstacles.

5. Street Inlets (Gullies).

These are openings on the road surface at the lowest point for draining rain water from roads, and admitting it into the underground storm water sewers (drains) or combined sewers.

These inlets are located along road sides at an interval of about 50 m. These inlets are connected to the nearby man holes by pipe lines. A street inlet is a simple concrete box having gratings or openings in vertical or horizontal direction.

6. Catch Basins (Catch Pits)

These are the street inlets provided with additional small settling basins. Grit, sand, debris etc., do settle in these basins, and their entry into the sewer is thus prevented.

Catch basins need periodical cleaning, as otherwise, the settled organic matter may decompose, producing foul odours, and may also become a breeding place for mosquitoes.

7. Flushing Tanks

Whenever, there are any chances of blockage of sewer pipes, such as in the case of sewers laid on flat gradients not producing self cleansing velocities, or near the dead end points of sewers, flushing devices are installed. These devices store water temporarily, and throw it into the sewer for the purpose of flushing and cleaning the sewer. Such devices are called flushing tanks.

8. Grease and Oil Traps

Grease and oil traps are those trap chambers which are constructed in sewerage system to remove oil an grease from the sewage before it enters into the sewer line. Such traps are located near the sources contributing grease and oil to the sewage. They are, therefore , generally located at places, such as, automobile repair workshops, garages, kitchens of hotels, oil and grease industries etc.,

The removal of oil and grease from the sewage before it enters into the sewer pipe is considered necessary because of the following reasons:

(1) The grease and oil, if allowed to enter the sewer, will stick to the sewer sides, thereby reducing the sewer capacity.

(2) The suspended matter in the sewage, also sticks to the sides of the sewer, due to sticky nature of oil and grease, thus further reducing sewer capacity.

(3) The presence of oil and grease in sewage makes the sewage treatment difficult.

9. Inverted Siphons

Whenever a sewer pipe has to be dropped below the gradient line for passing it beneath a valley, a road, a railway, a stream or any other depression in the earth's surface or where it passes beneath some other obstructions in the path, it is known as an inverted siphon. The sewage through such a pipe line will not flow under gravity, but will be flowing under pressure (as in the case of water pipe lines).

10. Storm Water Regulators

Storm water regulators are constructed in the combined sewerage systems, and permit the diversion of excess storm water into nearby stream or river. This excess sewage will be mainly composed of storm water, and therefore, it will not be foul in character.

12.6. MAINTENANCE OF SEWERS

Sewer maintenance generally involves their cleaning to keep them free from any clogging and to carry out repairs to the damaged portions, if any, so as to prolong their life and to ensure efficient functioning.

The proper maintenance of sewers in necessary in order to make the sewerage system function efficiently. The sewer maintenance generally includes:

(*i*) Frequent inspection and supervision,

(*ii*) Measuring rate of flow,

(*iii*) Cleaning and flushing,

(*iv*) Repairing the leaking joints or any other damaged portions; if any,

(*v*) Protecting them against their misuse,

(*vi*) Preventing explosions,

(*vii*) Removal or prevention of stoppages,

(*viii*) Ventilation of sewer gases,

(*ix*) Repair of appurtenances as and when required,

(*x*) Cleaning of street inlets, etc.,

The most general complaints received are about clogging. The sewers get blocked or clogged due to silting and deposition of debris, grit and other floating matter present in the sewage. This silting is more pronounced at low discharges, and in sewers laid on flatter gradients. Greasy and oily matters from kitchens or from such industries, further aid in clogging. Oil and grease traps should, therefore, be constructed at the sources of sources of such oil greasy sewage, and also along the sewer line.

Many a times, the general public also throw away their domestic wastes, and garbage, etc., into the manholes, which may find its way into the sewers, thus increasing the chances of their deposition and consequent blockage of the sewer pipes.

Inspite of all precautions taken to avoid their silting, sewers do silt, and, therefore, require frequent cleaning.

Inspection of Sewerage System

Inspection should be a routine practice. Besides, there should be emergency or special inspection when some trouble arises. Based on the inspection, maintenance programme with respect to following may be taken up :

(*i*) Ventilation of sewers,

(*ii*) Cleaning of sewers,

(*iii*) Roots penetration in the sewers,

(*iv*) Cleaning of street inlets and connection,

(*v*) Working of flushing tanks,

(*vi*) Repair work of sewers and appurtenances, and

(*vii*) Inspection and maintenance of pump house and treatment plants.

Equipment used in Maintenance

The equipment required for maintenance depend on the size and types of sewers serviced. A small city requires less and simples equipment than a large city. A fairly large city will require : trucks, power winches, sewer jetting machines, excavator, root cutters etc., The other manually operated or other minor equipment required may be : flexible steel rods fire hose, shovels, picks, hydrant, manhole tools, buckets, ropes, flash lights, split bamboos, safety lamps, hydrogen sulphide detectors, carbon monoxide detectors, traffic signs, flags, first aid kits, personal protection equipment (PPEs) like rubber boots, gloves, safety belts, ropes.

12.7. CLEANING OF SEWERS

Sewers should be periodically cleaned so as to avoid their clogging. For cleaning small sewers which cannot be entered into by manual labour, flushing operations are essential. A flexible fire hose with about 2.5 cm or less sized nozzle can be inserted down, and the water discharged under pressure. The water pressure stiffens the rubber hose pipe, and make it possible to push the hose in and out in the sewer, thus helping in flushing and cleaning of the sewer. Flushing can also be carried out by tipping larger volumes of water into the manholes.

Medium sized sewers are generally cleaned by using scraping instruments, which are dragged through the sewer. A commonly used device consists of a steel cylinder of 20 to 25 cm diameter and 60 to 75 cm long, and provided with a sharp cutting teeth on the open forward end. This type of a sewer cleaning rod called cane rodding is dragged in the sewer to and fro from the manholes, and pushed down the sewer until the obstruction is reached and removed.

Large sized sewers are entered into by men and generally cleaned by manual labour. However, sufficient precautions and care must be taken while entering the sewers, as otherwise, sometimes, foul, poisonous and explosive gases may cause serious hazards to the workers entering the sewer pipes.

Precautions While Entering Sewers

Various poisonous and explosive gases which and generally found present in sewers are : hydrogen sulphide (H_2S), carbon dioxide (CO_2), and methane (CH_4) alongwith petrol vapours. These gases are largely produced when the sewage gets stale and septic, which more commonly happens in hotter climates. In addition to the production of these hazardous gases, oxygen gets consumed by the organic matter and thus resulting in the shortage of oxygen inside the sewer, and may cause difficulty in breathing.

In order to avoid these problems, *ventilation of sewers* is generally adopted. In addition to the ventilation of sewers, the following precautions should also be taken while allowing the workers to enter the sewers.

1. Before entering a sewer through a manhole open the cover of the manhole in question as well as the covers of two others manholes, one on the upstream and the other on the downstream ; atleast half an hour in advance. This will help in obtaining some ventilation and exposure of the sewer to the atmospheric oxygen.

2. Tests should be carried out to detect the presence of any hazardous gases inside the sewer. The following tests may be carried out to detect their presence :

(*a*) H_2S gas may be detected by exposing a sheet of paper moistened with lead acetate for 5 minutes near the sewer entry. If the paper turns black, the presence of H_2S gas is indicated.

(*b*) Presence of CO_2 gas may be detected by lowering a miners safety lamp near the level of sewage in the manhole. If the flame extinguishes within 5 minutes, the presence of CO_2 gas is indicated.

(*c*) Presence of methane (CH_2) gas may be detected by lowering the minor's safety lamp in the upper layers of the sewer; methane, being lighter than air, is generally present in the upper layers of sewer. When the gas is present, it forms an explosive mixture with air, and the gauge cylinder of the safety lamp gets filled with the flame. The formation of flames will indicate the presence of methane gas inside the sewer.

When any of these gases are found to be present, special breathing apparatus fitted with hose masks should be used by the workers entering the sewers.

3. If the above hazardous gases are absent, a lighted lantern may be lowered down the manhole, so as to test the presence of oxygen. If it burns brilliantly, the sewer can be safe to enter.

4. The workers going down for inspection must be tied with ropes to their waists, and they may be held fast by the persons on top, so that they may be immediately pulled up in case of any risks. A responsible officer should preferably be present to direct the operation.

5. Any smoking or carrying of naked light inside the sewers are strictly prohibited.

QUESTIONS

1. Mention the qualities of a good material for constructing the sewers. Judging from this requirement discuss the suitability of
   (*a*) Cement concrete, (*b*) Stoneware, and
   (*c*) Reinforced cement concrete for sewers.
2. State the type of sewers used in sewerage used in sewage system, giving their use in sewerage system.
3. What are the factors considered while estimating the quantity of sewage of a town?
4. Explain as to how is storm water flow computed?
5. Explain the rational method of design of storm water sewers. Explain the relation between design rainfall intensity and the time of concentration.
6. Describe the laying of a sewer line in a trench.
7. Explain in brief about maintenance of sewers and their cleaning.
8. Write a procedure for estimating sewage and storm water flows.
9. How are sewers designed?
10. Explain in brief on:
    (*a*) Drop manhole (*b*) Storm water regulator
    (*c*) Catch pit (*d*) Flushing tanks.
11. Describe in order the various stages followed in the construction of sewers.
12. Write in detail about
    (*a*) Laying, jointing, and testing of sewers,
    (*b*) Consideration involved in determining the size of sewers.
13. Enumerate various forces that act on a sewer pipe laid underground.
14. Write a detailed note on 'manholes for sewers'.
15. Discuss the points you would keep in mind while deciding upon the programme for construction of sewers in a town.
16. Explain how will you test the newly laid sewer lines before bringing them into commission.
17. State different materials and shapes of sewers, and explain their merits and demerits.
18. Write short notes on :
    (*a*) Grease and oil trap, (*b*) Lamp-hole, and
    (*c*) Precautions taken during maintenance of sewers.

Sewage Pumping and Pumping Stations

13.1. INTRODUCTION

Sewage is required to be lifted up from a lower level to a higher level at various places in sewerage system. Sewage is lifted by pumps under the following circumstances :

(*i*) The sewage from localised low lying pockets in a city has to be pumped, so as to throw it up into the city's sewer pipes running at higher elevations.

(*ii*) When the area is flat, the laying of sewers at their designed gradients may involve deeper and deeper excavations (Fig. 13.1 (*a*)) in the forward direction of flow. In such circumstances, it may be advisable to lift the sewage at suitable intervals, and then to lay sewers at reasonable depth below the surface, as shown in Fig. 13.1 (*b*).

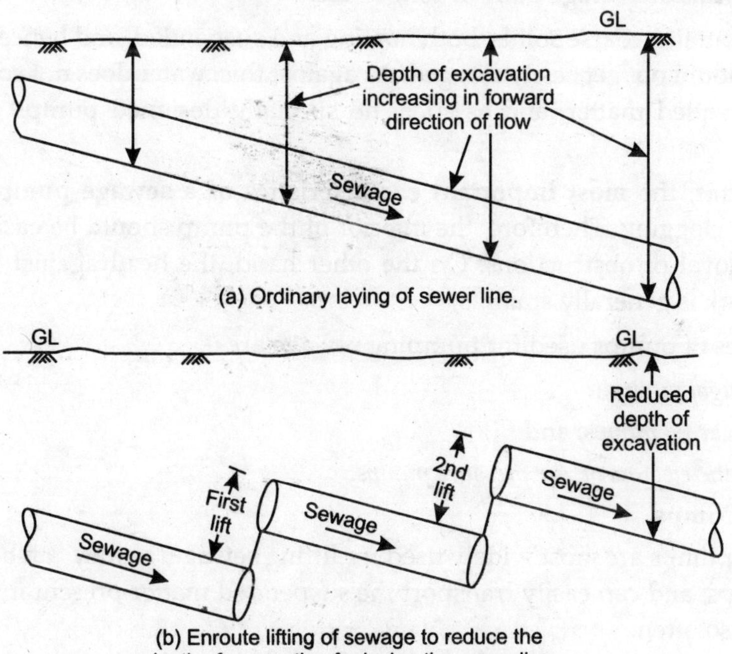

(a) Ordinary laying of sewer line.

(b) Enroute lifting of sewage to reduce the depth of excavation for laying the sewer line.

Fig. 13.1.

(*iii*) For disposing of the sewage of the basements of large commercial buildings, sewage may have to be pumped, as the street sewer may be higher than the level of the basement floor.

(*iv*) When the outfall sewer is lower than the level of the treatment plant, the sewage will have to be pumped up. Similarly, when the level of the treated sewage coming out from the treatment plant is lower than the level of the source of its disposal, it will have to be pumped before it can be disposed of.

(*v*) When a sewer has to go across a high ridge, then instead of driving a tunnel through the ridge, it is economical to pump the sewage into sewers the slope of the ridge at reasonable depth.

13.2. TYPES OF PUMPS

Various types of pumps that are commonly used for *lifting water* have already been discussed in Part I—"Water Supply Engineering" (Chapter 6). Before we discuss them here the types of pumps used for *lifting sewage*, it may be mentioned that there are two essential differences between the requirements of a pumping station in a water supply system and that in a sewage system :

1. Sewage cannot be stored except for short periods of time, and the flow is highly variable; hence in a sewage pumping station provision must be made for handling peak flows. In order to ensure this, two or more pumps, and in some cases two sources of power, are needed. As against this, water can be stored easily for future use, and as such, water supply pumps may be sized on the basis of average daily or longer use.

2. Sewage contains coarse solids, both floating and suspended, and hence requires special pumps which should not get easily clogged. As against this, water does not contain significant amount of suspended matter, and as such, no specially designed pumps are required for pumping water.

In view of this, the **most important characteristics of a sewage pump** is its ability to operate without clogging. Therefore, the interior of the pump should be easily accessible for cleaning or removal of obstructions. On the other hand, the head against which a sewage pump has to work is generally small.

Various types of pumps used for pumping sewage are :

1. *Centrifugal pumps* :
2. *Reciprocating pumps*; and
3. *Pneumatic ejectors or Air pressure pumps*.

1. Centrifugal Pumps.

Centrifugal pumps are most widely used for lifting sewage, as they can be easily installed in pits and sumps, and can easily transport the suspended matter present in sewage without getting clogged so often.

The horizontal axis type of centrifugal pumps are generally used. They are provided with either open or closed three vane type impellers. Open impellers are preferred for pumping sewage containing more solids. The clearance between the vanes is kept large enough so as to allow any solid matters entering the pump to pass out with the liquid sewage. This helps in preventing the clogging of pumps and damage to the rotor. Such pumps are, therefore, called **non-clog pumps.**

These centrifugal pumps, though, work at low efficiency (of the order of 30 to 60 per cent), but their use avoids the necessity of installing pre-screens for grit removal before pumping. However, where sewage has already been screened, ordinary centrifugal pumps of high efficiency may be used.

Other types of centrifugal pumps including submersible pumps have already been discussed in the chapter on "Pumps for Lifting Water."

2. Reciprocating Pumps.

Reciprocating pumps are generally not used these days, because of their high initial cost, difficulty in maintenance and greater wear and tear of their valves. However, they may sometimes be used for pumping difficult sludges and where large quantity is required to be pumped against low heads. They should, however, be amply protected by screen of 2 cm spacing.

Various types of reciprocating and displacements have already been discussed in the chapter on "Pumps for Lifting Water.

3. Pneumatic Ejector.

Pneumatic ejectors sometimes called *Air ejectors* or *Air Pressure pumps* are complete lift units in themselves, and are used for pumping smaller quantities of waste water, such as for rasing waste waters from the basements of building and thus discharging it into street sewers. Such a complete unit can be purchased ready made or fabricated for direct installation.

A pneumatic ejector, in general, consists of an air-tight tank into which the waste water flows by gravity, and out of which, the waste water is forced automatically whenever sufficient waste water has accumulated to raise a float, and thus opening the compressed air inlet valve.

The pneumatic ejectors are often used in pairs so that when one is empty, the other is filled.

Advantage of Air Ejectors.

1. There is no necessity of providing primary treatment arrangement for sewage, when air ejectors are employed.
2. These are automatic complete units, and require least supervision.
3. They are useful in conditions where there are chances of small capacity centrifugal pumps getting clogged, if used.
4. Such ejectors are not likely to be closed anywhere.
5. Compared to the construction of a pumping station, They are found to be economical at places where smaller quantities of waste waters are required to be lifted.

Disadvantages of Air Ejectors.

The only disadvantages of air-ejectors is that they have very low efficiency, of the order of 15 per cent or so.

13.3. CAPACITY OF PUMPS

It is desirable to install minimum of four pumps in a pumping station. Their capacities are divided so that one has capacity equal to minimum rate of flow, one is equal to the average rate of flow, and one is equal to the maximum rate of flow. It is further advisable that the combined capacities of two smaller pumps may be equal to the largest one. A standby pump with standby power should be provided with a capacity equal to the largest pumping unit. If the number and capacity of pumping units are properly choosen, the fluctuation in pumping rates are not considerable and frequent cutting-in and cutting-out in the cycle of pumping operation is avoided.

The guidelines for proper functioning of a sewage pumping station are as under :

1. Main pumps should be provided with atleast three pumping units having capacities such that if the largest unit is out of service for repairs the remaining units can pump the peak flow sewage.
2. Driving power should be available from atleast two sources.
3. The sewage pumping station should be provided with sufficient standy units.
4. Pumps are generally placed in the dry wells. This will help cleaning, maintenance and repair of pumps easy.
5. All pumps for raw sewage should be preceded by screens.
6. An emergency overflow or bypass should be installed wherever possible for the protection of the pumping station.
7. Suction and discharge pipes should not be less than 100 mm diameters.

13.4. SEWAGE PUMPING STATIONS

The building where pumps and other accessories are installed for lifting sewage is called a sewage pumping station. Such a building should be located at a carefully choosen site.

These pumping stations should preferably be located near a natural disposal unit, such as, a stream, a lake or a river, etc.,, for the disposal of treated sewage, but at the same time the site should be such that it is not liable to get flooded either due to seepage from the adjoining river stream or due to high floods in that river stream.

Elements of Pumping Stations

A typical sewage pumping station is shown in Fig. 13.2. Such an installation has the following major components :

1. Grit channel.
2. Coarse and fine screens.
3. Wet Well, called Sump well.
4. Dry well or Pumps room.

5. Motor room.
6. Rising mains; and
7. Other accessories starters, valves, flow recorders, etc.

The construction and functioning of these components is briefly discussed below :

1. Grit Channel.

The sewage entering the pumping station generally contains a lot of solid matter, such as, grit, gravel, sand, etc., in addition to the solids in suspension like feces, papers, rags, etc., It is necessary to remove as much of this material as possible before pumping, so as to minimise the wear and tear of the pump impeller and of the rising main.

The solid matter is, therefore, first of all, removed before the sewage enters the pump. In order to achieve this purpose, the sewage is first passed through grit channel (or grit chamber), where its velocity is considerably reduced. Due to the reduced velocity, the grit settles down in a pit and is removed by an endless chain to which perforated buckets are fixed. The chain is operated by power, and grit is dredged into a closed container, till it is removed and disposed of.

2. Screens.

After the sewage has been excluded from the solid grit matter by settling it down in the grit channel, it is passed through the screens; which help in extracting other matter, such as, excess of rags, sticks, papers, etc., Such matter, if not removed, may cause the choking of pump impellers.

However, use of screens may be avoided by using disintegrating pumps (as pointed out earlier).

Two types of screens commonly used are:– *coarse screens* and *fire screens*. Normally, in large installations, both these types of screens are provided together. The sewage is first passed through coarse screens to intercept solids like scrubbing brushes, blocks of wood, etc., The sewage is then passed through fine screens to intercept all except very fine particles of sewage. The screens are cleaned mechanically by rakes having fingers attached to either end of an endless roller chain running over sprockets.

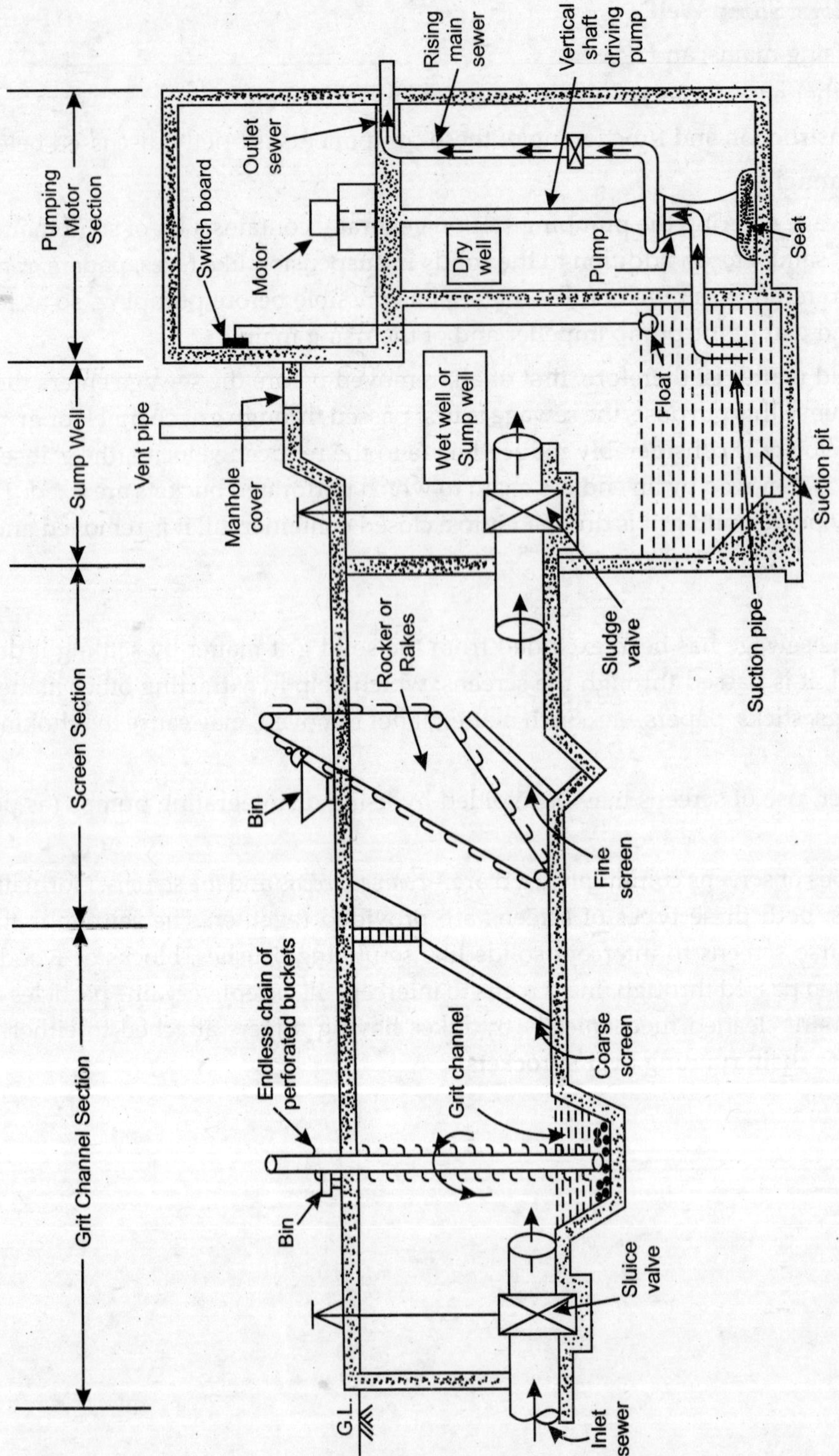

Fig. 13.2. A Typical Sewage Pumping Station.

3. Wet Well or Sump Well

The purpose of providing a sump well is to form a suction pit from which the pump may draw sewage through the suction pipe. It also acts as an equalising basin to minimise the load fluctuations on the pump. A float connected to a switch is provided in this sump well is such a way that when the sewage rises above the float level, the switch gets pressed and pump automatically starts functioning, thus pumping sewage. This designed level of the sewage in the sump well is kept above the pump level, which avoids the necessity of priming of the pump. The capacity of the sump well should be about 15 to 30 minutes of the peak flow.

4. Dry Well or Pump Room

In this room, pumps are installed, and the end of their suction pipe is placed in the suction pit or the wet well. The general layout of a pump room should include a repair workshop, office and store room, etc., Enough provision for future extension may also be made. The number and sizes of pumps required should be carefully worked out, and depends upon the quality of sewage likely to enter into the station.

The provision for stand-by units should also be made.

5. Motor Room

This room is situated above the pump room, and accommodates the electric motors to drive the pumps.

6. Rising Mains

The sewage, after being pumped, is taken to the high levelled gravity sewer through the rising mains. The rising main may be of cast iron or asbestos cement pressure pipes. To avoid siltation, the velocity of flow in the rising main should not be less than 0.75 m/sec. at any time of flow. To avoid excessive head loss, the maximum velocity of flow in the rising main is usually limited to 1.8 m/sec.

QUESTIONS

1. When is pumping of sewage required in a sewage system. How is it more difficult than the pumping of water in water supply schemes?
2. Draw a sketch of a sewage pumping station, showing all the necessary arrangements to pump sewage of a town of one lakh population at the outfall sewer.
3. Under what circumstances is pumping of sewage necessary? What types of pumps would you recommend for pumping sewage and why?
4. Write short notes on the following with respect to sewage plant.
   (*a*) Selection of type and capacities of the pump,
   (*b*) Number of pumps and standby capacity of the pumping units.
5. State the pumps usually employed for the pumping of sewage and sludge. How is capacity of the wet well in a sewage pumping station determined?
6. Differentiate between ejectors and pumps. Describe the working of an ejector.

Sewage Collection from Residential and Other Buildings

14.1. INTRODUCTION

The sewage produced in residential and other buildings is required to be removed and conveyed to the municipal sewers, as early as possible with minimum nuisance or hazard to health. For this purpose an efficient plumbing system is required to be constructed.

The plumbing system includes fixtures for collecting waste and foul matter let out through traps into branches, carry through the stack, convey through house drains and sewers and finally discharges to the public sewers.

The building drainage include water closets, urinals, traps, waste pipes, inspection chamber, house sewer lines and a number of other fittings and fixtures. In order to safeguard from future problems due to bad or improperly designed plumbing works, a properly designed and carefully constructed an efficient plumbing system in any building is necessary.

Various **sanitary fittings** provided, in buildings, *e.g. wash basins, bath tubs or bath rooms, kitchen sinks, urinals, latrines (with their flushing cisterns)*, etc., do contribute to the sewage, and hence a *system of pipes* should be laid and connected with the sanitary fittings in such a way that the sewage (or sullage) automatically flows by gravity through this system, and ultimately discharges into the municipal sewer (also known as public sewer).

14.2. REQUIREMENTS OF SANITARY PLUMBING SYSTEM

A plumbing system should be designed keeping following requirements (principles) into consideration :

1. Waste water from latrines, urinals, sinks, bathrooms, wash basins etc., should also flow by gravity, automatically with self cleaning velocity.
2. The number of sanitary installations should depend upon the number of persons using them and the type of building in which they are installed, *viz.* hotel, bus stand, cinema, hostel, factory, office, railway station or any other building.
3. All the pipes should be laid in the outer walls or underground. This enables easy access for repair and escaping of foul gases, if any,

4. The outfall house sewer should be at elevation so as to prevent back flow from municipal sewer.
5. All sharp bends and junctions, which tend to become dead-ends for collection of sewage solids, should be avoided except through manholes.
6. The entire system should be properly ventilated from the start to the end.
7. The plumbing system should contain enough number of *traps* (water seals), at suitable points for its efficient functioning or to avoid evolution of foul smells.
8. The *house sewer* should be disconnected from the *public sewer* by the provision of an ***intercepting trap***, so as to prevent foul gases from public sewer to enter the house sewer.
9. Joints of sewers should be water-tight.
10. Lateral sewers should be laid at sufficient slopes, so as to develop self cleaning velocities.
11. Adequate ventilation eliminates the foul gases.

14.3. TRAPS USED IN SANITARY PLUMBING

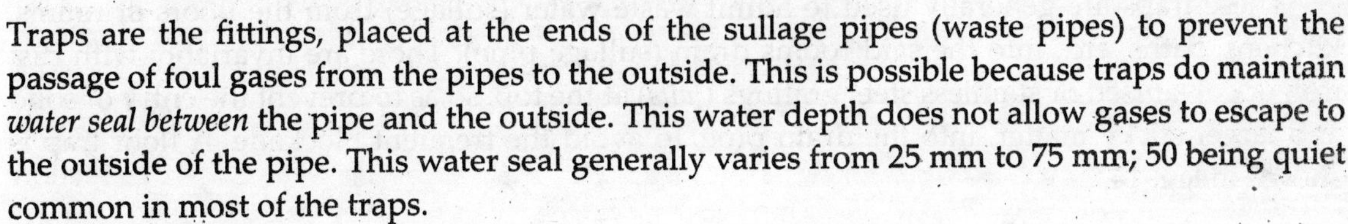

Traps are the fittings, placed at the ends of the sullage pipes (waste pipes) to prevent the passage of foul gases from the pipes to the outside. This is possible because traps do maintain *water seal between* the pipe and the outside. This water depth does not allow gases to escape to the outside of the pipe. This water seal generally varies from 25 mm to 75 mm; 50 being quiet common in most of the traps.

14.3.1. Types of Traps

Traps can be classified as under :

1. According to their shapes
2. According to their use

1. *Depending upon their* shapes, the traps are of three types, *i.e.*

(*i*) *P-trap;*

(*ii*) *Q-trap; and*

(*iii*) *S-trap.*

These three types of traps are shown in Fig. 14.1. A trap essentially consists of a U-tube, which retains water, acting as a seal, between the foul gases (inside the pipe) and the outside atmosphere. They are largely used for baths, sinks, and laboratories, where they are made with enlarged mouth, so that the waste pipe may be thoroughly flushed out.

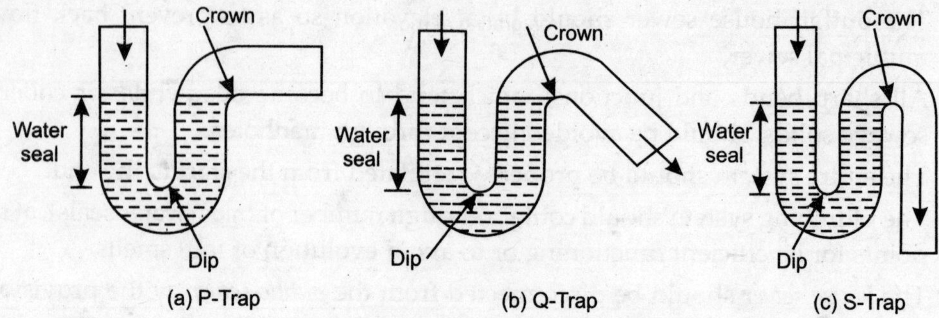

Fig. 14.1. P, Q and S Traps.

2. ***Depending upon their* use**, traps may again be of three types *i.e.*
   1. *Floor trap;*
   2. *Gully trap;*
   3. *Intercepting trap.*

These three traps are briefly described below :

1. Floor Traps

These traps are generally used to admit waste water (sullage) from the floors of rooms, kitchens, baths, etc., into the said rooms drain (sullage pipe). These are invariably with cast iron or galvanised or stainless steel gratings (*Jalls*) at the top, so as to prevent the entry of solid and larger sticky matter, into the drain pipe, to avoid the frequent blockade. A floor trap is shown in Fig. 14.2.

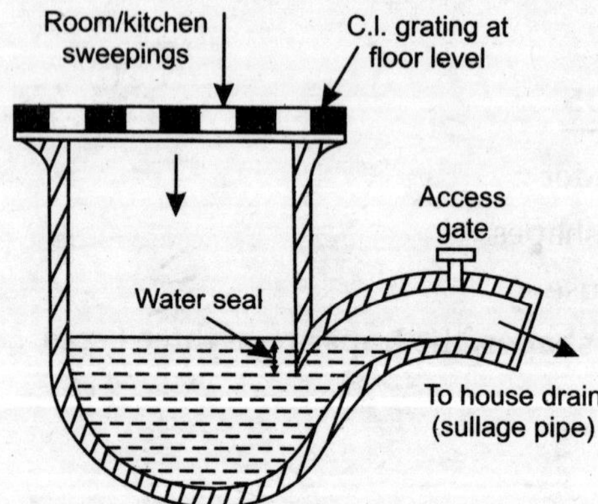

Fig. 14.2. Floor Trap.

2. Gully Traps

A gully trap is used to discharge waste water to sewer drain so that the foul air and insects from sewer do not escape to the outside. It is usually made of stoneware or cast iron.

The rain water pipes or sullage discharging into drains, are often connected to them through such traps. Gully traps may either have a S-trap or a P-trap. The water seal is usually 50 mm to 75 mm deep. The top of the trap is covered by a C.I. grating to exclude the entry of coarser materials to avoid blockade.

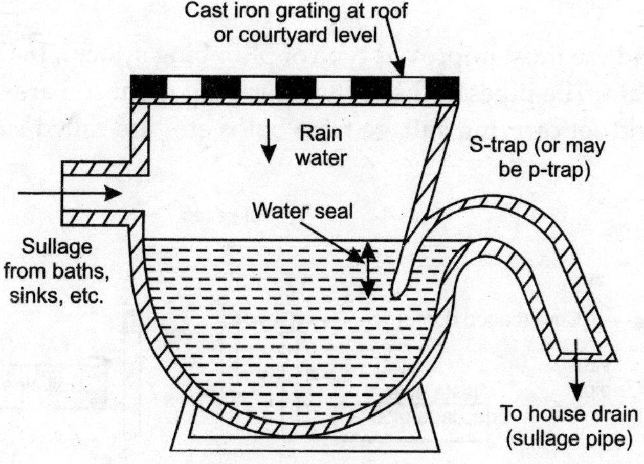

Fig. 14.3. Gully Trap.

3. Intercepting Traps

This trap (Fig. 14.4) is provided at the junction of a house sewer and a municipal sewer, so as to prevent the entry of the foul gases of the municipal sewer into the house drainage system. This trap at such a junction is often provided in a small man-hole constructed just near the house. This trap is provided near its top with an access gate or a plug, called *cleaning eye,* for removing silted matter from inside the trap incase of blockade. It has a high depth of water seal, say about 100 mm.

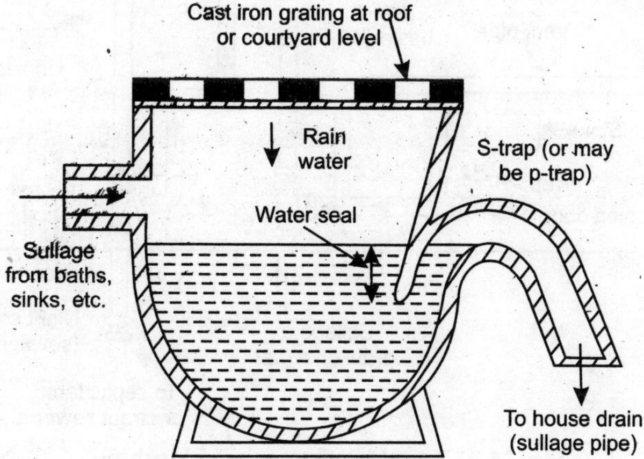

Fig. 14.4. An Intercepting Trap.

14.4. SYSTEMS OF PLUMBING

Following systems are adopted in plumbing of drainage work in buildings :

1. Two pipe system.
2. One pipe system.
3. Single stack system.
4. Partially ventilated single stack system.

1. Two Pipe System

This is the best and the most improved type of plumbing system. In this system, two seats of vertical pipes are laid. The pipes of the first set carrying night soil are called ***soil pipes,*** and the pipes of the second set carrying sullage from baths etc., are called ***sullage pipes or waste pipes.***

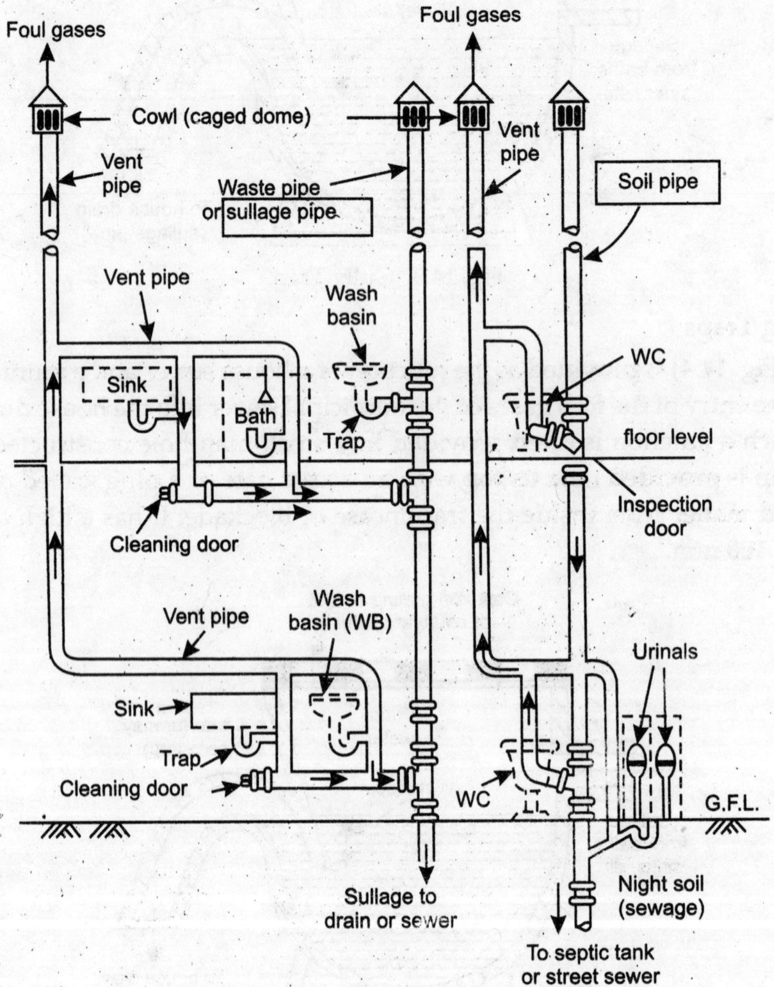

Fig. 14.5. Two Pipe System of Plumbing.

The soil fixtures, such as latrines and urinals, are, thus, all connected through branch pipes (laterals) to the *vertical soil pipe*; whereas the sludge fixtures such as baths, sinks, wash basins, etc., are all connected through branch pipes to the vertical *waste pipe*. The soil pipe as well as the waste pipe, are separately ventilated by providing separate vent pipes, as shown in Fig. 14.5. This arrangement, thus requires four pipes, and is quite costly.

2. One Pipe System

In this system, instead of using two separate pipes, only one main vertical pipe is provided, which collects the night soil as well as the sullage water from their respective fixtures through branch pipes. This main pipe is ventilated in itself by providing cowl at its top, and in addition to this, a separate vent pipe, however, is also provided, as shown in Fig. 14.6. This system, thus, has two pipes, instead or four pipes of the two pipe system.

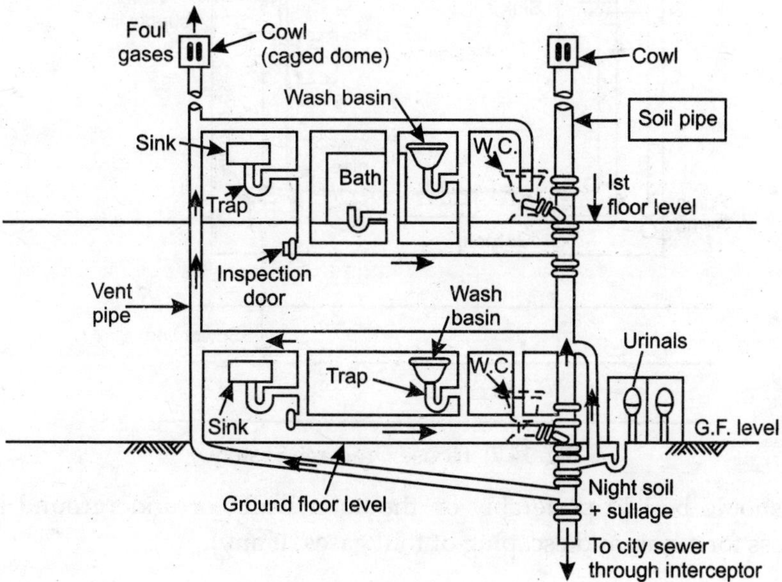

Fig. 14.6. One pipe system.

3. Single Stack System

This system is a single pipe system without providing any separate ventilation pipe. Hence. it uses only one pipe, which carries the sewage as well as the sullage, and is not provided with any separate vent pipe, except that it itself extended upto about 2 m higher than the roof level and provided with a cowl, for removal of foul gases. In a single stack system, we have to provide a discharge pipe of minimum diameter of 100 mm for the foul water drains or stack.

4. Partially Ventilated Single Stack or Single Pipe System

This is an improved form of single stack system in the sense that in this system, the traps of the water closets are separately ventilated by a separate vent pipe called ***relief vent pipe.*** This system, thus, uses two pipes as in a single pipe system, but the cost of branches, (laterals) is considerably reduced compared to single pipe system, because the sullage fixtures are not connected to the vent pipe.

14.5. SEWERAGE PLAN OF BUILDING (HOUSE DRAINAGE)

Figure 14.7 shows a general layout plan of a building water supply, waste water and rain water connections.

Waste water from latrines, urinals, sinks, bathrooms, wash basins, etc., should always flow by gravity, automatically with self cleaning velocity. The requirements for sanitary installations depend upon the number of persons using them and the type of building in which they are being installed, *viz.* a cinema hall, hotel, bus stand, hostel, or a factory or any other building.

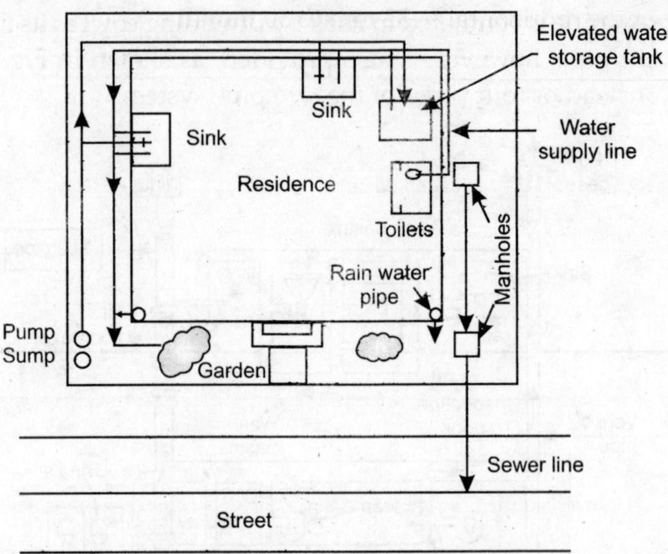

Fig. 14.7. House Drainage System.

All pipes should be laid preferably on the outer walls, or underground locations. This gives easy access for repair and escaping of foul gases (if any).

14.6. DESIGN OF SEWER PIPES

1. Quantity of Flow

The flow in the drainage pipes of a house sewerage system is irregular in nature and is not continuous. The average rate of flow is usually based on a water supply per capita per day. The maximum flow may be taken as three times this average.

2. Pipe Sizes and Gradients

The usual pipes sizes for house drains and sewers are 100 mm, 150 mm, and 230 mm, and in some extreme cases, 300 mm also. The lateral pipes should be laid with gradients sufficient to prevent any deposition of large amounts of solid matter to avoid blockade. The recommended gradients for sewers carrying maximum flow (3 × average flow) when running half full and the corresponding discharges, as per IS : 1742—1960, are given below :

Dia. of pipe sewer in mm	Gradient	Discharge in m^3/min (m^3/day)
100	1 in 35	0.18 (259)
150	1 in 65	0.42 (605)
230	1 in 120	0.93 (1339)
300	1 in 200	1.66 (2390)

The commonly adopted sizes for house drains, for average condition, are as follows :

1. Soil pipe—100 mm dia.
2. Sullage pipe (vertical)—75 mm dia.

3. Sullage pipe (horizontal)—32 mm to 50 mm dia.
4. Vent pipe—50 mm dia.

3. Sewerage Plans of Buildings

Generally it is mandatory to prepare detailed plans of house sewerage system and to get it approved from the municipalities. While preparing such plans, the following points need attention :

1. The site plan of the building should be drawn to a convenient scale, and position of gully traps (G.T.) should be marked on it.
2. The 'L'— section of the proposed sewer (drain) generally for drains larger than 150 mm dia should be drawn, indicating ground levels, invert levels, depth of cutting, sizes of manholes, sizes and gradient of pipes, etc., Position of public sewer should be marked on the site plan.
3. It is advisable to join house sewer to a manhole on public sewer line, if site conditions permit.
4. Detailed plans of different floors of building show the positions of floor traps, sanitary fittings, etc.,
5. House drains are laid as per approved sewerage plan of the building.
6. House drains are tested for water-tightness.

14.7. TESTING OF HOUSE SEWERS

After the house sewers are laid and the pipes fitted in position, they are usually tested to ascertain the water-tightness of the joints. The sewers (drains) are usually tested in sections between successive chambers, and the vertical pipes are tested in groups, each group having one common inspection chamber. Three tests, are commonly performed to test house drains; *viz.*

(*i*) Air test;
(*ii*) Hydraulic test or Water test; and
(*iii*) Smoke test;

1. Air Test

In this test, the section of the pipe to be tested is plugged at upper and lower ends, and air is forced into the pipe with the help of a pump, and all the joints are coated with soap solution. The escape of air from the leaky joints will then form bubbles at the joint, which can be repaired.

2. Hydraulic Test

In this test, the sewer is plugged at the lower end, and water is filled in the pipe through the upper end manhole to maintain a pressure of about 1 to 1.5 m head of water. If there is found to be no drop in water level in this manhole in about 10 minutes time, then it can be assumed that no joint is leaking. In case of drop in the water head, the leaking joints are repaired.

14.8. SANITARY FITTINGS AND OTHER ACCESSORIES

The *sanitary fillings* are the appliances used in a house plumbing system, to receive the human excreta and sullage waste of the house. Such fittings include :

(A) Soil Fittings

1. *Water Closets* (W.C.); and
2. *Urinals.*

(B) Wastewater (Wash water) Fittings

1. *Flushing cisterns of water closets;*
2. *Wash basins* (WB);
3. *Bath tubs; and*
4. *Kitchen sinks.*

All these fittings are briefly discussed below :

14.8.1. Water Closets (W.C.)

A water-closet is a water-flushed plumbing fixture, designed to receive the human excreta directly, and is connected to the soil pipe by means of a trap. Commonly used water-closets are :

(*i*) Indian type (squatting type).

(*ii*) Pedestal (or European) type.

These are discussed hereunder.

1. Indian Type W.C.

This type of W.C. is used in squatting position. The W.C. pan alongwith the trap is fixed flush with the floor of W.C. or slightly above. Two foot rests are provided on either side of the pan. The pan has inbuilt flushing rim having number of holes through which the flushing water from the cistern is discharged. The pan (seat) is fitted either with a P-trap, or S-trap, which is then connected to the soil pipe.

It is made of vitreous china or porcelain, and the pan (seat) and trap are in two different pieces and joined together, as shown in Fig. 14.8. The trap has an opening for fixing an *anti-siphonage* pipe or a vent pipe. The pan has a flushing rim to spread the flush water. The excreta does not fall directly into the trap, and needs to be forced into it and into the soil pipe, by flushing the pan with water released under pressure from the flushing cistern. The excreta may therefore become foul or get decomposed if not properly flushed. This W.C. requires 5-10 lit of water for proper flushing.

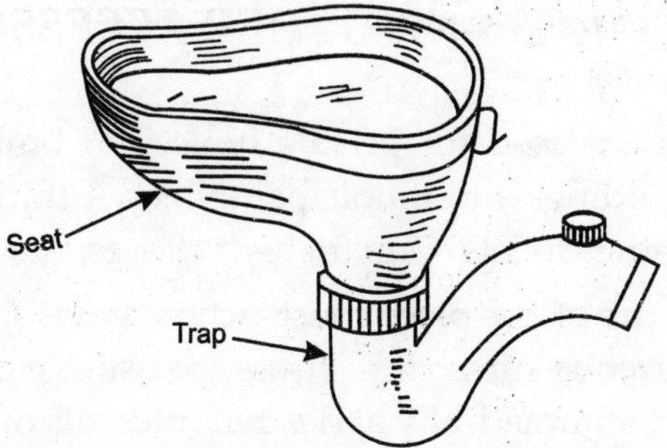

Fig. 14.8. Indian Type W.C. Pan and Trap.

2. European Type W.C.

This type of W.C. is used in sitting position over a plastic seat hinged to the appliance. This is a pedestal type of appliance with the pan and trap in a single piece. The European type W.C. is sometimes called wash-down type of W.C. In modern days, most of city homes are fitted with such European W.C.s. It is provided with a wide flushing rim and a 50 mm trap in one piece construction. It is provided with an inlet for connecting to the flushing pipe. The W.C. may be provided with a *P-trap* or a *S*-trap. These types of water-closets require less space then the Indian type, and can be located alongwith the bath in the same room as it won't look bad. They are normally flushed with a low level cistern. A perspective view of such a W.C. is shown in Fig. 14.9.

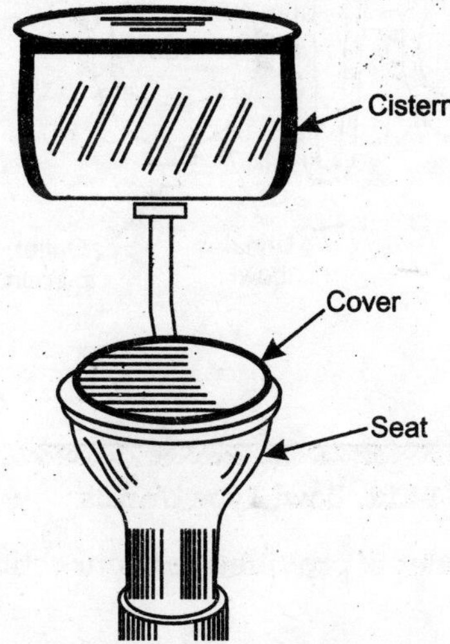

Fig. 14.9. Pedestal or European Type W.C.

Besides above, Anglo-Indian type of W.C. is also available, which can be used both in squatting position as well as in sitting position. This type of W.C. are not very common these days.

14.8.2. Urinals

The discharged from urinals is connected to the soil pipe. These are available mainly in following forms :

(*i*) The *bowl type or the basin type*; and

(*ii*) The *stall type* or the *slab type*.

The ***bowl type*** is generally used for private houses or buildings, and the *stall type* is generally used for public buildings and municipal toilets. A third type, called *squatting plate type*, is used specifically in ladies toilets. Only in high classes.

The bowl type of urinals are of one piece construction and is fixed on wall. These are made of glazed earthenware or vitreous china clay. These are usually provided with siphonic type of flushing cistern, which act automatically at desired intervals of time.

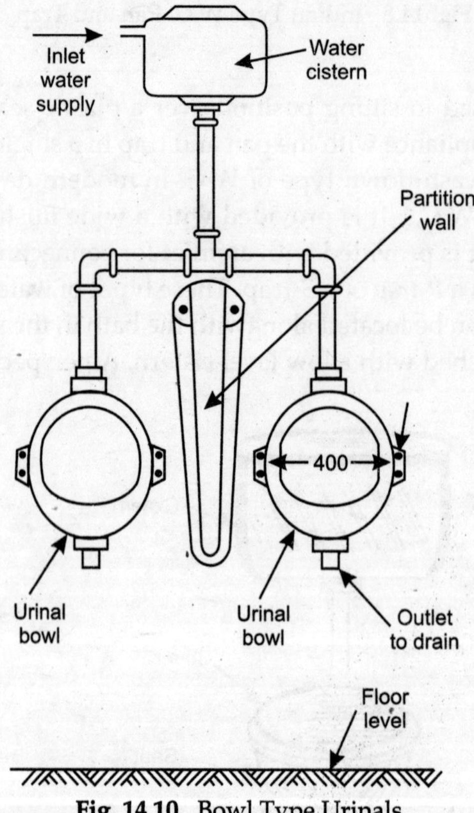

Fig. 14.10. Bowl Type Urinals.

At the bottom of the bowl an outlet is provided for connecting the urinal to the soil pipe through a floor trap.

The ***stall or slab type*** urinals are used in cinema house, restaurants, municipal toilets, railway stations, offices etc., This system consists of simple compartments made by raising side walls, and the walls infront is glazed, against which the urine is passed by males. At the bottom, a drain is made along the front wall in which the urine flows and is finally connected to the soil pipe through a floor trap. The flushing is provided in the usual manner, by a cistern.

14.8.3. Flushing Cistern

A flushing cistern is used for storage and discharge of water for flushing of contents from a

W.C. or urinal. A flushing cistern is a device which releases a fixed quantity of water, under pressure so as to flush and clean the pan trap of a W.C. or a urinal.

In Indian type water-closets, high level cisterns fixed at about 2 m above the floor level are generally provided; whereas, in European type water closets, low level cisterns fixed at about 60 cm above the floor level are generally provided. The high level flushing cisterns used in Indian W.C. normally made of cast iron and operated by pulling a chain. Whereas the low level cisterns are normally made of glazed china ware and are used western W.C.s. and are operated by turning a handle

14.8.4. Wash Basins

A wash basin is used for washing hands, face or brushing teeth etc., in a standing position. It is made of glazed earthen ware or vitreous china clay. The type of wash basin normally used in a house has an oval shaped bowl with an overflow slot at top. It has a provision for making holes for installing one, two or even three taps. Normally two pillar taps are provided one for cold water and the other for hot water.

It has a circular waste water hole at the bottom for its draining out from the basin. The basin is normally mounted on 2 angle irons fixed in the wall. The top of the wash basin should be kept at a height of about 75 to 80 cm from the floor level.

Sometimes, separate cold and hot connects are given and mixed water comes out through a mixer arrangement. The hotness of the coming water can be adjusted upto the required extent, by opening the cold and hot water taps to the desired extent.

14.8.5. Baths and Bath Tubs

Bath rooms often produce waste water, either when one takes a bath or washes clothes. In normal Indian homes bath is taken without a bath tub, either under a shower or by bucket and a mug. In high class buildings, however, bath tubs are provided. The tub can be filled with water, which can be drained out by opening the outlet plug. It is generally made of vitreous earthenware. Two taps are provided to obtain hot and cold supplies both. It is provided with an overflow pipe also, to drain excess water, above the full supply level.

The usual dimensions of a bath tub are 1.8 m length × 0.75 m width × 0.45 m depth near waste pipe. Overall height with feet being about 0.60 m.

14.8.6. Kitchen Sinks

Kitchen sinks are used to clean the utensils. The sinks are commonly used in kitchen, hospitals and laboratories. It is made of marble, vitreous china clay, stainless steel or plastic. Sinks may be made with or without overflow arrangement. It has a circular waste hole for fixing the metallic strainer in which the waste pipe is attached for conveying the discharge from sink to the floor trap.

QUESTIONS

1. Draw a neat sketch showing water and sanitary fittings required for a building.
2. Differentiate between single pipe and single stack plumbing systems.
3. Explain brief with sketches :
   (*a*) Traps
   (*b*) Flushing cistern
   (*c*) Sink
4. Write short note on :
   (*a*) W.C.
   (*b*) Wash basin
   (*c*) Automatic flushing tank
5. Explain one pipe and two-pipe systems in house plumbing
6. What are the principles adopted in providing an efficient and economic system of drainage in buildings?
7. Sketch an Indian type Water Closet, with associated fittings and fixtures. Explain its working.
8. What is meant by ventilation of house sewers, and how is it achieved? Also discuss the uses of antisiphonage pipes in multi storeyed blocks.
9. Show by sketches the urinals generally used in high class public buildings, and in municipal street toilets, in India.
10. Name and explain with sketches, the different types of traps used in house sewer constructions.
11. Describe different types of plumbing systems of for buildings. State their merits and demerits.
12. Write short note on the following :
    (*a*) Gully trap and S-trap.
    (*b*) One-pipe system of plumbing.
    (*c*) Soil pipe, waste pipe and vent pipe.
    (*d*) Testing of house drains.

Disposal of Sewage Effluents

15.1. INTRODUCTION

Sewage from individual blocks or dwelling units is collected internally in a colony or premises by radial feed system in which lateral sewers running underground receive their discharge directly from house drainage system. Then these laterals are connected to branch sewers and then to main sewer. Main sewers discharge into the trunk sewer which lead the entire sewage for treatment and disposal. Sewer systems are designed to cater for peak loads which include domestic discharges, industrial waste, storm flow and allowance for infiltration.

Large volumes of sewage is discharged into the water bodies either in untreated or in partially treated condition, thereby throwing heavy load of pollutants containing pathogens, oxygen depleting biodegradable organic matter, non-biodegradable organic chemicals like detergents, sediments and suspended inorganic matter etc.,

The discharge of sewage causes large scale pollution of water bodies. The Bureau of Indian Standards (BIS), and Central Pollution Control Board has prescribed and notified general standards for discharge of pollutants from effluents into water resources. Therefore, for the disposal of sewage into water bodies, proper treatment is necessary.

In view of the above, the study of the sources of disposal is important, because the amount of treatment required to be given to a sewage depends very much upon the source of disposal, its quality and capacity to tolerate the impurities present in the sewage effluents.

There are two general **methods of disposing** of the sewage effluents :

(*a*) *Dilution i.e.* disposal in water; and

(*b*) *Effluent Irrigation or Sewage Farming, i.e.* disposal on land.

Both these methods of disposal are described below :

15.2. DISPOSAL BY DILUTION

Disposal by dilution is the process whereby the treated sewage is discharged into a river stream, or a large body of water, such as a lake or sea. The discharged sewage, in due course of time, is purified by what is known as *self purification process of natural waters*. The degree and amount of treatment given to raw sewage before disposing it into the river-stream in question, depend upon

(*i*) The quality of raw sewage,

(*ii*) *The self purification capacity of the river stream and*

(*iii*) *The intended use of its water.*

The dilution method for disposing of the sewage can be adopted under the following conditions :

(*i*) When sewage is comparatively fresh (4 to 5 hr. old), and free from floating and settleable solids, (or are easily removed by primary treatment).

(*ii*) When the diluting water has a high dissolved oxygen (D.O.) content.

(*iii*) Where diluting waters are not used for the purpose of navigation or drinking water supply *for atleast some reasonable distance* on the downstream from the point of sewage disposal.

(*iv*) When the outfall sewer of the city or the treatment plant is situated near some natural waters having large volumes.

15.3. STANDARDS OF DILUTION FOR DISCHARGE OF SEWAGE INTO RIVERS

The ratio of the quantity of the diluting water to that of the sewage is known as the *dilution factor*; and depending upon this factor, the following standards and degrees of treatment required to be given to a particular sewage, have been laid.

Dilution factor	*Standards of purification required*
Above 500	No treatment is required. Raw sewage can be directly discharged into the volume of dilution water.
Between 300 to 500	Primary treatment such as plain sedimentation should be given to sewage , and the effluents should not contain suspended solids more than 150 ppm.
Between 150 to 300	Treatments such as sedimentation, screening and esssentially chemical precipitation are required. The sewage effluent should not contain suspended solids more than 60 ppm.
Less than 150	Complete thorough treatment should be given to sewage. The sewage effluent should not contain suspended solids more than 30 ppm, and its 5 days B.O.D. at 18.3°C should not exceed 20 ppm.

With the increasing pollution of surface streams due to indiscriminate discharge of domestic and industrial wastewaters, without bothering to look into the available dilution ratios, it has become imperative to limit the concentrations of various pollutants being discharged into the surface water sources along with the sewage and industrial effluents.

The Bureau of Indian Standards (BIS), previously known as Indian Standard Institution (ISI), has laid down its *guiding standards* for sewage effluents, vide IS 4764-1973, and for industrial effluents vide IS : 2490-1974.

When the industrial wastewaters are disposed of into public sewers, then also, their quality has to be checked, by following the standards prescribed by IS : 3306-1974 or the standards notified by the Government of India under Environment (Protection) Rules,1986.

15.4. SELF PURIFICATION OF NATURAL STREAM

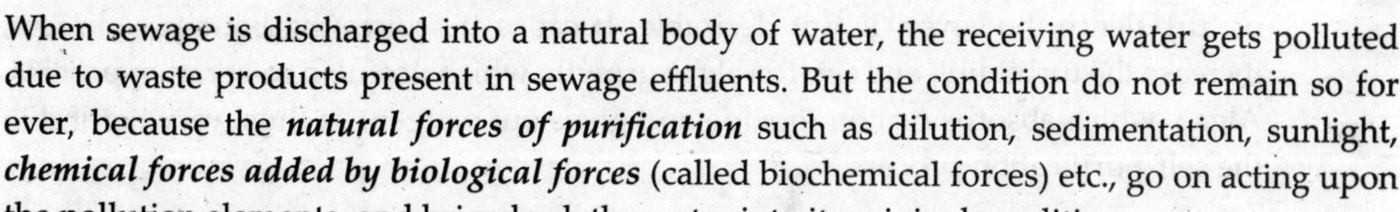

When sewage is discharged into a natural body of water, the receiving water gets polluted due to waste products present in sewage effluents. But the condition do not remain so for ever, because the ***natural forces of purification*** such as dilution, sedimentation, sunlight, ***chemical forces added by biological forces*** (called biochemical forces) etc., go on acting upon the pollution elements, and bring back the water into its original condition.

Aerobic bacteria utilise the pollutant material, in the presence of sunlight and oxygen available in surface water and decompose organic matter. River waters absorb oxygen from the atmosphere continuously.

This automatic purification of polluted water, in due course, is called the **self-purification phenomenon.** However, if the self-purification is not achieved successfully either due to too much of pollution discharged into it or due to other causes, the river water itself will get polluted, which in turn, may also pollute the sea where the river outfalls.

15.4.1. Factors of Self Purification

1. **Dilution.** In this process,volume of water is very high in comparison with quantity of sewage. This reduces the concentration of impurities. This also maintains more dissolved oxygen to satisfy the BOD of sewage.
2. **Dispersion.** Strong currents and turbulence in a stream cause dispersion of solids. Thorough mixing of sewage and water helps in decomposition of the sewage due to biochemical reaction.
3. **Sedimentation.** Heavier solids settle at a lower level due to gravitational forces. These are carried away by the stream owing to the high velocity.
4. **Temperature.** Warm and summer temperatures are good for sedimentation and biological activity.
5. **Sunlight.** The sunlight has a bleaching and stabilising effect of bacteria. The evolution of oxygen, by a process of *photosynthesis,* in river water due to sunlight will help in achieving self-purification through oxidation.
6. **Oxidation.** The oxidation of the organic matter present in the sewage effluent, will start as soon as the sewage outfalls into the river water containing dissolved oxygen. The deficiency of oxygen so created, will be filled up by the atmospheric oxygen.

The process of oxidation will continue till the organic matter has been completely oxidised.

7. **Reduction.** Reduction occurs due to hydrolysis of organic matter settled at the bottom either chemically or biologically. Anaerobic bacteria will help in splitting the complex organic constituents of sewage into liquids and gases, and thus pouring the way for their ultimate stabilization by oxidation.
8. **Turbulence.** In the body of water it helps in breaking the surface of the stream or lake, and helps in rapid re-aeration from the atmosphere. Thus, it helps in maintaining aerobic conditions in the river stream, and in keeping it clean. Too much of turbulence, however, is not desirable, because it scours the bottom sediment, increase the turbudity, and retards algae growth, which useful in re-aeration process. Algae, which absorbs carbon dioxide and gives out oxygen, is thus, very helpful in the self-purification process.

15.5. DISPOSAL OF SEWAGE IN LAKES

15.5.1. Lake Pollutants

Disposal of wastewaters in confined lakes or reservoirs, is much more harmful than its disposal in flowing streams and rivers. Water quality management in lakes entirely different from that in rivers. It is infact the **phosphorous** (a nutrient largely contained in industrial as well as domestic wastewaters), which seriously affects the water quality of lakes; and is hence considered as the **prime lake pollutant.**

Oxygen demanding wastes may be the other important lake pollutants. The *toxic chemicals* from industrial wastewaters may also sometimes very adversely effect some special classes of the lakes.

15.5.2. Stratification in Lakes

The water of a lake gets stratified during summers and winters. **During summer season,** the surface water of a lake gets heated up by sunlight and warm air. This warm water being lighter, remains in upper layers near the surface, until mixed downward by turbulence from winds, waves, boats and other forces. Since such turbulence extends only to a limited depth from below the water surface, the top layers of water in the lake become well mixed and aerobic. This warmer, well mixed and aerobic depth of water is called ***epilimnion zone.*** The lower depth, which remains cooler, poorly mixed and anaerobic, is called the ***hypolimnion zone.*** There may also exist an intermediate zone or a dividing line, called ***thermocline* or *metalimnion***, as shown in Fig. 15.1. (*a*).

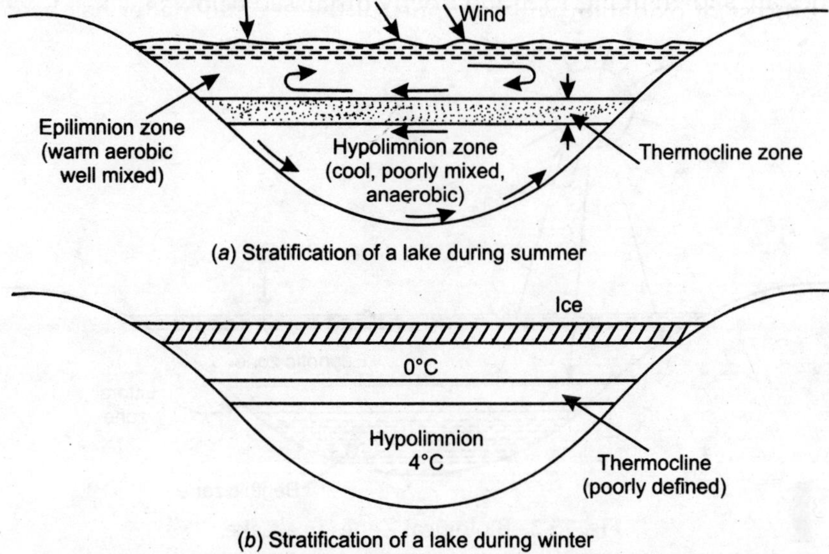

(*a*) Stratification of a lake during summer

(*b*) Stratification of a lake during winter

Fig.15.1. Stratification of Lakes.

The depth of epilimnion zone depends upon the size of the lake for the same temperature changes. It may be as little as 1 m in small lakes and may be as large as 20 m or more in large lakes. This depth also depends upon the storm activity in the spring when stratification is developing. A major storm at the right time will mix the warmer water to a substantial depth and thus create a deeper epilimnion zone than its normal depth. *Once formed, lake stratification is very stable, and can only be broken by exceedingly violent storms.*

With the onset of winter season, the epilimnion cools, until it is more dense than the hypolimnion. The surface water then sinks, causing '**overturning**'. The water of the hypolimnion rises to the surface, when it cools and again sinks. *The lake, thus becomes completely mixed,* making it quite aerobic. In regions of freezing temperatures, when the temperature drops below 4°C, the above process of overturning (*or turn over*) stops, because water is most dense at this temperature. Further cooling or freezing of the water surface results in winter stratification, as shown in Fig. 15.1. (*b*).

With the passing winters and commencement of spring season, the surface water again warms up and overturns, and lake becomes completely mixed. The lakes in regions of temperature climate will, therefore, have atleast one, if not two, cycles of stratification and *turn-over* every year.

15.5.3. Biological Zones in Lakes

Lakes have been found to exhibit distinct zones of biological activity, largely determined by the availability of light and oxygen. The most important biological zones are :

(*i*) *Euphotic zone;*

(*ii*) *Littoral zone*; and

(*iii*) *Benthic zone.*

These zones are shown in Fig 15.2. and briefly discussed below :

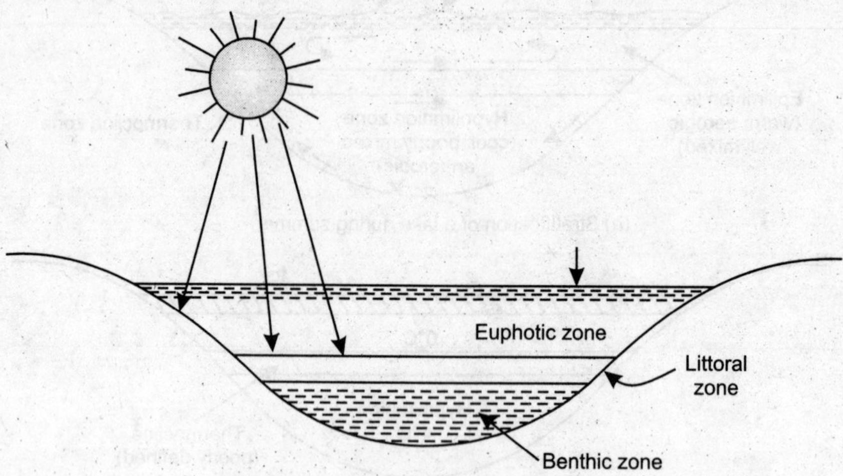

Fig. 15.2. Biological Zones in a Lake.

1. Euphotic Zone

The upper layer of lake water through which sunlight can penetrate, is called the euphotic zone. All plant growth occurs in this zone. In deep water, algae grow as the most important plants, while rooted plants grow in shallow water near the shore.

The depth of the euphotic zone is reduced by the turbidity, which blocks sunlight penetration. In most lakes, the turbidity is due to algae growth; although colour and suspended clays may substantially reduce sunlight penetration in some lakes. It is important to note that the bottom of euphotic zone only coincides with the thermocline.

2. Littoral Zone

The shallow water near the shore, in which rooted plants grow, is called the littoral zone. The extent of littoral zone depend on the slope of the lake bottom, and the depths of the euphotic zone. The littoral zone cannot extend deeper than the euphotic zone as shown in Fig. 15.2.

3. Benthic Zone

The bottom sediments in a lake comprise what is called the *benthic zone*. As the organisms living in the overlying water die, they settle down to the bottom, where they are decomposed by the organisms living in the benthic zone. Bacteria are always present in this zone, the presence of higher life forms, such as worms, insects etc., however, depends upon the availability of oxygen at the lake bottom.

15.6. DISPOSAL OF SEWAGE IN SEA WATER

Sewage solids, when thrown into sea water, chemically react with the dissolved matter of sea water, resulting in precipitating some of the sewage solids, *giving a milky appearance to the sea water and forming* **sludge banks.** These sludge banks are undesirable, as they are likely to produce offensive hydrogen sulphide gas by reacting with the sulphate rich water of the sea.

This is because, sea water normally contains 20% less oxygen than that contained in fresh water of a river stream. Moreover, sea water normally contains a large amount of dissolved matter. As such, the capacity of sea water to absorb sewage solids is not as high as that of fresh water of a stream.

Since the sea contains too large volumes of water, most of these deficiencies get removed, provided the sewage is taken deep into the sea and away from the coast line.

To prevent the spreading of sewage on the sea shore, the sewage should preferably be disposed off during low tides.

The following points should be kept in mind while discharging sewage into the sea to avoid marine pollution :

(*i*) The sewage should be discharged in deep sea only.

(*ii*) In order to mix the sewage properly with sea water, the sewage should be released at a minimum depth of 3 to 5 metres below the water level, and by taking it sufficiently inside (about $1\frac{1}{2}$ km) from the shore line. This measure will prevent the sewage from accumulating on the shore, and thus preventing nuisance to baths and recreation centres on the shores.

(*iii*) Before deciding the position of outfall point, the sea currents, wind direction, velocity, etc., should be thoroughly studied. The point of disposal should be such that the sewage is taken away from the shore by the winds, and not brought back near to the shore.

The discharge of wastewaters into the sea should, however be controlled in respect of the quality of effluents, by adhering to the prescribed standards. The standards prescribed by BIS under IS : 1968—1976 are shown in Table below.

BIS (ISI) Standards for Wastewater Effluents Discharged in Marine Coasts (sea water).

S. No.	*Constituent Pollutant contained in the Wastewater Effluent*	*Tolerance Limit*
(1)	(2)	(3)
1.	BOD	100 mg/l
2.	COD	250 mg/l
3.	pH value	5.5 to 9.0
4.	Total suspended solids	100 mg/l
5.	Oil and grease	20 mg/l
6.	Fluorides (as F)	15 mg/l
7.	Ammoniacal Nitrogen (as N)	50 mg/l

15.7. DISPOSAL OF SEWAGE ON LAND FOR IRRIGATION

In this process, the sewage effluent (treated or diluted) is allowed to flow over cultivable land. The percolating water may either join the water-table, or is collected below a system of under-drains. This method can then be used for irrigating crops. This method, in addition to disposing of the sewage, may help in increasing crop yields (by 33% or so) as the sewage generally contains a lot of fertilising minerals and other elements (like nitrogen, phosphate, potash etc.,) which help in increasing the yield. This type of irrigation is known as broad irrigation or sewage farming. The sewage effluent before being used as irrigation water, must be made safe.

As mentioned above, when raw or partly treated sewage is applied to the land, a part of it evaporates, and the remaining portion percolates through the ground soil. While percolating through the soil, the suspended particles present in the sewage are caught in the soil voids. If proper aeration of these voids is maintained, the organic sewage solids caught in these voids get oxidised by aerobic process. Such aeration and aerobic conditions will more likely prevail, if the soil is sufficiently porous and permeable (such as sands and porous loams).

However, if the land is made up of heavy, sticky and fine grained materials (such as clay, etc.,), the void spaces will soon get choked up, and thus resulting in non-aeration of these voids. This will lead to the evolution of foul gases, and excessive clogging may also result in ugly ponding of sewage over the farm land, where mosquitoes may breed in large number, causing further nuisance.

Application of too strong or too heavy load of sewage will also similarly result in the quick formation of anaerobic conditions. The greater is the sewage load, more likely it will be for the soil to get clogged. Hence, if the sewage load is reduced either by diluting it or by pre-treating it, it may also possible to avoid the clogging of the soil pores. The degree of treatment required depend upon the type of the soil of the land. If the soil, to be irrigated, is sandy and porous, the sewage effluents may require lesser treatment, as compared to the case where the soil is less porous and sticky.

Sewage Sickness of Land

When sewage is supplied continuously on a piece of land, the soil pores or voids may get filled up and clogged with sewage solids retained in them. The time taken for such a clogging will, of course, depend upon the type of soil and the load present in the sewage. After clogging of these voids, free circulation of air will be prevented. This results in evolution of foul gases like hydrogen sulphide, carbon dioxide, and methane. This phenomenon, of soil getting clogged, is known as ***sewage sickness of land.***

Preventive Measure for Sewage Sickness of Land

1. **Primary Treatment of Sewage.** Sewage should be disposed of after primary treatment, such as screening, grit removal, and sedimentation.

2. **Choice of Land.** The piece of land used for sewage disposal should be sandy or loamy. Clayey lands should be avoided.
3. **Under-drainage of Soil.** The land on which the sewage is disposed of, should be provided with a system of under-drains (*i.e.* open joined pipes) laid below.
4. **Giving the Rest to Land.** Some extra land should be kept as reserve and stand-by for diverting the sewage during the period the first land is at rest. During the rest period, the land should be thoroughly ploughed so that it gets broken up and aerated.
5. **Rotation of Crops.** Different crops in rotation should be grown instead of growing only single type of crop.
6. **Applying in Thin Layers.** The sewage should be applied in thin layers. Greater depths of sewage on a land results in its clogging.

15.8. DISPOSAL BY DILUTION V/S LAND DISPOSAL

1. For disposal by dilution, large volumes of natural clean waters are required; whereas, for land disposal, large areas of land, preferably with sandy soils, are required.
2. The cost of land, in land disposal method, is generally very high in cities and urban areas. Whereas, in the rural areas, where the sewage produced is less, and land being available at less cost, land disposal method is opted.
3. In land disposal method, when cost of land is high, good management is required. Dilution method of disposal is a simple method, and does not require too much of management.
4. Dilution method of sewage disposal requires nil or small head pumping, because rivers flow through the lowest contour, while land disposal requires high head pumping, thus, making the land disposal method costlier.
5. Land disposal saves the inland rivers from getting polluted by sewage and returns the fertilising elements, which were once drawn it.

15.9. LOW-COST SANITATION IN RURAL AREAS

Rural sanitation is different from sanitation in urban areas. One of the main difference is the flow of wastewater in the streets, which is the main cause of unhygienic and insanitary conditions. Disposal of human waste is a big problem in a rural areas due to absence of underground sewage system. There is a great need in rural areas to properly dispose off sewage and storm water to avoid health hazard to rural population.

Sullage in rural areas is generally taken along side drains flowing along houses, and then finally discharged into some open surface drain or just left unattended into some outside pond, depression etc., The human excreta however, is burried underground through the *latrines* (*privies*) of different designs.

Where comparatively large amount of water is available due to the availability of an organized municipal supply, *septic tanks* are constructed.

15.9.1. Sewage Disposal in Rural Areas

In rural areas, where there is no provision for water carriage system, various types of latrines (privies) may be provided. Most common types of latrines are :

(*i*) Pit latrines

(*ii*) Borehole latrines, and

(*iii*) Aqua privies.

1. Pit Latrines. In this system, excreta falls in a hole in the ground. The size of the pit is 130 × 100 × 150 cm. Its depth may vary from 130 cm to 280 cm depending upon the groundwater table. The seat and superstructure are constructed over it. Sometimes, lime is applied frequently to reduce the odour and check insects. When the pit is the about three-fourth full, the squatting seat and the superstructure are removed. The pit is covered with about 60 cm of local earth and abandoned.

Once the pit is filled, the seat and the superstructure are constructed over a new pit nearby.

2. Borehole Latrines. These are suitable for sparsely populated areas and are generally constructed 4 m and 30 m away from the houses and wells respectively. These are similar to pit latrines and consists of a pit 30 to 40 cm in diameter, 4 to 8 m deep, on the top of which is fitted a squatting plate made of cement concrete (Fig. 15.3). The bore hole is generally provided away from the superstructure. The seat is connected to the bore hole.

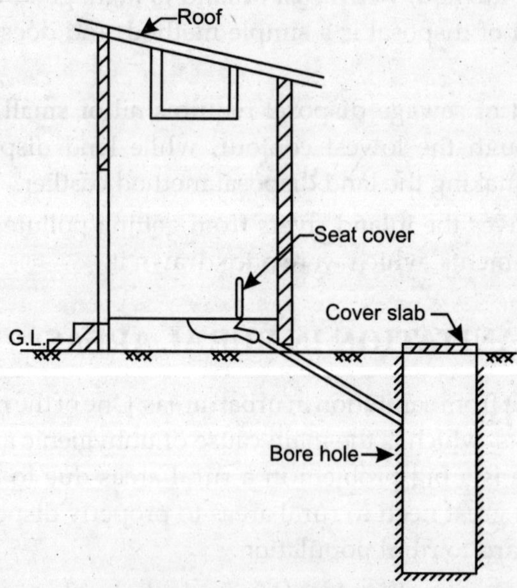

Fig. 15.3. Bore Hole Latrine.

The trap provided under the seat reduces odour and fly nuisance. When the bore hole is filled, it is covered with a thick layer of soil. A new bore hole is dug nearby and connected to the seat. The bore can be used as compost for agriculture.

3. Aqua Privies (Latrines). These are also termed as wet latrines. An aqua privy consists of three chambers as shown in Fig. 15.4. The end of the pan is taken about 10 cm deep in water to provide a water seal. Night soil falls smoothly into the first chamber and it is decomposed under aerobic conditions. The vent pipe in the second chamber is used to release the foul gases, formed during decomposition, to the atmosphere. In the third chamber, aerobic bacteria attack the organic matter. The effluent from the second chamber is taken to the third chamber through a pipe whose outlet is situated near the bottom of the chamber. The third chamber is small and fixed with clinkers. The effluent, while rising through the clinkers, gets purified. It can be used for farming and gardening. Aqua privy is the most useful method of disposal of night soil, without the help of water carriage system. It is economical hygienic.

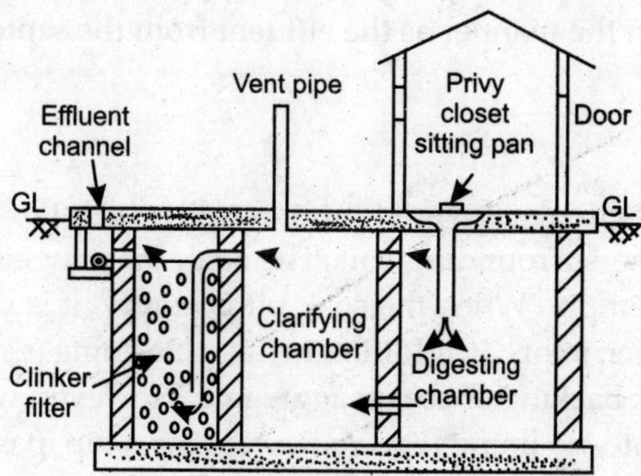

Fig. 15.4. Aqua Privy.

15.9.2. Septic Tanks

A septic tanks may be defined as a primary sedimentation tank, with a longer detention period (12 to 36 hours, against a period of 2 hrs. in an ordinary sedimentation tank) and with the extra provisions for digestion of the settled sludge.

In a septic tank (Fig. 15.5), sewage and sludge are detained for 12 to 24 hours in a closed water tight chamber. Sedimentation and digestion are carried out in this septic tank under anaerobic activity. Foul gases escape through a vent shaft. The digested sludge is periodically removed once in a year or two, and disposed. The effluent from the septic tank is highly offensive (200 mg/L organic solids and 100 mg/L BOD) and hence disposed into ground through soak pits (Fig. 15.6) or dispersion trenches. Septic tanks should be commissioned with well-digested sludge. Septic tanks are very useful for individual houses, hostels and small groups of populations.

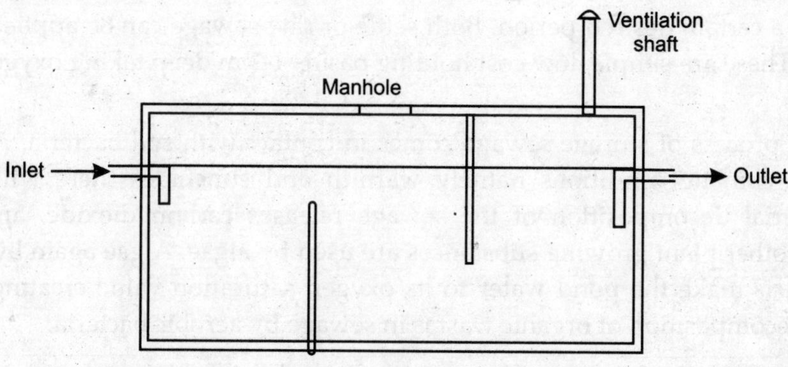

Fig. 15.5. Septic Tank.

A septic tank, is thus, a horizontal continuous flow type of a sedimentation tank, directly admitting raw sewage, and removing about 60 to 70% of the dissolved matter from it. Septic tanks are generally provided in areas where sewers have not been laid and for catering to the sanitary disposal of sewage produced from isolated communities and various public institutions etc.,

If only water closets are connected to the septic tank, the sewage flow will be about 40-70 litres/capita/day. But when sullage is also discharged into the septic tank, the sewage flow may be as much 90-150 litres/capita/day. Where a large quantity of sludge is expected, it may separately be disposed of in the manner as the effluent from the septic tank.

15.9.3. Soak Pits

A *soak pit or seepage pit* is a circular covered pit, through which the effluent is allowed to be soaked or absorbed into the surrounding soil. The soak pit may either be filled with stone aggregate or may be kept empty. When the soak pit is empty, it is lined with brick, stone or concrete blocks with dry open joints. In addition to this, this lining is supported below the inlet level by atleast 7.5 cm thick backing of coarse aggregate. However, when the soak pit is filled with stone or brick aggregate, no lining required except for a top masonry ring constructed to prevent damage by flooding of the pit by surface run off.

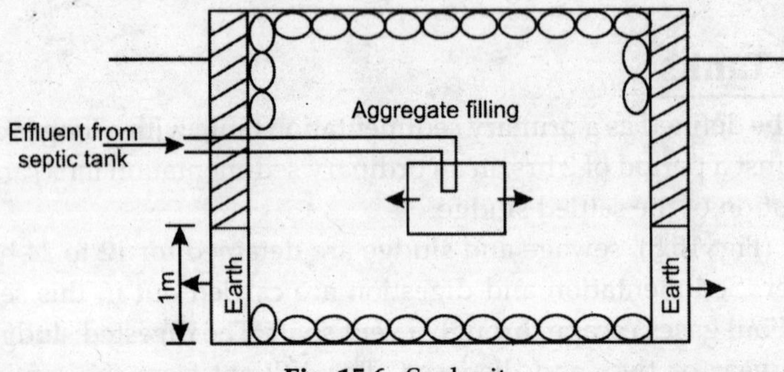

Fig. 15.6. Soak pit.

15.10. OXIDATION PONDS

These are shallow earthen basins excavated in the ground which retain the sewage or wastewater for a certain desired period. Both settle or raw sewage can be applied and stored in these tanks. These are simple, low cost holding basins 1-2 m deep taking oxygen (air) from the atmosphere.

During the process of storage sewage comes in contact with soil bacteria. Again, under the favourable climatic conditions namely warmth and sunshines there will be growth of algae. Bacterial decomposition of the sewage releases carbon dioxide, ammonia etc., Ammonia and other plant growing substances are used by algae. Algae again by the process of photosynthesis make the pond water to its oxygen saturation value creating favourable condition for decomposition of organic wastes in sewage by aerobic bacteria.

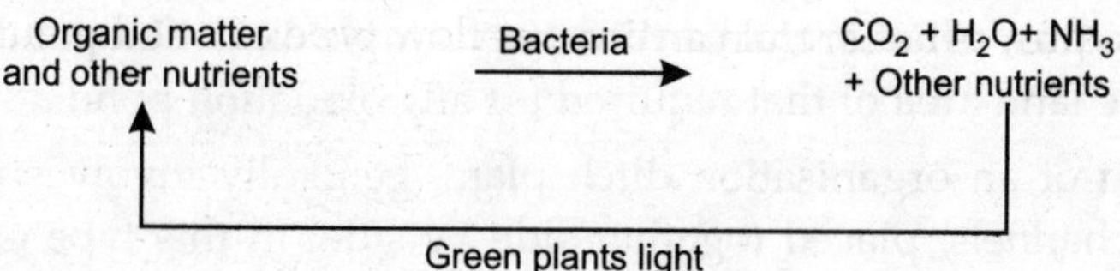

The essential feature of the oxidation pond is that bacteria and algae live together (symbosis) for mutual benefits. CO_2 and the nutrients are used by algae (green) and O_2 is released in daylight.

$$CO_2 + H_2O \longrightarrow CH_2O + O_2$$

Various processes which take place in an oxidation pond (Fig. 15.7.) are sedimentation, oxidation, digestion, photosynthesis, evaporation and seepage.

During the designed eletention period in an oxidation pond, the organic solids will either be oxidized or will settle down. The average sludge accumulation observed in several ponds is about 2 to 5 cm depth per year. At a maximum of this rate, sludge removal is required, once in about 6 years for 1.2 m deep ponds.

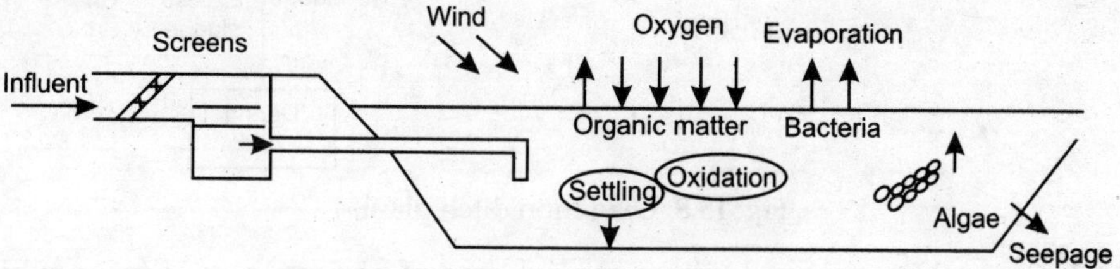

Fig. 15.7. Oxidation Pond.

Design Considerations for Oxidation Ponds

1. Detention period is usually 10-40 days depending upon latitude of the place (temperature and sunlight availability).
2. Organic loading may be taken from 300—150 kg/hectare/day depending upon the climate and location of the place.
3. The pond depth is taken as 1—2 m. Rectangular ponds of length to breadth ratio of 2 may be considered.

As mosquito and odour nuisance is expected, these ponds are kept away from residential areas, considering the wind direction.

15.10.1. Oxidation Ditches

This method is suitable for giving complete treatment of sewage from a small community. It is similar to extended aeration type treatment unit with mechanical aeration system. These units are long, narrow channel like structures having length 150 to 1000 m, width 2 to 5 m, and depth 2 to 4 m. The aeration is done with the help of cylindrical cage rotors 60 to 75 cm in diameter. This system has a detention time of 12 to 24 hours, and an air supply of 1.5 to 2.0 kg of O_2 per kg of applied B.O.D.

Such a plant, also proves better than an ordinary simple oxidation pond, as it requires only about one-fifth of the land area of that required for any oxidation pond.

The construction of an organisation ditch plant, generally involves the construction of a number of ditch channels, placed together side by side. In this type of plant, the quality of effluent obtained is quite good, with suspended solids removal at about 95% and BOD removal at about 98%.The method can be advantageously used for treating small quantities of wastewaters at places where sufficient land is not available constructing large sized oxidation ponds.

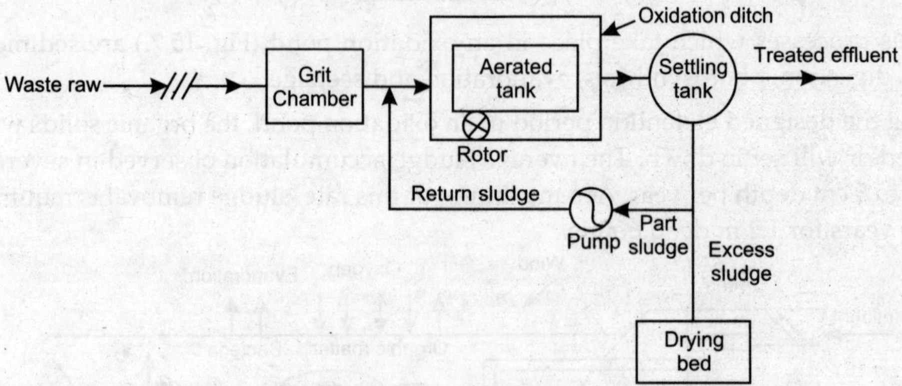

Fig. 15.8. Oxidation ditch plant.

Each ditch channel is equipped with a special type of horizontal axis rotor, which serves the purpose of agitating and circulating the sewage and thereby oxygenating the same. The rotor not only gives oxygen supply to the sewage, but also helps in keeping the sewage-solids in suspension (which acts as carriers of aerobic bacteria), by imparting sewage a velocity of more than 0.03 m per sec. The thoroughly aerated sewage is then settled in a settling tank by stopping the rotors for about 2 hours and then taking out the clear supernatant liquour at the time of no flow. A part of the settled sludge is used for recirculation, and the excess settled sludge, which is well stabilised due to long detention periods (characterised by low BOD) can be easily dried on sand beds, and disposed, off suitably.

QUESTIONS

1. Describe briefly the various methods adopted in final disposal of wastewater sludges.

2. Explain the construction and working principles of :

(*a*) Septic tank. (*b*) Soak pit.

(*c*) Oxidation ponds.

3. Describe the activated sludge treatment process.

4. Describe the construction and working of :

(*a*) Oxidation ditch.

(*b*) Method of wastewater disposal on land.

5. Describe the theory of disposal by oxidation ponds. State its design criteria. Describe its suitability for adoption as a means of disposal of sewage of a town.

6. Explain self-purification of streams and indicate how sun-light helps in self-purification.
7. The methods of natural treatment of sewage farming oxidation pond and dilution. Give briefly the details of these methods and state under which conditions you would recommend them.
8. Distinguish between oxidation pond and oxidation ditch.
9. What is the primary source of oxygen in oxidation ponds?
10. Discuss the suitability of oxidation ponds for sewage treatment in our country.
11. Discuss the control measures to prevent river pollution.
12. Enumerate the two general methods for sewage disposal.
13. What are the factors affecting self-purification of polluted streams?
14. Write a brief note on the use of sewage effluent for irrigation.
15. Explain the difference between the dilution process, if the wastewater effluents are disposed in stream waste and sea water.
16. Explain clearly the methods, problems, and limitations of land disposal of sewage.
17. Give a list of methods available for treatment of sewage for both rural and urban conditions.
18. Mention various methods of disposal of effluent from septic tank. Describe one of them in detail.
19. Enumerate and discuss various types of privies that are used in conservancy system of sanitation.
20. Discuss various methods of human excreta disposal in areas where sewers have not been laid.

Treatment of Sewage (Domestic Wastewaters)

16.1. INTRODUCTION

The sewage, also known as domestic wastewaters are required to be treated for the purpose of converting a raw harmful sewage into an acceptable final effluent and also to dispose of the solids removed in the process. It is therefore, necessary first to determine the characteristics of the raw sewage and the required characteristics of the effluent, or the required degree of treatment, before proceeding with the design of the sewage treatment plant.

The method of treatment depend on several considerations like, the characteristics of effluent sewage, condition of available disposal points, effluent standards desired etc.,

The sewage contains human excreta, and wastewaters from bathrooms containing wastewaters from cloth washings, like soaps, detergents, etc., The sewage contains about 99% water and 1% solids. Out of all the solids present in sewage, about 70% are organic and 30% are inorganic. The inorganic solids contain grit, salts, and metals in varying proportions. In several cases, large volumes of sewage is discharged into the water bodies either in untreated or in partially treated condition, thereby throwing heavy load of pollutants containing pathogens; oxygen depleting biodegradable organic matter; non-biodegradable organic chemicals like detergents; sediments and suspended inorganic matter; etc.,

The discharge of sewage causes large scale pollution of water bodies. Laws have, therefore, been enacted, laying down norms and standards, for different pollutants which prescribe permissible limits for the maximum allowable concentrations of these pollutants in the discharged sewage.

In order to control the quantum of pollutants being discharged into the water bodies, the Central Pollution Control Board has notified general standards for discharge of pollutants from effluents into water sources.

The domestic wastewaters (sewage) as well as industrial wastewaters (industrial effluents) need to be treated in a suitable treatment facility, before they can be discharged into a river, or at some other suitable place. The degree of treatment depends upon the composition of the wastewater and the specified standards for the source of disposal. These standards will decide the maximum limit on the BOD, suspended solid (SS), and other pollutants, which may be permitted in the treated wastewater, before the same is allowed to be discharged into the said source of disposal. Different treatment units and their combinations are, infact, employed to treat different type of wastewaters, as to bring down their polluting contents.

However, before we describe the treatment units, we will describe the guiding standards recommended by the Bureau of Indian Standards (BIS) for discharge of sewage and industrial effluents into water bodies and no land.

Objectives of Sewage Treatment

The purposes of sewage treatment are :

1. To prevent waterborne diseases.
2. To remove organic solids from sewage, and to convert solids into stable products by biological decomposition.
3. To dispose treated effluents effectively and safely.
4. To safeguard water sources from pollution.
5. To prevent nuisance and offensives odours.

16.2. BIS GUIDING STANDARDS FOR DISCHARGE OF SEWAGE AND INDUSTRIAL EFFLUENTS

The guiding standards prescribed by the Bureau of Indian Standards (BIS) for the discharge of sewage and industrial effluents in water bodies are given in Table 16.1, and the standards for discharge of these effluents on land are given in Table 16.2. For both these cases, the standards have also been laid down under the Environment (Protection) Ac, 1986.

Table 16.1. BIS (ISI) Standard for Discharge of Sewage and Industrial Effluents in Surface Water Sources and Public Sewers

S. No.	*Characteristic of the Effluent*	*Tolerance limit for Sewage Effluent discharged into Surface Water, as per IS 4764-1973*	*Tolerance Limit for Industrial effluents discharged into*	
			Inland surface waters, as per IS 2490-1974	*Public sewers as per IS 3306-1974*
(1)	(2)	(3)	(4)	(5)
1.	BOD	20mg/L	30 mg/L	500 mg/L
2.	COD	—	250 mg/L	—
3.	pH value	—	5.5 to 9.0	5.5 to 9.0
4.	Total Suspended Solids (TSS)	30 mg/L	100 mg/L	600 mg/L

5.	Temperature	—	40°C	45°C
6.	Oil and grease	—	10 mg/L	100 mg/L
7.	Phenolic compounds (as Phenol)	—	1 mg/L	5 mg/L
8.	Cyanides (as CN)	—	0.2 mg/L	2 mg/L
9.	Sulphides (as S)	—	2 mg/L	—
10.	Fluorides (as F)	—	2 mg/L	—
11.	Total residual chlorine	—	1 mg/L	—
12.	Insecticides		Zero	—
13.	Arsenic (as As)	—	0.2 mg/L	—
14.	Cadmium	—	2 mg/L	—
15.	Chromium, hexavalent (as Cr)	—	0.1 mg/L	2 mg/L
16.	Copper	—	3 mg/L	3 mg/L
17.	Lead	—	0.1 mg/L	1 mg/L
18.	Mercury	—	0.01 mg/L	—
19.	Nickel		3 mg/L	2 mg/L
20.	Selenium	—	0.05 mg/L	—
21.	Zinc	—	5 mg/L	15 mg/L
22.	Chlorides (as Cl)	—	—	600 mg/L

Table 16.2. BIS (ISI) Standards of Wastewater Effluents to be Discharged on Land for Irrigation

S. No.	Characteristics / Constituent of Effluent wastewater	Tolerance limit as per IS : 3307-1965
(1)	(2)	(3)
1.	BOD at 20°C	500 mg/L*
2.	pH value	5.5 to 9.0
3.	Total Dissolved Solid (TDS)	2100 mg/L
4.	Oil and greases	30 mg/L
5.	Chlorides (as Cl)	600 mg/L
6.	Boron	2 mg/L
7.	Sulphates	1000 mg/L

8.	Percentage of Sodium with respect to total constent of Sodium, Calcium, Magnesium and Pottasium	60%
9.	Radioactive Materials (*i*) α-emitters (*ii*) β-emitters	 10^{-9} μC/mL 10^{-8} μC/mL

16.3. CLASSIFICATION OF SEWAGE TREATMENT

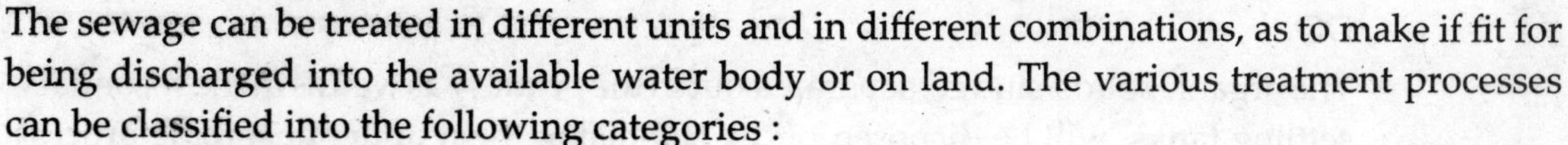

The sewage can be treated in different units and in different combinations, as to make if fit for being discharged into the available water body or on land. The various treatment processes can be classified into the following categories :

1. *Preliminary treatment*
2. *Primary treatment*
3. *Secondary treat*
4. *Tertiary treatment*

Sometimes, the preliminary and primary treatments are classified together, as primary treatment.

The above mentioned treatment processes are discussed below :

1. Preliminary Treatment

Preliminary treatment consists of separating the floating materials (like dead animals, tree branches, papers, pieces of rags, wood, etc.,), and also the heavy settleable inorganic solids. It also helps in removing the oils and greases, etc., from the sewage. This treatment reduces the *BOD* of the wastewater, by about 15 to 30%. The processes used are : **Screening** for removing floating papers, rags, clothes, etc.,; **Grit chambers** for removing grit and sand; and **Skimming tanks** for removing oils and greases.

2. Primary Treatment

Primary treatment consists in removing large suspended organic solids by sedimentation in **Settling tanks** or **sedimentation tanks.** The liquid effluent from primary treatment, often contains a large amount of suspended organic material, and has a high BOD.

The organic solids, which are separated out in the sedimentation tank (in primary treatment), are often stabilised by anaerobic decomposition in a *digestion tank* or are *incinerated.* The residue is used for landfills or soil conditioners.

3. Secondary Treatment

Secondary treatment is further treatment of the effluent, coming from the primary sedimentation tank. This is generally accomplished through biological decomposition of organic matter, which can be carried out either under aerobic or anaerobic conditions. In these biological units, bacteria will decompose the fine organic matter, to produce clearer effluent.

The treatment reactors, in which the organic matter is decomposed (oxidised) by aerobic bacteria are known as ***aerobic biological units;*** and may consist of:

(*i*) Filters (*intermittent sand filters as well as trickling filters*);

(*ii*) **Aeration tanks,** with the feed of *recycled* **activated sludge** (*i.e.* **the sludge,** which is settled in secondary sedimentation tank, receiving effluents from the aeration tank);

(*iii*) **Oxidation ponds and Aerated lagoons.**

Since all these aerobic units, generally make use of primary settled sewage, they are easily classified as secondary units.

The effluent from the secondary biological treatment usually contain a little BOD (5 to 10% of the original), and may even contain several milligrams per litre of DO.

The organic solids/sludge, separated out in the primary as well as in the secondary settling tanks, will be disposed off by stabilising them **under anaerobic** process in a **Sludge digestion tank.**

4. The Tertiary Treatment (Final or Advance Treatment)

This treatment consists in removing the organic load left after the secondary treatment, and particularly to kill the pathogenic bacteria. *This treatment, which is normally carried out, while using the river stream for collecting water for reuse or for water supplies*. It may, however, sometimes be adopted, when the outfall of sewage is very near to the water intake of some nearby town. ***The sewage treatment is, therefore, usually confined upto secondary treatment only.***

The various treatment units, which may be used for treating sewage, and the extent of BOD, solids and bacteria removed by them, have been summarised in Section 16.11.

Schematic Outline of Sewage Treatment

As discussed above, the primary treatment of sewage is carried out in various stages through screens, the grit chamber, the skimming tank, the primary sedimentation and the detritous tank. Secondary (biological) treatment consists of trickling filters, an activated sludge treatment process, the oxidation ponds, the sludge digester, the sludge drying beds, etc., An advanced treatment is given for the removal of specific characteristics from the sewage, as a special case. Chemical precipitation and chlorination are adopted only when they are required. Finally, the treated waste waters are disposed on land or water bodies.

The plan of treatment is shown in Fig. 16.1. The treatment (purification) process in each unit is given below in brief :

1. *Screen chamber* removes dead animals, branches of trees, logs of wood, rags and other coarse floating materials.
2. *Grit chamber* removes heavy settleable inorganic solids such as sand.
3. *Skimming tank* removes oil and grease.
4. *Sedimentation tanks* are designed to separate solids from liquids under gravity and remove suspended settleable organic solids (about 60%) and BOD load of 30% associated with solids removal.

5. *Trickling filters* are devices that depend on micro-organisms for removing the settleable, non-settleable, and dissolved organic solids (70–90%) and BOD load (65%–70%).
6. *Activated sludge treatment* method utilises additional oxygen and allows greater contact between wastewater and micro-organisms for the removal of organic load (80–90%) and BOD reduction (70%–80%).

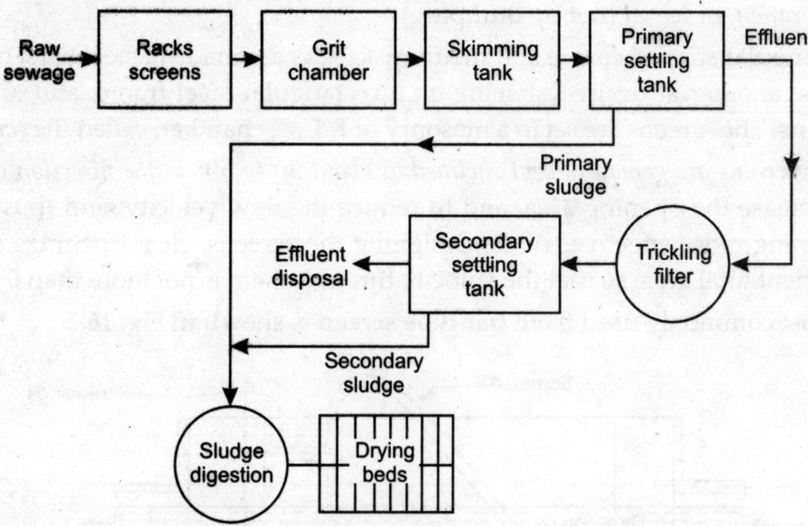

Fig. 16.1. Layout of Sewage Treatment Plant.

Different types of treatment plants contain different combinations of treatment units. These combinations can also be changed, and some other combinations made, depending upon the local needs. The topography and geology will influence the choice of the units to be adopted.

Now, all these treatment units are being discussed in detail.

16.4. SCREENING

A screen is the first unit in wastewater (sewage) treatment. These are provided to retain floating matter and coarse solids. Screens consist of parallel bars, rods, gratings, wire mesh or perforated plates. The screens are kept ahead of the grit chamber either in a separate chamber or in the same grit chamber.

Screening is the very first operation carried out at any sewage treatment plant, and consists of passing the sewage through different types of screens, *so as to trap and remove the floating matter*, such as pieces of cloth, paper, wood, cork, hair, fibre, kitchen refuse, fecal solids, etc., present in sewage. These floating materials , if not removed, will choke the pipes, or adversely affect the working of the sewage pumps. Thus, the main idea of providing screens is to protect the pumps and other equipments from the following damages due to the floating matter of the sewage.

Types of Screens

Depending upon the size of the openings, screens may be classified as coarse screens, medium screens, and fine screens.

1. **Coarse screens** (also known as Racks) have spacing between the bars (*i.e.* opening size) as about 50 mm or more. These screens do help in removing large floating objects from sewage. The material separated by coarse screens, usually consists of rags, wood, paper, etc.,, which may be disposed off by *incineration, burial,* or *dumping.*
2. **Medium screens** have spacing between bars as about 6 to 40 mm. The screenings usually contain some quantity of organic material, which may be disposed off by *incineration,* or *burial* (not by dumping).

   Rectangular shaped coarse and medium screens are made of steel bars, fixed parallel to one another at desired spacing on a rectangular steel frame, and are called **bar screens.** The screens are set in a masonry or R.C.C. chamber, called the *screen chamber.*

   These screens are generally kept inclined at about 30 to 60° to the direction of flow, so as to increase the opening area, and to reduce the flow velocity; and thus making the screening more effective. While designing the screens, clear openings should have sufficient total area, so that the velocity through them is not more than 0.8 to 1 m/sec.

   A most commonly used **fixed bar type screen** is shown in Fig. 16.2.

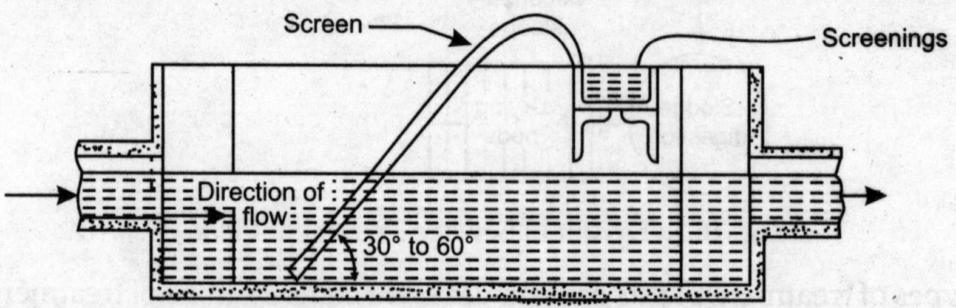

Fig. 16.2. Fixed Bar Type Coarse or Medium Screen.

3. **Fine Screens** have perforations of 1.5 mm to 3 mm in size. The installation of these screens prove very effective, and they remove as much as 20% of the suspended solids from sewage. These screens, however, get clogged very often, and need frequent cleaning. They are, therefore, used only for treating the industrial wastewaters, or for treating those municipal wastewaters, which are associated with heavy amounts of industrial wastewaters. These screens will considerably reduce the load on further treatment units.

   Wire mesh are generally used for constructing fine screens. The metal used should be resistant to rust and corrosion. The fine screens may be **disc** or **drum** type and are operated continuously be electric motors, and is cleaned by a cone brown.

16.5. GRIT CHAMBERS OR GRIT CHANNELS

The grit chambers or grit channels, as they are usually called, are the sedimentation basins placed usually after the fine screens and before the primary sedimentation tank. The grit chamber removes the inorganic grit, such as sand, gravel, and other mineral matter that has a nominal diameter of 0.15 to 0.20 mm, and specific gravity more than 2.5.

In this, the flow passes through the channel in a horizontal direction. The channel is designed to give a horizontal straight line flow velocity, which is kept constant over varying discharge. (about 0.3 m/s). The constant velocity is achieved by providing a *velocity control section, such as a proportional flow weir* at the effluent end of a rectangular channel; or by providing a *parabolic (or V) shaped channel*, having varying width with depth.

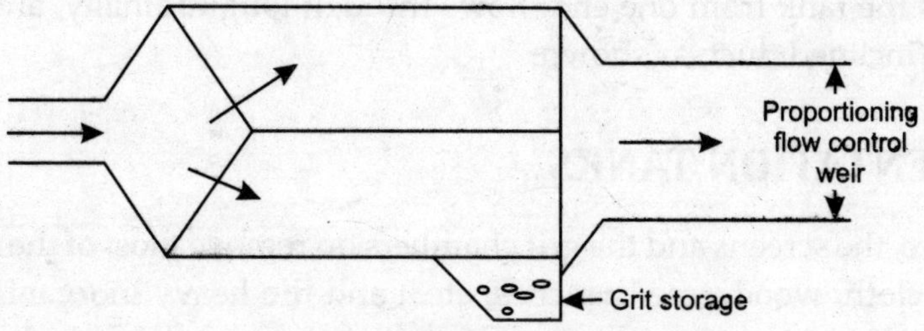

Fig. 16.3. Grit Chamber.

Fig 16.3. shows the line sketch of a simple **rectangular grit channel** (provided with a proportioning flow weir at the effluent end), which is often used in the treatment of sewage.

16.6. SKIMMING TANKS

Skimming tanks are employed for removing oils and greases from the sewage, and are placed before the sedimentation tanks. They are used where sewage contains too much of grease or oils, which include fats, waxes, soaps, fatty acids, etc., These greasy materials may enter into the sewage from the kitchens of restaurants and, houses, from motor garrages, oil refineries, soap and candle factories, etc.,

These oil and greasy materials may be removed in a skimming tank, in which air is blown by an aerating device through the bottom. The rising air tends to coagulate and solidify the grease, and cause it to rise to the surface (being pushed in separate compartments), from where it is removed.

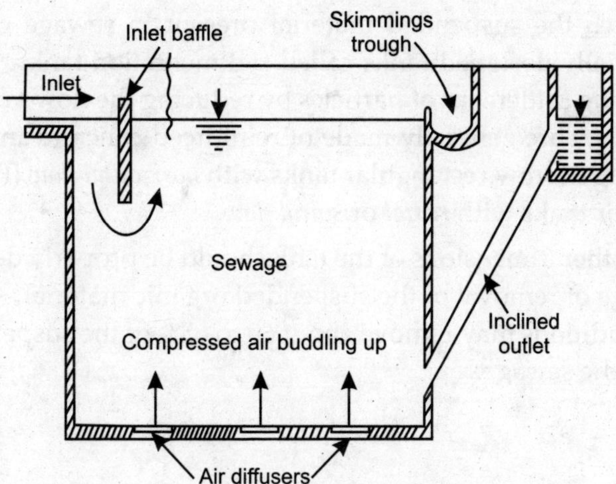

Fig. 16.4. Skimming Tank.

The typical details of a skimming tank are shown in Fig 16.4. It consists of a long trough shaped structure divided into two or three lateral compartments by means of vertical baffle walls (having slots in them) for a short distance below the sewage surface, as shown. The baffle walls help in pushing the rising coagulated greasy material into the side compartments (called stilling compartments). The rise of oils and grease (as they lighter than water) is brought about by blowing compressed air into the sewage from the bottom of the tank. The collected greasy materials are removed (*i.e.* skimmed off) and disposed off either by burning or burial.

Sewage enters the tank from one end, flows through longitudinally, and finally goes out through a narrow inclined duct, as shown.

16.7. SEDIMENTATION TANKS

As discussed above, the screens and the grit chambers do remove most of the floating materials (like paper, rags, cloth, wood, tree branches, etc.,) and the heavy inorganic settleable solids from the sewage. However, a part of the suspended *organic solids* which are too heavy to be removed as floating matters, and too light to be removed by grit chambers (designed to remove only the heavy inorganic solids of the size more than 0.15–0.2 mm, and of sp. gravity 2.65), are generally removed by the sedimentation tanks.

In a complete sewage treatment, the sedimentation is carried out twice; once before the biological treatment (*i.e. primary sedimentation*) and once after the biological treatment (*i.e. secondary sedimentation*). When *chemical coagulants* are also used for flocculating the organic matter during the process of sedimentation, the process is called *chemical precipitation*, or *sedimentation aided with coagulation*. This is generally not used in modern days.

Other sewage treatment units which work on the principle of sedimentation are : *Septic tanks*, Imhoff tanks etc.,

Primary Sedimentation Tanks

The clarification of sewage by the process of 'sedimentation' can be affected by providing conditions under which the suspended material present in sewage can settle out. This is brought about in specially designed tanks called sedimentation tanks. Sedimentation basins are designed for effecting settlement of particles by reducing the flow velocity or by detaining the sewage in them. They are generally made of reinforced concrete and may be rectangular or circular in plan. Long narrow rectangular tanks with *horizontal flow* (Fig. 16.5) are generally preferred to the circular tanks with *radial* or *spiral flow*.

The capacity and other dimensions of the tank should be properly designed, so as to effect a fairly high percentage of removal of the suspended organic material. A plain sedimentation tank under normal conditions may remove about 60 to 65% of the suspended solids, and 30 to 35% of the BOD from the sewage.

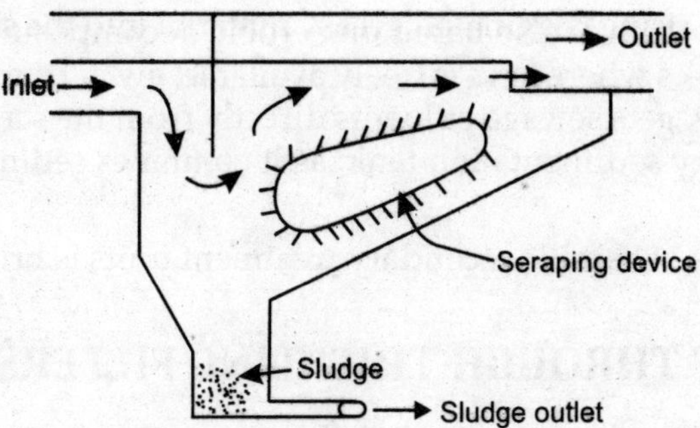

Fig. 16.5. Primary Sedimentation Tank.

Chemical Precipitation and Coagulation

The chemical precipitation, as also mentioned above, is also known as "sedimentation aided with coagulation" or "type II Sedimentation". Very fine suspended particles, present in wastewaters, which cannot be removed in plan sedimentation, may sometimes, be settled by increasing their size by changing them into flocculated particles. For this purpose, a chemical compound (like ferric chloride, ferric sulphate, alum, chlorinated coppersas etc.) called ***coagulant***, is added to the wastewater, which on through mixing forms a gelatinous precipitate called ***floc***. The fine mud particles and other colloidal matter present in wastewater get absorbed in these flocs, forming the bigger sized flocculated particles. The process of addition and mixing of chemical is called ***coagulation***. The coagulated sewage is then made of pass through sedimentation tank, where the flocculated particles settled down, and get removed.

The process of coagulation of sewage is similar to that of water as described thoroughly in the chapter on "Purification of water" in part-1 of this book. Since the coagulation of sewage is seldom adopted in modern days, we are not repeating the entire coagulation process here again.

16.8. SECONDARY TREATMENT UNITS

The effluent the primary sedimentation tank contains about 60 to 80% of the unstable organic matter originally present in sewage. This organic matter is oxidised/decomposed, and removed in a secondary treatment unit through biological treatment, with the help of aerobic or anaerobic bacterias. The aerobic bacterial treatment is preferred, since anaerobic decomposition is not only slow (having 1/3rd speed of aerobic decomposition) but also produces bad smells due to generation of methane (CH_4) gas.

The two treatment units, which are commonly used for the **aerobic secondary treatment**, are :

1. *Trickling filters*; and
2. *Activated sludge plant*.

The **trickling filters** are generally used for small and medium sized sewage treatment plants (STPs); while activated sludge treatment is usually adopted for large sized plants for metro cities, etc., Both these secondary units are aerobic units, and, hence, use aerobic bacteria for oxidation/decomposition of the organic matter contained in sewage. In such an aerobic bacterial treatment unit, the main aim is to produce suitable environment, which may help in stimulating the growth of aerobic bacteria.

A third type of unit, which is also based on aerobic decomposition is an ***oxidation pond***, which may be used at places where land is freely available, away from the main residences. An oxidation pond, generally, gets sewage effluents directly from the screens or the grit channels, without the use of primary sedimentation tank, as it combines sedimentation with biological oxidation.

The use of these types of aerobic secondary treatment units is briefly described below :

16.9. TREATMENT THROUGH TRICKLING FILTERS

The **high rate trickling filters** are now almost universally adopted for giving secondary treatment to sewage. These filters, also called as *percolating filters or sprinkling filters*, consist of tanks of coarser filtering media, over which the sewage is allowed to sprinkle or trickle down, by means of *spray nozzles* or *rotary distributors*. The percolating sewage is collected at the bottom of the tank through a well designed *under-drainage system*.

The purification of the sewage is brought about mainly by the aerobic bacteria, which form a bacterial film around the particles of the filtering media. In order to ensure the large scale growth of the aerobic bacteria, sufficient quantity of oxygen is supplied by providing suitable ventilation facilities in the body of the filter; and also to some extent by the intermittent functioning of the filter.

The effluent obtained from the filter is taken to the secondary sedimentation tank for settling out the organic matter, oxidised while passing down the filter.

The most commonly used circular filter tanks, are provided with **rotary distributors** having a number of distributing arms (generally four arms are used) as shown in Fig. 16.6. These distributors rotate around a central support either by an electric motor, on more generally by the force of reaction on the sprays.

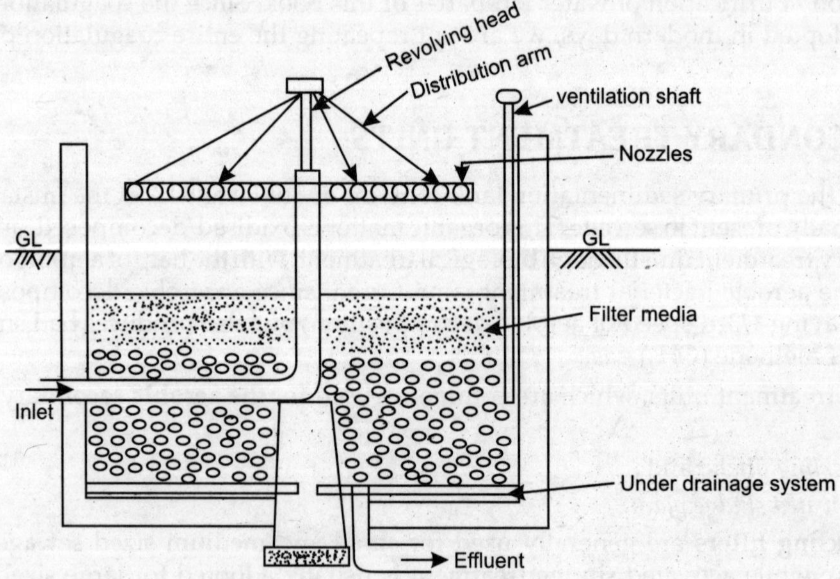

Fig. 16.6. Trickling Filter.

The **filtering media** consists of coarser materials like cubically broken stones or slag, free from dust and small pieces. The size of the material used may vary between 25 to 75 mm.

Usually, stones from rocks like granite or limestone may be used for this purpose. The depth of the filtering media may vary between 2 to 3 metres. The filtering material may be placed in layers; with coarsest stone used near the bottom, and finer material towards the top. The walls of the filter tank are provided with openings for circulation of air, all through.

A satisfactory ventilation is achieved when properly designed **under drains,** having adequate openings, are provided under the filter bed. Besides ensuring satisfactory drainage, such drains, will also ensure satisfactory ventilation and aeration of the filter bed. Vitrified clay blocks (Fig. 16.7) are generally used as under-drains. These blocks have top openings of such size that the stone can be placed directly on them, and yet furnish flow channels with sufficient capacity for the heaviest hydraulic loading. These blocks are laid on a reinforced concrete floor, (about 10 to 15 cm thick) which is sloped gently (at about 1 in 300) towards the main effluent rectangular channel. This main **effluent channel** may be provided adjoining the central column of the distributor, as shown in Fig. 16.6.

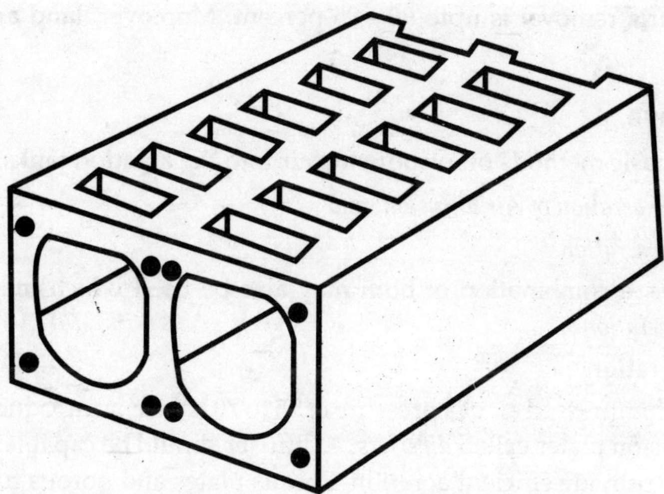

Fig. 16.7. Under-drain Block for Standard Trickling Filters.

Secondary Settling Tanks for Trickling Filters.

It was pointed out earlier that a filter does remove *only a very small percentage* of the finally divided suspended organic matter present in sewage, while *the majority of this organic matter undergoes a change of character, only* due to the biological oxidation and nitrification taking place in the filter. Due to this aerobic oxidation and nitrification that takes place in the filter, the organic solids are converted into coagulated suspended mass, which is heavier and bulkier, and would thus settle down by gravity, if allowed to do so in sedimentation tanks. The effluent of the filter is, therefore, passed through a sedimentation tank, called *Secondary clarifier* or **Secondary settling tank** *or* **Humus tank.**

16.10. ACTIVATED SLUDGE TREATMENT

The activated sludge process is an excellent method of treating the sewage coming out from the primary clarifier. The sewage effluent from primary sedimentation tank, is mixed with 20 to 30 per cent of own volume of **activated sludge,** which contains a large concentration of highly active aerobic micro-organisms. The mixture enters an aeration tank, where the micro-organisms (coated around the sludge solids) and the sewage, are intimately mixed together, with a large quantity of air for about 4 to 8 hours.

Under these conditions, the moving organisms will oxidise the organic matter, and the suspended and colloidal matter tend to coagulate and form a precipitate, which settles down readily in the secondary settling tank. The settled sludge (containing micro-organisms) called **activated sludge,** is then recycled to the head of the aeration tank, to be mixed again with the sewage being treated. New activated sludge is continuously being produced by this process, and a portion of it being utilised and sent back to the aeration tank, whereas the excess portion is disposed off properly alongwith the sludge collected during primary treatment, after digestion.

The effluent obtained from a properly operated activated sludge plant is of high quality, usually having a lower BOD than that of a trickling filter plant. BOD removal is upto 80—95 per cent, and bacteria removal is upto 90—95 percent. Moreover, land area required is also quite less.

Methods of Aeration

There are two basic methods of introducing air into the aeration tanks, *i.e.*

1. *Diffused air aeration of Air diffusion; and*
2. *Mechanical aeration*
3. Sometimes, a combination of both may also be used which may then be called as *combined aeration*.

1. Diffused Air Aeration

In this system, compressed air at a pressure of 35 to 70 kN/m^2 is introduced into the aeration tank, through diffusion plates called *diffusers*. A diffuser should be capable of diffusing air into small bubbles as to provide efficient aeration. Porous plates and porous tubes made of quartz or crystalline alumina are generally used as diffusers.

2. Mechanical Aeration

In this system, the sewage is thoroughly agitated by means of mechanical devices like paddles, so as to use the atmospheric air, rather than introducing compressed outside air. The thorough agitation of sewage brings it into intimate contact with the atmospheric air, as to cause sufficient aeration. The aeration period depends on the mechanical process adopted for agitation, and usually varies from 6 to 8 hours. The quantity of returned sludge in mechanically aerated aeration tanks is usually about 25 to 30 percent of the flow of sewage.

3. Combined Aeration

In this system, the diffused air aeration as well as mechanical stirring are combined together in a single unit. In this type of an aerator, the aeration of sewage is achieved by diffusing air through bottom diffuser plates, as well as by rotating paddles @ 10 to 12 r.p.m. Spiral motion, so set up, brings about the required aeration.

16.11. SEWAGE TREATMENT UNITS, THEIR FUNCTIONS, AND EFFICIENCIES AT A GLANCE

S. No.	Type of Treatment	Purification effected	Process or unit employed	BOD removal as percentage of original	Removal of suspended and dissolved solids as percentage of original	Removal of bacterial load as percentage of original	Disposal of residuals
(1)	(2)	(3)	(4)	(5)	(6)	(7)	(8)
1.	Preliminary Treatment	(*a*) Removal of *floating materials* like dead animals, tree branches, pieces of rags, wood and another large sized floating materials.	Coarse and **screens** of different designs	5—10	2—20	10—20	Screening can be' disposed off, easily, either by burial or burning.
		(*b*) Removal of *heavy settleable inorganic solids.*	**Grit chambers** or *Detritus tanks*	10—20	20—40	10—20	The grit can be easily disposed off either by burning or burial for raising low areas.
		(*c*) Removal of *fats and greases.*	**Skimming tanks**	20—30	20—40	10—20	The skimming contain unstable volatile organic materials, and have to be disposed off, by first stabilising them indigestion tanks by anaerobic process.

2.	Primary Treatment	Removal of suspended settleable organic solids	(1) **Sedimentation tanks**	30–35	60–65	25–75	Sludge containing organic material has to be stabilised first in digestion tanks, and the digested material is then used as a manure
3.	Secondary or Biological Treatment	Removal of fine suspended non settleable solids and colloidals, including dissolved organic matter	**2. Conventional low rate filters, followed and preceded by sedimentation.**	90–95	70–92	90–95	Sludge containing organic material has to be stabilised in digestion tanks and the residue is used as manure
			or *(ii)* **Modern high rate trickling filters,** followed and preceded by sedimentation	65–95	65–90	80–95	As above
			or **3. Activated treatment in** *Aeration tank and secondary tank*	75–95	85–90	90–98	As above
			or *(iv)* **Oxidation ponds**	85–90	85–90	90–98	Effluents are generally disposed off by using them for irrigation
4.	**Final or Tretiary Treatment**	**Removal of Pathogens and the remaining very fine dissolved organic matter**	*Chlorination* (seldom used for sewage treatment)	100%	100%	100%	–

16.12. OXIDATION PONDS AND STABILISATION PONDS

As dissolved in previous chapter, stabilisation ponds are open *flowing-through* earthen basins, specially designed and constructed to treat sewage and biodegradable industrial wastewaters. Since these ponds are open tanks, exposed to the sun and atmosphere, they generally remain aerobic. The digestion of organic matter is brought about by aerobic bacteria, which remain suspended in the sewage.

Such ponds provide comparatively long detention periods, extending from a few days to several days, during which time, the wastes get stabilised by the action of natural forces.

Stabilisation ponds may be classified as *aerobic, facultative,* or *anaerobic,* depending upon the mechanism of purification. In a *totally aerobic pond,* the stabilisation of wastes is brought about by aerobic bacteria, which flourish in the presence of oxygen. The oxygen demand of such bacteria in such a pond is met by the combined action of algae and other micro-organisms (such as bacteria and protozoa*), called **algal photosynthesis**, or **algal-symbiosis**. In this symbiosis, the algae (which are microscopic plants) while growing in the presence of sunlight, produce oxygen by the action of *photosynthesis*; and this oxygen is utilised by the bacteria for oxidising the waste organic matter.

In an *anaerobic pond,* however, the stabilisation of wastes is mainly brought about by the usual anaerobic conversion of organic wastes to carbon dioxide, methane, and gaseous end products, with eruption of foul odours and pungent smells.

In a *facultative pond,* the upper layers work under aerobic conditions, while the anaerobic conditions prevail in the bottom layers. The *upper aerobic layer of the pond acts as a good check against the evaluation of the foul odours from such a pond*. The treatment affected from such a pond is comparable to that of conventional secondary treatment processes.

The totally aerobic ponds will have to be very small depths, such as below 0.5 m, and still then, may require occasional stirring, to prevent anaerobic conditions in the settled sludge. *It is, therefore practically difficult to construct and use such purely aerobic ponds.* The facultative ponds, with depths varying between 1.0 m to 1.5 m, are thus most widely used for treatment of sewage. Deeper ponds in the depth range 0.25 to 4 m may also sometimes be constructed to work anaerobically.

Stabilisation ponds have been used to treat wastewaters for many years, especially for small communities.

The term **oxidation pond** was originally referred to that stabilisation pond which received partially treated sewage; whereas the pond that received raw sewage was used to be called a **sewage location**; but in recent years, the term *oxidation* pond has been widely used as a collective term for all years of ponds, and most particularly the *facultative stabilisation ponds*.

The results of the oxidation pond treatment are: the oxidation of the original organic matter and the production of algae, which are discharged with the effluent. This results in a net reduction in BOD, since the algae are more stable than the organic matter in wastewaters, and degrade slowly in the river stream into which the effluent is discharged.

The effluent should be entered at the centre of the pond above or below the liquid surface for a uniform mixing of the influent with the oxygen saturated pond liquor. The outlet should provide withdrawal of effluent from different depths of pond.

16.13. MECHANICALLY AERATED LAGOONS

A mechanically aerated lagoon is a deeper oxidation pond, with oxygen introduced by mechanical aerators rather than relying on the photosynthetic oxygen production alone. *As these ponds are deeper than the oxidation ponds, and as they are artificially aerated, less detention time and areas are required.*

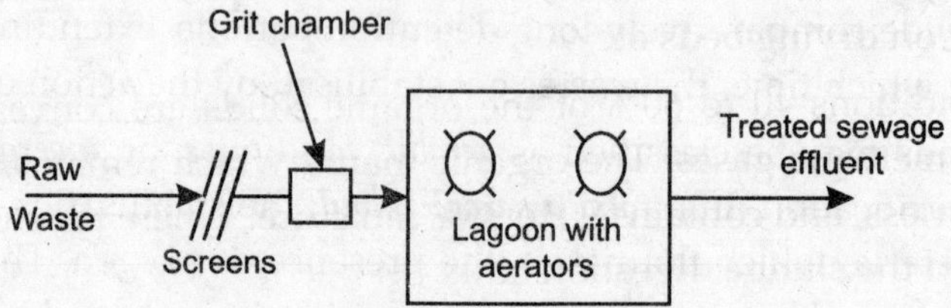

Fig. 16.8. Flowsheet for a mechanically.

The depth of these basins, normally ranges between 2.4 to 3.6 m, and detention time between 4 to 10 hours. The land area required is about 5 to 10% of that required for an equivalent oxidation pond. With the detention period of 4 to 10 hours, the efficiency obtained ranges between 60 to 90%. The aerated lagoons are frequently used for treating industrial wastewaters as well as city sewage. A simple flow chart of such a lagoon is shown in Fig. 16.8.

16.14. SLUDGE TREATMENT AND DISPOSAL

Different treatment units, as discussed above, give out sludge. The sludge so produced contain objectionable odour, objectionable organic substances, pathogenic organisms etc., and are unsafe to public health. Therefore, due care is required in treatment and safe disposal of the sludge.

The sludge produced from various sewage treatment units are :

1. **Pre-sedimentation sludge** also known as primary or raw sludge, is produced by pre-sedimentation tanks.
2. **Activated sludge** produced is collected through secondary settling tanks. The quantity of sludge thus produced is huge and its handling, treatment and disposal become a big task.
3. **Trickling filter sludge** produced from trickling filter unit is collected through humus tank or secondary settling tank.
4. **Chemical precipitation sludge** have higher solid content due to higher suspended solid removal and floc accumulation. It may be fed to the digester alongwith raw sludge. It may be conditioned, filtered and dried.

   The main problem posed by the sludge is its high moisture content, about 95%. The sludge, therefore, becomes very bulky, and difficult to handle. If the moisture content of the sludge is reduced, its volume will go on decreasing.

16.14.1 Sludge Digestion Process

Sludge treatment process is known as *sludge digestion*. Sludge if disposed of without treatment, the organic matter may decompose, producing foul gases and a lot of nuisance, pollution, and health hazards. In order to avoid such pollutions, the sludge is first of all, stabilized by decomposing the organic matter under controlled *anaerobic conditions* and then disposed of suitably after drying on drying beds etc.,

In anaerobic conditions 40 to 60% of the organic solids are converted by bacteria into carbon dioxide and methane gases. The organic matter which remains, in chemically stable and practically odourless, and contains 90 to 95% moisture. Where the sludge digestic process is carried out is called the sludge digestion tank.

Well digested sludge contains nitrogen and phosphates for use as a fertilize. Methane gas is used as a fuel. Sludge volume is reduced as gases escape and liquefaction also takes place. Water separates out from well digested sludge.

Following are the main *advantages of sludge digestion.*

1. A portion of the solids will be liquefied and gasified thereby reducing sludge volume to be handled.
2. Volatile solid contents will be reduced by 55 to 75% and overall solid reduction may be 35 to 45%.
3. Moisture content is reduced by about 5%.
4. Nature of digested sludge becomes such that it gives up its water content more easily.
5. It makes the sludge fit for further drying easily and economically.
6. Digested sludge is used as a fertilizer and the gases produced by digestion may be used as a fuel.

Effect of Temperatures

The process of digestion is greatly influenced by temperature; rate of digestion being more at higher temperature and *vice versa*.

Effect of pH Value

During the digestion process, a lot of volatile organic acids are formed in the breakdown of organic material. These volatile acids are then converted into methane gas. If the methane gas is not formed properly, an accumulation of volatile acids may occur, causing pH to drop to a value as low as 5.0, which will suppress further bacterial action. Hence, during digestion, care must be taken to keep the acidity well under control, so that the pH during digester start-up does not go below 6.5 or so, and thus to see that alkaline conditions (with optimum pH about 7.2 to 7.4) may prevail ultimately, in the final stage of digestion.

16.14.2. Sludge Digestion Tank or Digesters

Digesters are used for the sludge digestion under controlled conditions. Digesters is a hopper bottomed concrete tank circular is shape. Generally, the tank is partially underground and partly above ground. However, it should be above groundwater table. The sludge digesters are of two types; normally.

(*a*) Conventional digester

(*b*) High rate digesters

In *conventional digesters* the sludge is fed approximately at the mid height and digestion is allowed to take place. In this, there is no extra arrangement for mixing of the sludge and raising the temperature, therefore, the rate of digestion is slow.

In high rate digesters, the sludge is continuously added and vigorously mixed. The digester is also heated to maintain maximum activity in the mesophilic region.

Sludge digestion tank as shown in Fig. 16.9, is about 10-20 m diameter and 6-8 m depth is provided with outlets and devices for;

(*i*) The collection of gases,

(*ii*) The collection of supernant liquor,

(*iii*) The collection of digested sludge

(*iv*) Breaking the scum

(*v*) Scraping the sludge

(*vi*) Mixing the contents

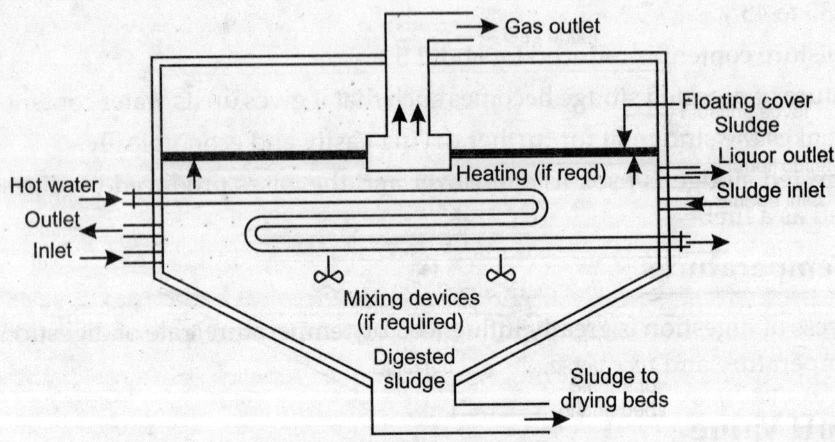

Fig. 16.9. Sludge digestion tank.

It may have fixed roof with separate gas holding device or a floating type of roof. 0.016 m^3/capita (1/m^3 /kg of volatile solids) is the amount of gas production. Volume of a digestion tank may be calculated at 0.033 to 0.060 m^3/capita or from the relation

$$V = [V_f - \frac{2}{3}(V_f - V_d)]\, t$$

where

V = Volume of the digestor (m^3)

V_f = Volume of fresh sludge (m^3/day)

V_d = Volume of digested sludge (m^3/day)

t = digestion period in days

16.14.3. Disposal of Digested Sludge

The digested sludge from the digestion tank contains a lot of water, and is, therefore, first of all, *dewatered or dried up*, before further disposal either by burning or dumping. In India the water of sludge is removed by drying the sludge on drying beds; whereas, in western countries, dewatering is generally done by *vacuum filters* or by using *high speed centrifuges.*

Sludge Drying Beds

The digested sludge is applied on a well drained bed. From this bed water is drained and perolates, and also evaporates leaving the solids to dry for sometime. The dried sludge is used as a manure or soil conditioner. This method is used where adequate land is available. The major portion of the liquid drains off in a few hours after which air drying commences by evaporation.

A sludge drying bed consists of a bottom layer of a gravel of uniform size over which is laid a bed of clean sand. Open jointed tile under-drains are laid in the gravel layer to provide effective drainage as the liquid percolates through the sand and gravel.

QUESTIONS

1. Enlist the treatment units which are employed in primary treatment of domestic wastewaters. Describe anyone such unit drawing a neat sketch.
2. Describe the aim of the following sewage treatment methods :
   (*i*) Screen, (*ii*) Sedimentation,
   (*iii*) Grit chamber, (*iv*) Activated sludge process.
3. Write short notes on the following :
   (*a*) Skimming tank (*b*) Plain sedimentation tank
   (*c*) Grit chamber (*d*) Screen
4. Explain two important units/processes which are commonly used for treatment of domestic wastewaters.
5. Explain the following :
   (*a*) Trickling filters (*b*) Activated sludge process
   (*c*) Aeration tanks
6. Explain the use of secondary settling tanks in treatment of sewage.
7. Explain as to how do "high rate trickling filters" differ from "conventional trickling filters"?
8. Give by a line diagram the outline of the method of purifying sewage in different steps from outfall sewer to the effluent discharge points in the river for
   (*i*) Medium sized city, and (*ii*) For a metro city.
9. Distinguish between primary treatment and secondary treatment. Draw the flow diagram for sewage treatment using activated sludge process.

10. What is the object of sewage treatment? Explain how would you decide the type of treatment to be given for a particular sewage.
11. With a flow diagram, explain the functions of each unit in primary treatment.
12. Explain the working principles of;
    (*a*) Oxidation ponds (*b*) Trickling filters.
    (*c*) Septic tank.
13. Explain the process of sludge digestion and state the factors affecting same.
14. Describe briefly the various methods adopted in final disposal of waste water sludges.

Treatment of Industrial Wastewaters

17.1. INTRODUCTION

The industrial wastewaters usually contain a lot of pollutants, and do have a great potential for polluting the water bodies into which such industrial effluents are let off. The nature and composition of industrial wastewaters usually vary from industry to industry and even within the same industry, depending upon the raw materials used and the technologies adopted by the industry.

The industrial wastewater may contain pollutants of different types, ranging from simple nutrients and biodegradable organic matter to complex toxic heavy metals and dangerous volatile organic chemicals (VOCs). Major polluting industries are :

1. Dairy plants.
2. Cane sugar and distilleries.
3. Fertiliser industry.
4. Paper industry.
5. Refineries and Petrochemicals.
6. Pharmaceuticals.
7. Steel plants.
8. Tanneries.
9. Textile industry.
10. Electroplating and metal finishing.
11. Pesticides manufacturing industry.

The wastewaters from industries like dairies, distilleries, tanneries, sugar factories, textiles, etc., contain a lot of biodegradable organic matter, and can be treated by using comparatively simpler biological treatment methods, which are also used for treating domestic sewage. Certain other industrial wastewaters may, however, be more difficult to treat. For example, metal plating industries release substantial quantities of toxic heavy metals and cyanides in their wastewaters. The chemical industry release wastewaters with highly variable composition, which are often acidic or alkaline in nature. Wastewaters from industries like fertilizers, oil-refineries, petrochemicals, pharmaceuticals, etc., contain several pollutants, and need elaborate treatment processes. Industries, producing and using volatile synthetic organic chemicals, produce wastewaters, which contain highly dangerous VOCs.

Solid, liquid and gaseous effluents from various industries are generally disposed of into the nearby environment. Almost all industries use water for some process or the other and finally discharge the wastewaters. The pollutants generated by industries include;

(*i*) Acids, (*ii*) Alkalies (*iii*) Suspended solids,

(*iv*) Toxic organic and inorganic chemicals

(*v*) Non-toxic organic and inorganic chemicals

(*vi*) Materials producing colours, turbidity, odour and taste

(*vii*) Oil and grease (*viii*) Radioactive material

(*ix*) No-biodegradable substances.

These pollutants should undergo a purification process within the factory boundaries and suitable measures be taken while disposing them.

Wastewaters obtained from industries are generally much more polluted than the domestic or even commercial wastewaters. Still, several industrialists try to discharge their effluents into our natural river streams, through unauthorised direct discharges. Such a tendency on the part of the industries pollute the entire river water to a grave extent, thereby making its purification almost on impossible task.

Sometimes, the industries discharge their polluted wastewaters into municipal sewers, thereby making the task of treating that municipal sewage, a very difficult and a costly exercise. Polluted Yamuna is a glaring example of industrial pollutants discharge in the river. The industries are prevented by legal laws, from discharging their untreated effluents. It, therefore, becomes necessary for the industries to treat their wastewater in their individual treatment plants, before discharging their effluents either on land or lakes or rivers, or in municipal sewers.

17.2. HARMFUL EFFECTS OF INDUSTRIAL WASTEWATERS

Industrial wastewaters have following main undesirable characteristics:

1. Soluble organics cause depletion of oxygen in river water.
2. Suspended solids impair normal aquatic life.
3. These cause undesirable tastes and odours.
4. Heavy metals, cyanides and toxic substances are harmful to living beings.
5. Colour and turbidity create unaesthetic conditions in water bodies.
6. Hot water released from factories act on solubility and affect bacteriological action.
7. Acids and alkalies affect aquatic life.
8. Inorganic materials may give rise to various problems such as hardness and corrosion.
9. Some harmful chemicals released from industrial wastewaters are : arsenic, cadmium, chromium, lead, mercury, cyanide, nitrogen, iron, zinc, metallic sulphates, chlorides, nitrates, fluorides, phosphates, and sulphide.

In view of the above, a decision is required to be taken whether the domestic sewage and industrial waste should be treated separately or combined :

The industrial waste may be mixed with domestic sewage for combined treatment before ultimate disposal into stream, in following circumstances :

(*i*) When domestic and industrial waste mixture is homogeneous in composition and is uniform in flow rate,

(*ii*) The mixture has less suspended solids and BOD,

(*iii*) The mixture is near neutral zone of pH, and

(*iv*) The mixture is free from toxic chemical content,

(*v*) The industrial waste can be mixed with domestic sewage for further treatment, after partial treatment is given to industrial wastes.

17.3. METHODS OF TREATING INDUSTRIAL WASTEWATERS

Industrial wastewaters usually contain several chemical pollutants and toxic substances in too large proportions. *The characteristics of the produced wastewater vary from industry to industry, and also vary from process even in the same industry.* Such industrial wastewaters cannot always be treated easily by the normal methods of treating domestic wastewaters, and certain specially designed methods or 'sequence of methods' may be necessary.

The normal biological treatment method for sewage are dependent upon the bacterial activity within the sewage; and as the toxic chemicals present in the industrial wastewaters may destroy the bacterial activity, these normal methods may not suffice, unless medified and/or supplement by additional techniques.

In order to achieve this aim, it is generally necessary to isolate and remove the troubling pollutants from the wastewaters, before subjecting them to usual treatment sequence of treatment processes *The sequence of treatment processes adopted should also be such as to help generate useful by products.* This will help economise the pollution-control measures, and encourage the industries to develop such treatment plants. Depending upon the *quantum, concentration, toxicity, and presence of non-biodegradable organics* in an industrial wastewater, its treatment may consist of any one or more of the following processes :

1. *Equalisation;*
2. *Neutralisation;*
3. *Physical treatment*
4. *Chemical treatment;* and
5. *Biological treatment*

These processes are briefly discussed below :

17.4. EQUALISATION OF INDUSTRIAL WASTEWATERS

Equalisation consists of holding the wastewater for some pre-determined time in a continuously mixed basin, to produce a uniform wastewater. Such an arrangement is necessary when the wastewater produced by the industry varies in characteristics and quantity over the entire day.

Fluctuations or variations in the input quality affect the purification process. Therefore, consistency in the quality should be maintained by storing the liquid wastes for a certain period of time and mixing suitably in a tank. From an equalisation tank or pond, uniform nature of the influent is fed into the treatment units.

17.5. NEUTRALISATION

Neutralisation means neutralising the excessive acidity or alkalinity of the particular wastewater, by adding alkali or acid, respectively, to the wastewater. This may be achieved either in the *equalisation tank*, where possible, or a separate *neutralisation tank*.

Most of the biological treatment plants function effectively within a range of 6 to 9 pH. Thus, industrial wastes of acidic or alkaline nature are neutralised in a tank or pond. Sometime wastewaters from nearby units are mixed to obtain neutrality. For example, acid wastes may be passed through beds of limestone, or waste fuel gas, as carbonic acid from boiler units, may be allowed to react with caustic wastes to obtain a pH value of about seven.

In case, industrial wastes and domestic sewage are mixed for combined treatment, separate storage tanks help to properly proportion and regulate the quality and quantity of wastes into a treatment plant. In such cases, industrial wastes may be partially treated and disposed into domestic sewage for further treatment.

17.6. PHYSICAL TREATMENT OF INDUSTRIAL WASTEWATERS

Physical treatment consists of separating the suspended inorganic matter by physical processes, like sedimentation and floatation.

1. ***Sedimentation.*** Sedimentation is employed to separate the heavier settleable solids, when the wastewater contains a high percentage of such heavy inorganic solids.
2. ***Floatation.*** Floatation consists of certain of fine air bubbles in the waste tank, by introduction of air into the tank from the bottom. The rising air bubbles, attach themselves to the fine suspended particles, increasing their buoyancy, and finally lifting them to liquid surface for consequent removal by skimming.
3. ***Inorganic Dissolved Salts.*** These are removed by
   (*i*) An iron exchange (demineralisation) process,
   (*ii*) Evaporation units,
   (*iii*) Dialysis, or
   (*iv*) Algae use (oxidation pond) depending upon the requirement.
4. ***Removal of Organic Dissolved solids.*** This operation in a waste treatment process utilise the biological activities. The micro-organisms, as already discussed, are sensitive to temperature, pH value, oxygen levels, presence of toxic chemicals etc., in industrial wastes. Trickling filters, and activated sludge units are used for oxidising organic matter and removing BOD. Anaerobic sludge digestion helps in reducing the solid volume, imparting drainability and converting solids into liquids and gases.

Dewatering is done through drying beds of coarse gravel as discussed in previous chapter.

5. ***Elutriation.*** It is a washing operation, which reduces alkalinity and eliminates odours generated from waste solids.

17.7. CHEMICAL TREATMENT OF INDUSTRIAL WASTEWATERS

Chemical treatment is often necessary before the biological treatment, through sometimes, it may not be required at all. Sometimes, it may however, serve as the final stage of treatment. The chemical treatment is used to recover the dissolved organic matter from the wastewater, and may consists of one or more of the following processes :

(i) *Reverse osmosis or hyper filtration;*

(ii) *Electrodialysis;*

(iii) *Chemical oxidation;*

(iv) *Chemical coagulation or chemical precipitation;* and

(v) *Adsorption;*

All these processes are briefly pointed out below :

1. *Reverse Osmosis.*

In reverse osmosis treatment process, the wastewater containing dissolved salts are filtered through semipermeable membranes at a pressure higher than the osmotic pressure. Such a treatment requires pre-treatments like : (*a*) *activated carbon adsorption;* or (*b*) *chemical precipitation* followed by some kind of *filtration*.

2. Electrodialysis.

In electrodialysis treatment process, dissolved salts from wastewater are separated by passing an electric current through the wastewater tank, installed with 'ion exchange membranes'.

3. *Chemical Oxidation.*

In chemical oxidation, chemicals like *chlorine* and *ozone* are used to reduce substances like *ammonia* and *cyanide*, etc., from the wastewater, in addition to reducing BOD load on biological treatments.

4. *Chemical Coagulation.*

Chemical coagulation, as we know, is adopted in treating raw water supplies, and helps in sedimentation of unsettleable micro and colloidal impurities, which get absorbed in the gelatinous flocs, formed by the chemical reactions between coagulants, or between the coagulant and the alkalies present in raw water.

However, in treatment of industrial waste-waters, such chemicals are generally added to help out the precipitation of dissolved and troublesome impurities, such as *phosphorus* or *phosphoric compounds*, etc., This process is then popularly known as **chemical precipitation.**

The *alum*, or ferric salts like *ferric chloride* or *ferric sulphate* may be used to directly *precipitate out phosphorus*. Sometimes, *lime* is also used to precipitate out phosphorus.

Chemical precipitation, through introduces additional chemicals in wastewater and produces larger quantity of sludge, yet is generally adopted these days for removing nutrients like *nitrogen from* specific industrial wastewaters. Toxic metals may also sometimes be precipitated out by using and developing appropriate chemical processes.

5. *Adsorption.*

Adsorption is generally used to remove non-biodegradable organics (like synthetic detergents), colour and odour from the wastewater. This process involve passing the pre-treated wastewater through the beds of activated carbon.

17.8. BIOLOGICAL TREATMENT OF INDUSTRIAL WASTEWATERS 

Biological treatment of industrial wastewater is necessary, when they contain larger quantities of biodegradable substances. Such biological treatment may be used.

The criteria for selecting a particular *conventional biological treatment*, may differ for different types of industrial wastewaters. The system parameters for a given type of industrial wastewater may be determined by laboratory experiments.

Conventional biological treatment may consists of : (*a*) *aerobic units* like : cheaper *stabilisation ponds* or *aerated lagoons*, or costlier units like *tricking filters* or *activated sludge plant*; and (*b*) *anaerobic units* like *anaerobic lagoons* and *septic tanks*.

It has also been observed that most of the industrial wastewaters do not certain sufficient nutrients for micro-biological growth; and hence nutrients like *urea* (containing nitrogen), *super phosphate* (containing phosphorus) etc., may have to be added to the reactors.

Acute *Toxic wastewaters* must be handled with special care and specific treatments. Say for example, toxic wastes, like cyanides, formaldehydes, phenols, etc., may need acclimatised growth of special type of bacteria. Toxic metals, like copper, zinc, chromium, etc., may need pre-separation chemically, as otherwise, they may interfere with the biological oxidation by tying upto the enzymes, essentially required for microbial growth.

In conclusion, it can be stated that the selection of a sequence of particular treatment processes depends on the characteristics of the wastewaters, and also upon the permissible requirements of the effluents. *The treatment processes needed for different industries will, therefore, generally vary from one industry to another*.

Use of simple biological methods have succeeded in industries, like *fruit processing, dairies, slaughter houses, and textiles*. Physical and chemical treatments prior to biological treatment, for separation of troubling pollutants, like chromium, arsenic, cyanide, mercury, several nitrogenous substances, etc.,, is very necessary and important for industries *like fertilizers, dyes and pigments, pesticides, electroplating, paper and pulp, etc., Suitable pre-treatments to the wastewaters of such industries, before subjecting them to the biological treatment, is the prime requirement for designing and planning the treatment plant for a particular industry*.

QUESTIONS

1. List out the harmful effects due to disposal of industrial wastewaters without adequate treatment.
2. How will you decide about combining industrial and domestic wastes before treatment and disposal?
3. Write a short note on the treatment of industrial wastewaters.
4. List out various methods of treating industrial wastewaters. Explain the 'equalisation' and 'neutralisation' methods.
5. Explain chemical treatment methods of industrial wastewaters.
6. Explain the physical treatment methods of industrial wastewaters.
7. Explain biological treatment methods for industrial wastewaters.

Solid Waste and Its Management

18.1. INTRODUCTION

Solid Waste is a material, which is thrown away or set aside as worthless and is viewed as a discarded material which has no consumer value to the person abandoning it. Though the waste may be in the form of solid, liquid or even gas, here only **Solid Waste** is discussed.

Solid waste is defined as," The unwanted or useless solid materials generated from human activities in Residential, Industrial or Commercial areas."

18.2. WASTE CATEGORIZATION

Waste Categorization is the process by which the composition of different waste sources are analyzed. Waste categorization plays an important part in any treatment of waste which is proposed.

Solid waste can be classified in the following categories:

1. Municipal Solid Waste (MSW).

Municipal solid waste consists of household waste, construction and demolition debris, sanitation residue and commercial units waste. The term **garbage** is generally used to putrescible or bio-degradable food waste. **Rubbish** is used to denote non-putrescible or non-biodegradable materials – Both these (Garbage and Rubbish) together have kitchen food wastes, paper, cloth, plastic, glass, metals, masonry, packing material etc., It may also include ash, cinder, dead animals, abandoned vehicles, pipes, sludge etc., Latest addition to it being e-waste.

As per Ministry of Environmental and Forest (MoEF)," Municipal solid waste (MSW) includes commercial and residential waste generated in municipal or notified areas in either solid or semi-solid form excluding industrial hazardous wastes but including treated bio-medical wastes."

2. Industrial Solid Waste (ISW).

In majority of cases it is termed as' hazardous waste' as they may contain toxic substances. Hazardous industrial waste may be ignitable, corrosive, reactive, or radioactive. All industrial waste is not hazardous.

3. **Bio-Medical Waste or Hospital Waste.**

It is any kind of waste containing infectious (or potentially infectious) materials. This may include **Sharps,** soiled waste, disposables, anatomical waste, cultures, discarded medicines, chemical wastes etc.,, usually in the form of disposable syringes, swaps, bandages, human excreta etc., Obviously, these can be serious threat to human health, if not properly treated before disposal.

4. **Agricultural Waste:**

The waste generated during various processes of crop production and livestock rearing is called agricultural waste. This includes **Plant Remains** like husk, straw, cotton and tobacco waste, coconut waste, nutshells; **Animal waste** includes animal excreta (Gobar) and **Processing waste** as food crops are processed to extract grain. In this category huge amount of waste is generated particularly in villages.

5. **Nuclear Waste.**

Radioactive waste from the nuclear power and weapon industries has recently become a matter of great concern. The most important is the **spent nuclear fuel,** which remains active even for hundreds of years. Parts of cell phones, computers etc., are also source of nuclear waste.

Sometimes radioactive leak and accidents in nuclear plants or willful explosion of nuclear devices cause greatest of hazards.

Composition of Municipal Solid Waste

For proper solid waste management, it is absolutely necessary to know the composition of solid waste in any area. It must be realised that there is considerable variation in composition of solid waste in different areas. The composition also changes with time. For example: in rural communities there is more food and other organic waste, in cities paper, plastics and inorganic waste is considerable. Earlier in waste there was more ash which is now at reduced low level.

18.3. COMPOSITION OF MUNICIPAL SOLID WASTE IN INDIA

Type of waste	% of Total
Biodegradable	51
Plastic	10
Paper	7
Other (Textile, glass, metal, silt, street, sweeping, inert)	32

Source: Task Force on Waste to Energy Planning Commission 2014

The character and composition of refuse (waste) depends upon various factors like geographical location of the area, its weather, food habits of the people and socio-economic condition of the community.

According to one estimate the typical composition of solid waste in a city like Patna is:

(a) *Compostable solid waste:*

(i)	Vegetable/leaves	40%
(ii)	Paper (waste)	1%
(iii)	Miscellaneous	11.5%
	Total	**52.5%**

(b) *Recyclable waste:*

(iv)	Glass/Ceramics	2%
(v)	Plastic	4%
(vi)	News paper/ Packing Material	4%
(vii)	Metals	1%
	Total	**11%**

(c) *Landfill waste*

(viii)	Ash/Soil/ street sweeping	36.5%
	Grand Total	**100.0%**

Source : M.Sc. Dissertation, Patna University 2005.

Here it must be realised that the estimate varies from surveyor to surveyor, size and randomness of the sample.

18.4. MUNICIPAL SOLID WASTE GENERATION IN INDIA

India is one of the mega underdeveloped countries of the world with respect to population status and dynamics. The growing rate of population is a direct factor responsible for enhancement of municipal solid waste generation.

According to the Press Information Bureau, India generates 62 million tonnes of waste (mixed waste containing both recyclable and non-recyclable waste) every year, with an annual growth rate of 4% (PIB 2016). 62 million tonnes annually averages out to about 450 grams of waste per person per day. However, there is a lot of variability in per capita waste (MSW) generation in India. Daily household municipal solid waste generation ranges from 170 grams per person per day in small towns to 620 grams per person per day in large cities.

Following is the list of top MSW generating cities in India 2016*

S.N.	City	Population in Millions	MSW Tonnes per day
1.	Delhi	19.1	9620
2.	Mumbai	20.0	8600
3.	Kolkata	14.7	6000
4.	Chennai	10.1	5000
5.	Bangluru	10.4	4200

6.	Hyderabad	9.1	4000
7.	Ahmedabad	7.5	2500
8.	Pune	5.8	2300
9.	Surat	5.8	1680
10.	Kanpur	3.0	1500
11.	Lucknow	3.3	1200
12.	Nagpur	2.7	1000
13.	Jaipur	3.5	1000
14.	Ludhiana	1.7	850
15.	Indore	2.5	850
16.	Coimbatore	2.6	850
17.	Agra	2.0	790

* Urban Agglomerations

Sources : State Pollution Control Boards, Municipal Corporations, U.N. Population estimates. ICRIER, New Delhi, 4th January 2019.

18.4.1. Hazardous Waste Generation in India

One study estimates that nearly 7.5 million tonnes of hazardous waste is generated in India annually. Out of this recyclable hazardous waste is about 3.35 million tonnes. Though waste as a whole is a problematic chapter for India, hazardous waste in particular is an area of concern.

Another area of concern is the **plastic waste**. The Central Pollution Control Board (CPCB) in 2015 estimated that average plastic waste generation in the country is around 6.92% of municipal solid waste. An estimate puts India's plastic waste generation at 26,000 tonnes every day.

Nuclear waste is the latest of hazardous waste group. The amount of such waste generated in India is around 4 tonnes per Giga watt (1000 MW) for one year electricity generation.

Bio-medical waste is also a separate category of hazardous waste. As per a report by Associated Chambers of Commerce and Industry (2018), the total quantity of **medical waste** generated in India is 550 Tonnes per day (TPD) and the figure is rapidly rising.

18.5. ENVIRONMENTAL AND OTHER IMPACTS OF MUNICIPAL SOLID WASTE

The impacts associated with municipal solid waste are divided into three categories: environmental, economic and land-use-related.

Environmental Effects:

(i) **Hazardous Gas Emissions:** Landfills emit large amount of methane, a powerful greenhouse gas. In addition to its effect on ozone layer, methane is highly combustible gas that may be responsible for explosion hazards in and around landfills.

(ii) **Water Quality Contamination:** When leachate seep into the groundwater, it can be a source of many contaminants, specifically organic compounds that may decrease the oxidation-reduction potential and increase the mobility of toxic metals.

Economic Effects:

(iii) **Siting Resistance:** No one wants to live near a landfill and as urbanization spreads, it becomes more difficult to find land that is suitable for dumping and amenable to the surrounding population.

(iv) **Disposal Costs:**Unlike recycling, which requires reprocessing used materials, or composting, which requires intensive sorting, landfill dumping needs far less money and effort. This superior efficiency is a major reason that land filling dominates the waste disposal methods, even when they may be more environmentally sound.

Land Use Related Effects:

When siting a landfill following issues have to be addressed.

(v) **Airport safety:** Landfills attract birds, which can threaten aircraft.

(vi) **Floodplains:** If a landfill is located in a flood plain care must be taken to ensure that its contents do not flow with floodwater.

(vii) **Wetland:** wetlands should always be avoided when locating landfills.

(viii) **Unstable Areas:** Landfills should not be allowed in areas threatened by mudslides or other forms of earth movement.

18.6. MUNICIPAL SOLID WASTE MANAGEMENT

Nature could take care of disposal of waste by diluting, decomposition and recycling material within its absorbing capacity. Presently most of the cities have exhausted this capacity.

With the rapid pace of urbanization huge amount of solid waste is generated every day and disposal of this unwanted material is a monumental problem with multidimensions.

Solid waste management is "collecting, treating, and disposing of solid material that is discarded because it has served its purpose or is no longer useful." Improper and unscientific disposal of municipal solid waste (MSW) can create unsanitary conditions, and these conditions in turn can lead to pollution of the **environment** and to outbreaks of vector-borne disease – that is, diseases spread by rodents and insects.

The tasks of solid-waste management present complex technical challenges. They also pose a wide variety of administrative, economic, and social problems-that must be managed and solved.

Principles of Waste Management: Waste management or waste disposal include activities and actions required to manage waste from its inception to its final disposal. This includes the **collection, transport, treatment** and **disposal** of waste, together with monitoring and regulation of the waste management process.

18.6.1. Existing Waste Management Practices

Following chart shows existing Municipal Solid Waste Management (MSWM) system is India.

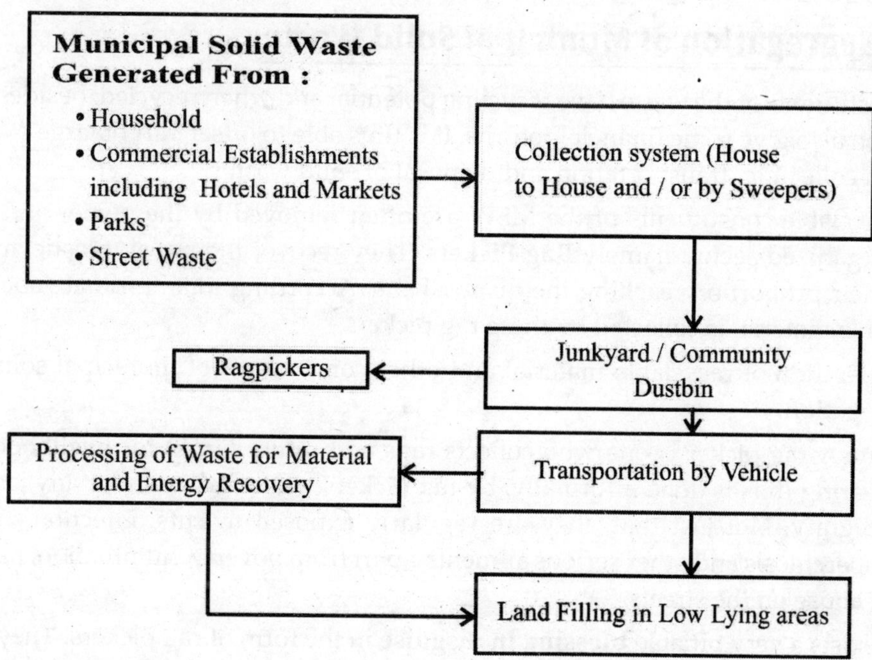

Fig.18.1. The flow chart of existing MSWN system is India

Source: Singh et. al. 2011.

For proper solid waste management, it is absolutely necessary to know the amount of waste generated and its composition. Municipal solid waste is a heterogeneous mixture of various constituents. Loose control over MSW may result in multiple environmental problems.

One of the concepts that was introduced in solid waste management was that of **'hierarchy of waste management'** also known as the 3Rs (Reduce, Reuse and Recycle). The 3R concept is a significant and innovative move towards cost reduction and profit maximization, whilst satisfying the people with safe and quality service.

Waste collection: It is a major part of the process of waste management. Primary collection of waste is the second essential step of solid waste management activity. Primary collection system is necessary to ensure that waste stored at source is collected regularly and it is not disposed of on the streets, drains, water bodies etc.,

In India, the system of primary collection of waste is practically non-existent, as the system of proper storage of waste at source is yet to be developed. Streets are generally treated as receptacles of waste and the primary collection is done, by and large, through street sweeping.

An appropriate system of primary collection of waste to be so designed that it synchronizes with storage of waste at source as well as waste **storage depots** facility ensuring that the waste once collected reaches the processing or disposal site.

Methods of Primary Collection of Waste:

Door step collection of waste can be done through **Containerized handcarts** affixed with a bell or a whistle provided to the person collecting the waste. Instead of a handcart a tricycle or a motorized vehicle may be more useful as these can cover larger areas and can save time also.

18.6.2. Segregation of Municipal Solid Waste

Certain constituents of the waste have recycling potential and when recycled, besides reducing the pollution also give some financial returns. It is thus able to offset part of large expenditure incurred on collection, transportation and disposal of solid waste.

The recyclable constituents of the MSW are often removed by the owner and part of it by the unorganized sector; namely **Rag Pickers**. They recover the paper, plastic, metals etc., reducing their proportion reaching the disposal site. According to an estimate about 50% of the recyclable material is collected by these rag pickers.

The separation of recyclable material and other components of municipal solid waste is called segregation.

Rag Pickers: A rag picker is one who collects **rags** and refuse (junk) for livelihood. A lot of garbage clearing thus is done informally, by rag pickers who work without any job security, salary or dignity. Not just that, they are regularly exposed to cuts, infection, respiratory diseases, tuberculosis and other serious ailments-apart from poverty, humiliation, harassment, and sexual abuse on the streets.

There exists a very pitiable **blessing in disguise** in the form of rag pickers. They are a blot on any civil society. A study in Patna University revealed that among the rag pickers there are more than half female rag pickers, most of whom are between the age of 5 and 15. They are exposed to various kinds of hazards frequently sexually assaulted. Their woes are multiplied when hospital waste is mixed with municipal waste.

18.6.3. Ideal Method of Segregation of Municipal Solid Waste

Segregation at source is perhaps the ideal situation and it should be encouraged. Household waste should be separated daily into different dustbins for the different categories of waste such as **Wet** and **Dry Waste** which should be disposed of separately.

Wet waste, which consist of leftover food stuff, vegetables, peels etc., The dry waste consist of cans, aluminium foils, plastics, metals, glass and paper which could be recycled. Waste can also be segregated as.

(1) **Bio-Degradable Waste:** It includes organic waste, *e.g.* kitchen waste, vegetables, fruits, flowers, leaves and paper.

(2) **Non-Bio-Degradable Waste:** This can be further categorised as.

(*a*) **Recyclable Waste:** Plastic, Paper, Glass, Metals etc., (*b*) **Toxic Waste:** Paints, chemicals, bulbs, cells, fertilizer and pesticide containers, spray cans, expired medicines etc.

(*c*) **Soiled Waste:** Hospital waste such as cloth/cotton soiled with blood, syringes, diapers etc.

18.6.4. Colour Coding of Waste Bins

Bins of different colours are placed at strategic points. Normally following colour coding is practiced.

Organic waste	Green
Glass	Yellow
Paper	White

Metals	Grey
Plastic	Blue
Hazardous	Red

18.7. SOLID WASTE REDUCTION, REUSE AND RECYCLING

In every aspect of human life unwanted materials are generated and then discarded simply because they are considered to be **wastes**. However, it is rightly said, **"Waste is not waste until we waste it."**

The Waste Hierarchy and the '3R$_s$'

Basically, the 3R concept is a sequence of steps on how to manage waste properly. The waste hierarchy sets out a hierarchy of options for managing waste in terms of what is best for the environment. It indicates an order of preference for action to reduce and further manage waste, and is usually presented diagramatically in the form of a pyramid. The aim of the waste hierarchy is "to extract the maximum practical benefits from products and to generate the minimum amount of waste.

Waste Reduction

At the top of the hierarchy is the waste reduction. There are many ways of reducing the amount of waste produced at home. These include – buy products that have less packing, make use of reusable rather than disposable items-like refillable containers, using own shopping bags, minimize food scraps or feed these to animals, repair rather than discarding household gadgets.

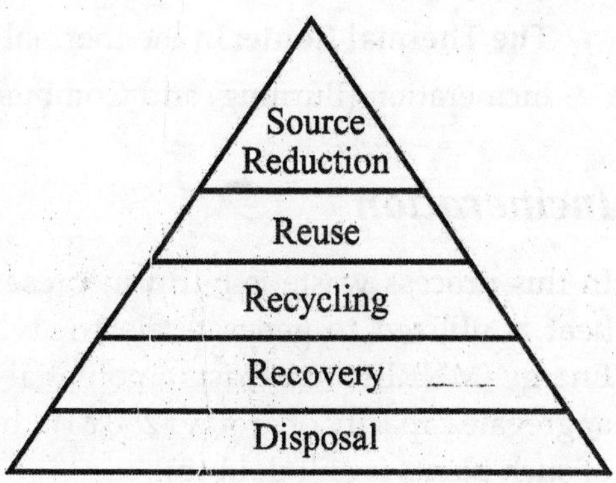

Industrial waste can be reduced to a large extent by using appropriate and modern technology.

Waste Reuse

This is the second option in the waste hierarchy. Reuse is defined as, "using a waste product without further transformation or without changing its shape or original nature." Different types of solid wastes can be reused, such as bottles, old clothes, books etc., Reuse results in lesser amount of waste.

There are several advantages of reducing and reusing waste. Reuse can be very helpful to poor and disadvantaged people who cannot afford to buy new goods. By reuse of material need for production of new goods is reduced which results in saving of energy, water and other resources. Environmental pollution also declines.

18.7.1. Waste Recycling

Recycling of waste means that the material is reprocessed before being used as new product. Recycling option depends upon the type of waste. For example paper can be used to make new paper or to be used for printing or packaging. Plastic and metals can be used to produce similar products.

The organic components of MSW have inherent calorific value (1620-2340 kcal/kg); which can be used to produce electricity and can also be converted to a manure.

The processing of the waste to reduce the pollution potential and also to recover some **value** can be achieved either through **Biological Route** or the **Thermal Route**.

The Biological Route: In the biological route mainly there are two processes:

Aerobic stabilization (Composting) and Anaerobic Processes (bio-methanation)

which are being used

Aerobic Stabilization: In aerobic stabilization decomposition of organic matter occurs in presence of oxygen used by microbes. The orgasnic fraction of waste thus yields a final product which can be used as an organic manure (called compost).

Anaerobic Process: It is known as **biomethanation** where the organic matter after proper segregation and size reduction is mixed with water and allowed to degrade under controlled anaerobic (in absence of oxygen) condition. The generated **bio-gas** having good value of 4500-5000 kcal/kg is used as a source of energy and digested residue as **compost**.

The Thermal Route: In the thermal route also two processes are commonly adopted.

Incineration (Burning) and Combustion (Pyrolysis).

Incineration

In this process waste is burnt in presence of excess amount of oxygen/air and the released heat is utilized to generate electricity. According to the Ministry of New and Renewable Energy (MNRE) there exists a potential of 1700 MW from urban waste. Around 92 plants with aggregate capacity of around 250 MW have been set up in India. However, in India the success of such plants is still doubtful.

Pyrolysis: The second process in the thermal route involves **combustion** of the material in absence of air or in oxygen deficient atmosphere. This process of pyrolysis results in three different products: **gas, liquid** and **char** each of which have certain calorific value. A fair portion of energy can be recovered in the process.

18.7.2. Composting

It is the most popular and proper method of disposal of organic portion of MSW as it sets up ideal condition for decomposition and return of organic materials (containing nutrients) to the soil as conditioner and fertilizer.

Composting is the decomposition of organic matter in the refuse by micro-organisms primarily bacteria, fungi, insects and worms. These micro-organisms that decompose organic matter require oxygen and water, and produce compost, carbon dioxide, heat and water.

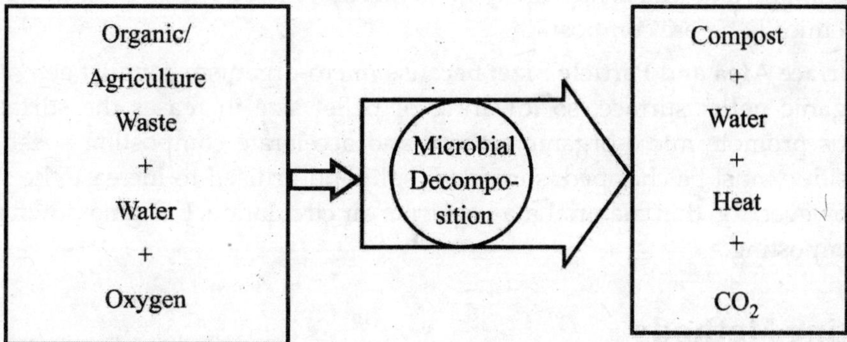

Fig. 18.2. The Basic Composting Process.

Tiny invertebrates like mites, millipedes, sow-bugs, earth worms and snails are the primary agents of physical decay.

Factors Affecting composting

Composting is a complicated bio-chemical reaction process is influenced by many factors as described below,

(1) **Carbon/Nitrogen Ratio (C/N Ratio):** If C/N Ratio is not correct this may seriously limit the composting process. The recommended ratio is 30 parts carbon to 1 part nitrogen by weight. (Normal range is 25:1 to 40:1). Softwood shavings, sawdust and straws are good source of carbon while manure are good for nitrogen supply.

(2) **Moisture Content:** Nutrients can be absorbed by micro-organism only when material is blended with water. The best water content for composting raw material is usually around 50 to 60% by weights. Lower (below 30%) moisture would adversely affect the composting process. Higher moisture content (above 60%) could result in leaching of nutrients, cut in air circulation and stench generation.

(3) **Oxygen and Aeration:** For successful composting sufficient oxygen is required. The ideal oxygen concentration is 18%. Oxygen requirement is greatest in the initial period. Increased porosity facilitates aeration. Wind may also help in increasing aeration.

(4) **Temperature:** At relatively higher temperature, the rate of decomposition is faster. Pile temperature between 32°C and 60°C indicate rapid composting. In the initial stages of composting, its temperature is almost the same as of surroundings. After one or two days, the temperature rises to 50°C to 65°C by the action of **mesophilic** bacteria, and remains so for few weeks. After this it again drops to ambient air temperature. At this temperature compost is said to be **mature**.

(5) **pH of material:** Generally the most appropriate pH value for micro-organism is neutral or weak alkali. The optimum pH value for efficient composting lies between 6.5 and 7.5. If pH falls below 6 (acidic), the bacteria becomes sluggish.

(6) **Nutrients Present:** Adequate level of phosphorus and potassium are main nutrients required in composting process which are normally available in the organic waste.

(7) **Presence of Toxic Substances:** Heavy metals like mangnese, copper, zinc, nickle, chromium and lead which are toxic to **thermophilic** bacteria should be immobilized chemically before composting.

(8) **Surface Area and Particle Size:** Because micro-organism conduct activities through organic **pellet** surface, so lowering of pellet size increases the surface area and thus promote micro-organic activity and accelerate composting speed. Therefore, residue must be chopped, shredded, split and bruised to increase the surface area. However, too thin material may obstruct air circulation, bringing down the speed of composting.

Composting Methods

The composting process is started by blending together different types of wastes and creating favourable condition for microbial activity. Following four methods are in practice.

(1) **Aerated Static Pile (ASP) Method:** The blended admixture is usually placed over perforated piping to facilitate air circulation in the form of a pile. It requires minimal turning and aeration. Moisture content has to be maintained.

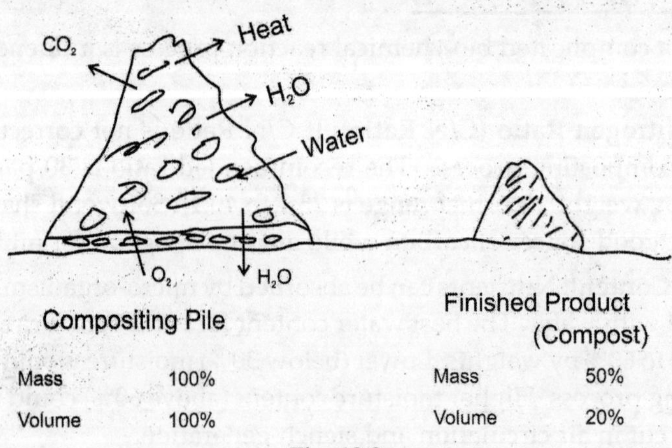

Fig. 18.3.

(2) **Open Windrow Method:** Organic matter is piled in long rows (windrows). Normally this method is suitable for large volumes of composting. These rows are generally

turned to improve porosity, air circulation, remove extra water and redistribute cooler and hotter portions of the pile.

(3) **Enclosed Channel Method:** This refers to composting systems in which composting takes place in channels made of Cement Concrete walls having holes. Floor is bedded with coconut shells to improve aeration. Mixing of organic waste is provided with automated turner. Additional aeration is given by mechanical system.

(4) **In-vessel Composting Method:** In this method the composting material is confined within a building, container or vessel. It can consist of metal or plastic tanks or concrete bunkers. Turning is done by mechanical devices.

(5) **Trench Composting:** If previous methods are too unsightly alternative method of trench composting is used. A **compost pit** is created. The depth of the pit can be about 30 cm, while the bottom area will depend upon the quantity of the organic matter available. The pit is covered to avoid accidents.

18.7.3. Vermicomposting

Breaking down of the organic material through the use of worms, bacteria and fungi is known as **vermicomposting**. The end product of vermicomposting is a substance called vermicompost or worm castings. Also known as **worm feces** this compost is very rich in nutrients.

The earthworm species (composting worms) most often used are **red wigglers** (Eisenia Fetida or Eisenea Andrei). A byproduct is a shiny liquid called **Vermiwash** which is very rich in nutrients.

Vermicomposting Process

A number of steps are involved for preparation of good quality vermicompost. Following steps are followed.

(1) A pit is dug of the size depending upon the available compostable material.

(2) A black polythene sheet is spread on the pit surface uniformly.

(3) About 7.5 cm layer of gravel mixed with sand is spread over polythene sheet.

(4) Above this, a layer of coconut coir is spread uniformly to prepare **vermibed**.

(5) Now the waste materials are spread over the coir to a depth of 15 to 20 cm mixed with cowdung slurry with microbes @ 300 g/ton.

(6) Above the waste materials a well rotted cowdung is spread from which about 200 earth worms species **Eisenia Foetida** are released.

(7) Each pit is covered with gunny bags to avoid evaporation loss.

(8) Water is sprinkled periodically to maintain moisture content of about 60% for survival of organism.

Advantages of Vermi-composting

As compared to conventional composting, it provides a way to compost organic matter more quickly. Vermi-compost is nutrient-rich organic fertilizer and soil conditioner that is relatively easy for plants to absorb.

Besides solid vermi-compost, vermi-composting also has a bye product in the form liquid manure or **vermiwash** which is very rich in nutrients. A 1.7 times growth in plant mass has been claimed due to the use of vermi-compost. It is also known as **Black gold.**

Disadvantages of Vermi-composting

- Vermi-composting requires careful management.
- It encourages pathogens more than normal composting.
- Vermi-composting requires low temperature to be maintained so the worms do not lose moisture and dry out.
- It requires more space.

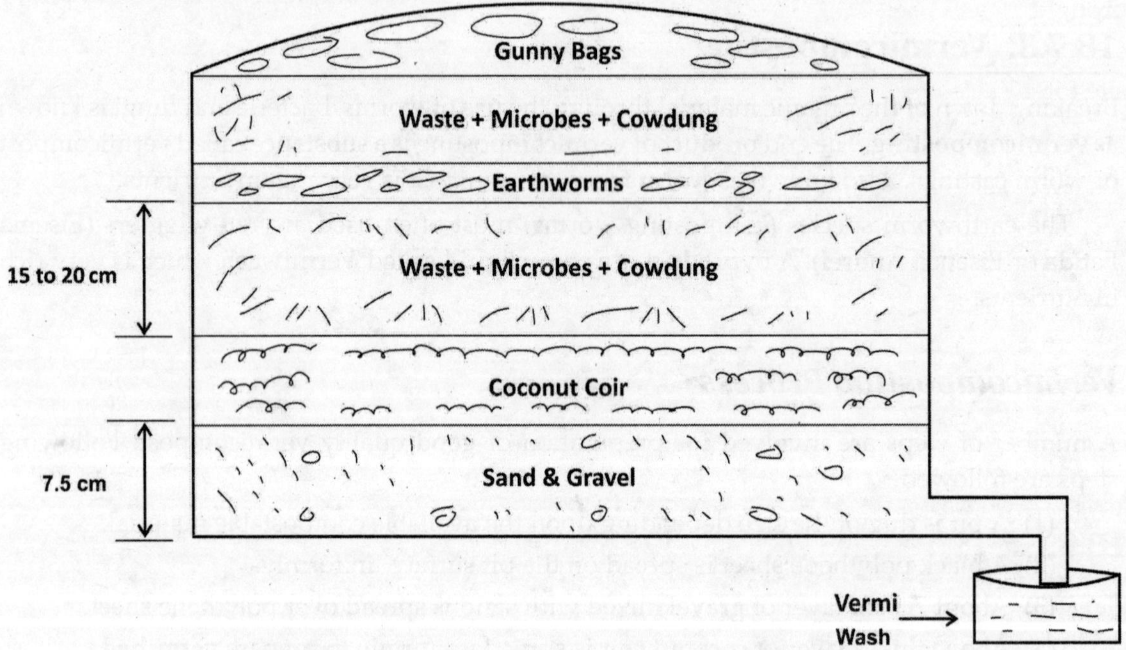

Fig. 18.4. Typical pit of vermi-composting size: 3m×4m×1m.

18.7.4. Land Filling by Solid Waste

A landfill is an engineered method for land disposal of solid waste. The Municipal Solid Waste Landfill (MSWLF) is a **discrete** area where after reduction and recycling, residue is dumped.

Landfilling is one of the major municipal solid waste disposal methods practiced worldwide. Though it is considered most cost-effective means of waste disposal, but poor management practices specially in developing countries likes India are the major causes of environmental pollution. Toxic gas emissions from landfills pose a serious threat to the environment as well as human health.

Land fills also generate a toxic soup known as **leachate** which is highly toxic and causes land and groundwater pollution.

Methods of land filling: There are three general methods of landfills, which are : area method, trench method and ramp or slope method.

The **area method** is best suited for flat or gently sloping areas where some land depressions are available. The waste material is spread, compacted and then covered with a layer of compacted soil every day.

The trench method consists of an excavated trench (a long narrow ditch) into which the solid waste are spread, compacted and covered. This method is suitable for nearly level ground where groundwater table is well below the ground surface. The soil excavated is used to cover the waste layers.

The ramp or slope method is the combination method where the deposition, covering, and compacting are done on a slope. The soil for covering is recovered from the site itself or brought from adjacent areas.

Criteria for selection of Landfill site: Following criteria should be followed while selecting site for landfill of solid waste.

(1) Land area and volume should be sufficient for projected need for several years.

(2) There must be a suitable buffer zone between the land fill site and population.

(3) Areas with steep gradient should not be selected.

(4) The groundwater table should be reasonably below the ground surface at the site.

(5) Water supply units should be well away from the boundaries of landfill.

(6) No major powers transmission lines, sewers, water supply lines etc., should be crossing the site.

(7) Ecologically sensitive areas should be avoided.

(8) An EIA should be done before finalizing the land fill site.

18.8. HAZARDOUS AND BIO-MEDICAL WASTE MANAGEMENT

Hazardous waste is the waste that posses substantial or potential threat to public health or the environment. It can be in solid, liquid gaseous or sludge form. They are generated primarily by chemical production, manufacturing and other industrial activities. Hazardous waste can also come out of hospitals, clinics and nursing homes.

They may cause damage during inadequate storage, transportation, treatment or disposal operations. Improper hazardous waste storage or disposal frequently contaminates surface and groundwater supplies.

People living in houses built near old abandoned waste disposal sites may be in particular vulnerable position.

Hazardous Waste Characteristics: Hazardous wastes are classified on the basis of their biological, chemical and physical properties. These properties generate materials, that are toxic, reactive, ignitable, corrosive, infectious, or radioactive.

Toxic Wastes are poisonous, even in very small or trace amounts. They may also cause chronic effects, slowly causing irreparable harm. Some of these are Carcinogenic, causing cancer after long exposure. Other are Mutagenic, causing major biological changes in the offsprings of exposed humans and wildlife. They may have acute effects causing violent silkness or even death.

Reactive Wastes are chemically unstable and react violently with air or water. They cause explosions or form toxic vapours.

Ignitable Wastes: These burn at relatively low temperature and may cause a fire hazard.

Corrosive Wastes: These include strong acidic or alkaline substances. They destroy solid material and living tissues upon contact by chemical reaction.

Infectious Wastes: These include used bandages, hypodermic needles and other materials thrown out of hospital, clinics and nursing homes.

Radioactive Waste: These emit ionizing energy which may harm living organisms. As some radioactive materials can persist in the environment for thousands of years before fully decaying, there is much concern over the control of these wastes. However, the handling and disposal of radioactive waste is not a responsibility of local municipal government due to complexity of the problem.

18.8.1. Transport of Hazardous Waste

Hazardous waste generated at a particular site often requires transport to an approved Treatment, Storage or Disposal Facility (TSDF). Because of possible threats to public safety and the environment transport has to be given special attention. Proper labelling, transport, and tracking of all hazardous waste should be strictly followed.

Transport Vehicles: Generally the hazardous waste is transported by trucks on highways. Small amounts may also be carried by inland waterways or railways. Highway shipment is most common because road vehicles can easily reach the collection sites as well as TSDFs.

These trucks are specially designed made of steel or aluminum alloy having standard dimensions. The capacity of these trucks can be upto 34,000 litres. The waste is put in containers of 200 litres capacity.

Treatment, Storage and Disposal of Hazardous Waste

While reduction and recycling are desirable options, they are not regarded as the final solution. There will always be a need for treatment and for safe storage or disposal of some amount of hazardous waste.

Hazardous waste can be treated by chemical, thermal, biological and physical methods.

Chemical Methods include ion-exchange, precipitation, oxidation, reduction and neutralization, which not only detoxify organic waste but also can destroy them. Special equipments are designed for burning waste. These include fluidized-bed incinerator, multiple-hearth furnace, rotary kiln and liquid-injection incinerator. One major problem associated with burning of waste is increased air pollution.

Biological treatment is also an option. One of the methods is land farming. In this technique the waste is carefully mixed with surface soil and microbes are added to metabolize the waste. Microbes can also be used to stabilize already contaminated sites. Process is called Biomediation. All the above three methods change the molecular form of the waste material.

Physical treatment on the other hand, concentrates, solidifies or reduces the volume of the waste. The process includes evaporation, sedimentation, flotation and filtration. Solidification is another physical process which includes Encapsulating the waste. The solid mass thus got is resistant to leaching. Waste can also be mixed with lime, flyash and water to form hard product.

18.8.2. Surface Storage and Land Disposal

Hazardous waste that are not destroyed by any of the above methods need to be disposed of properly. For most of these residual waste, land disposal is the ultimate destination. The two basic methods of land disposal are: (1) Land Filling and (2) Underground injection.

Prior to land disposal surface storage or containment systems are often employed as temporary solution. These include – open waste piles made over an impervious base, and should be protected from wind dispersion.

Secure Landfills: Land filling of hazardous waste is regulated more stringently than land filling of municipal solid waste. Hazardous waste must be deposited in the secure landfills, which provide atleast 3 m separation from the groundwater table for this two impermeable liners and leachate collection system have to be provided.

18.9. LEGISLATION TO MANAGE SOLID WASTES

Management of municipal solid waste is essentially the duty and responsibility of municipal committees and corporations in India, and the constitution and powers of municipal corporations are a subject included in the 'State List' in the Seventh Schedule of the Indian Constitution. Each State has its own Municipal Act, by and large, these Acts have been framed on similar lines. The municipal corporations have a wide range of subjects to look after : health and sanitation, primary education, construction and maintenance of roads and bridges, health-care, registration of births and deaths, inspection and licensing of food, and so on.

Most of the Municipal Acts have all the necessary legal provisions to (*i*) take action against the civic officials, if they fail to provide a clean and safe environment to the general public; and (*ii*) to penalize those who spoil the environment; yet in the absence of proper enforcement of all these laws, municipal and industrial waste have been disposed off with little regard to the environment, resulting in serious environmental problems.

In order to exercise some control on the private sector and the municipal bodies, as to force them to ensure safer disposal of their hazardous wastes and municipal solid wastes, the GoI has enacted the following Rules under the Environment (Protection) Act, 1986 :

(*i*) Hazardous Wastes (*Management and Handling*) Rules, 1989

(*ii*) Municipal Solid Wastes (*Management and Handling*) Rules, 2000.

A brief description of both these laws us given below :

1. Hazardous Wastes (Management and Handling Rules), 1989

Under the Environmental Protection Act, 1986, these rules were notified by GoI on **28.07.1989.** These rules have subsequently been amended twice, vide amendment notifications dated 6.1.2000 and 23.5.2003. *Schedule 1 to these rules* list the various hazardous wastes. Different kinds of waste substances have been categorised class A to class E, depending upon their reducing hazard, in *Schedule 2 to these rules. Schedule 3 to the rules* further lists various types of hazardous wastes, the imports and exports of which have some sorts of restrictions.

Under these Rules, the hazardous wastes shall be collected, treated, stored, and disposed off only in such facilities as may be authorised by SPCB or SPCC. An occupier or recycler handling hazardous wastes, not having hazardous waste treatment and disposal facility of his own, can be allowed to become a member of a *common treatment, storage and disposal facility* existing in the area. Such an occupier will send his wastes to this common facility for proper treatment and disposal, as otherwise, his authorisation is liable to be cancelled. These rules further provide guidelines for packing, labelling and transport of hazardous wastes. The occupier or operator of any waste facility is further supposed to identify, design and set up proper treatment and disposal site/facility. Guidelines are also given for imports and exports of hazardous wastes. Provision of imposition of fines for violations of these rules is also made in these rules. *However, no provision is made for any kind of imprisonments under these rules for violations of the rules.*

2. The Municipal Solid Wastes (Management and Handling) Rules, 2000

Under *the Environment (Protection) Act, 1986,* the Govt. of India has also notified these rules on **3.10.2000,** to provide some kind of responsibility on the municipal bodies for proper collection and disposal of municipal solid wastes. *Schedule 1 to these rules* provided target dates to municipal authorities for :

(*i*) Setting up waste processing and disposal facilities (31.12.2003);

(*ii*) Improving existing landfill sites as per provisions of these rules (31.12.2001);

(*iii*) Identifying landfill sites for future use and making these sites ready for operation (31.12.2002); and

(*iv*) For constant monitoring of the performance of waste processing and disposal facilities (once in six months). Although all the target dates have elapsed but no compliance is visible.

Under these Rules, Secretary Urban Development of the concerned State is the nodal agency—who is given the overall responsibility for enforcement of the provisions of these rules in metro cities. The Dy. Commissioners or the District Magistrates of the concerned districts are further responsible for implementing the provisions of the rules in their respective districts. The CPCB, SPCBs and SPCCs shall monitor the compliance of the standards laid down in these rules regarding ground water, ambient air, leachate quality, compost quality, including incineration standards, as specified under these rules. The SPCBs and SPCCs shall also grant authorisation to the municipal authorities for setting up any way waste processing disposal facility, including landfills.

In Schedule II, these rules further lay down guidelines on collection, segregation, storage, transportation, processing and disposal of MSW. Specifications for landfill sites are given in

Schedule III, whereby standards for ground water quality and ambient air quality in the vicinity of landfill sites are also laid down. *Schedule IV* to these rules, provide standards for composition of compost (to ensure its safe use); standards for disposal of treated leachates in public sewers and for land disposal (to avoid contamination of disposal sources); and operating as well as emission standards for incinerators.

The Rules appear to be like a BIS Code, than an anti-pollution law. No remedial measures or punitive actions are specified in these rules for violations of the rules, or for not making necessary arrangements for safe collection and disposal of the society's solid wastes.

QUESTIONS

1. Define 'municipal solid waste' and explain the necessary of its safe disposal.
2. Explain the term 'refuse', and give its composition and classification. Describe various methods employed for the collection and disposal of the refuse.
3. Describe and explain the importance of 'recycling and reuse' is disposal of solid waste (MSW) of a society. Which items are generally recycled from the MSW?
4. Write a brief note on the status of collection and transportation of municipal solid waste (MSW) in India.
5. Enumerate various methods used for disposal of MSW. Explain two widely adopted methods in India.
6. What do you understand by 'sanitary landfilling' and how it is practiced? Also explain the digestion process involved in such landfills.
7. Explain the term 'composting'. Describe the techniques adopted for compostings in India.
8. Differentiate between refuse and garbage. Mention the most significant property of the city refuse which guides the adoption of each of the following methods of refuse disposal :

   (*a*) Incineration (*b*) Sanitary landfill

   (*c*) Composting

[Ans. (*i*) High-calorific value, (*ii*) High density, (*iii*) High organic garbage content].

9. Discuss the method being adopted in India for collection and disposal of refuse.
10. How had the solid wastes of a society been disposed of in the sea? Why and how the practice has been phased out?
11. Differentiate between the hazardous and non-hazardous industrial solid wastes.
12. How are the industrial solid wastes disposed off in India, and what are their drawbacks?
13. Write a detailed essay on the existing laws and legislations of India in regard to disposal of solid wastes. What are the limitations and shortcomings of these laws?
14. Explain the salient provisions of the Municipal Solid Wastes (Management and Handling) Rules 2000, indicating their limitations.

15. Explain the impacts of improper disposal of solid wastes on human and the environment in general.
16. Explain the importance of 'recycling and reuse' in disposal of solid waste of a society.
17. Discuss different types and sources of solid wastes.
18. Discuss various effects on health due to improper disposal of solid wastes.
19. Write an essay on "solid wastes from construction activities."
20. Discuss various effects of solid waste on air, water and soil.
21. Write an essay on "air pollution due to solid wastes."
22. Discuss various aspects of water pollution due to solid wastes.
23. What is the 'Hazardous solid waste'? Explain in detail.
24. Discuss different 'waste management techniques'.
25. Write short note on :
    (*a*) Soil pollution due to solid wastes.
    (*b*) Waste to Energy.
26. What is vermi-composting? How it is different from composting?

PART – 3

Air Pollution Engineering and Noise Pollution

Air Pollution, Causes and Effects

19.1. INTRODUCTION

Before discussing the 'air pollution' we must understand about 'Environment' and 'Atmosphere". Therefore, in first few paragraphs, we shall concentrate on these two terms, as air is one of the most important constituents of man's environment.

The word **'environment'** is taken from the Greek word 'Environ' which means surroundings or outside. In simple words, environment refers to surroundings in which man lives and works. Man has very close and intimate relationship with environment, *i.e.*, surroundings.

F.J. Monkhouse defined environment as, "the wholesum of the surrounding external conditions within which an organism, a community or an object exists". Thus in simple words, we can say that 'environment' is a surrounding external conditions influencing development or growth of the people, animals or plants, living or working conditions etc.,

Environment supplies required food, nutrient and energy, and other living conditions that may influence the organisms directly or indirectly. Environment is a complex of several things (light, water, soil, temperature, air etc.,) which surrounds an organism.

Generally speaking the environment is equated with nature wherein physical components of the planet earth *viz*. land, air, water etc., support and affect life in biosphere. In the beginning, the environment was considered to be consisted of only physical aspects .of planet earth and biotic communities, but with the passage of time and advancement of society man extended his environment through his social, economic and political functions.

Our immediate concern is the quality of space we live in, the air we breathe, the food-we eat, the water we drink, and the resources we draw from the environment to support our economy. Environment is an inseparable whole and is constituted by the interacting systems of physical, biological and cultural elements which are interlinked individually as well as collectively in many ways. Here,

(*a*) Physical elements (space, water bodies, climate, soils, rocks and minerals) determine the variable character of the human habitat, its opportunities as well as limitations.

(*b*) Biological elements (plants, animals, micro-organism and man) constitute the biosphere.

(*c*) Cultural elements (economic, social and political) are essential man-made features.

On the basis of basic structure the environment may be divided into two basic types *e.g.*, physical or *abiotic* (non-living) environment, and (living) environment.

Atmosphere contains life-giving gases, like oxygen for man and animal, and carbon dioxide for plants. It also acts like a greenhouse and therefore, keeps The Earth warmer than it would otherwise be. Thus, the atmosphere acts like a blanket for the earth and regulates the heat balance. It also protects us from the harmful ultraviolet radiation of the sun.

The atmosphere serves as a storehouse for water vapour, which leads to precipitation. The composition of the atmosphere constitutes is taken as the composition of pure dry air, which constitutes mainly of nitrogen (78%) and oxygen (21%) that together make up 99% of the air by volume.

Air, water and surroundings are polluted by various activities of man, like industries, energy conversion plants, infrastructural projects etc., We need clean air, water, atmosphere and surroundings for good health and better quality of living. Various process without pollution control provisions affect the environment adversely. These have certain degrading effect on the environment.

The large-scale emission of CO, CO_2, NO_2, SO_2, particulate matter, heat etc., into the atmosphere results in local pollution and global environmental degradation effects, such as:

- Adverse effects on human health as discussed in the earlier chapters.
- Acid rains harming the agriculture, forests, animals, human health, aquatic life, drinking water in the lakes etc.,
- Global warming, melting of polar snow caps, increased level of ocean, environmental disturbance etc.,

Normally, the nature has self cleaning processes like photosynthesis, water cycle, seasons, winds etc., resulting in clean environment and ecological balance. The extraction, processing, conversion and use of energy are accompanied by degradation of the environment.

Due to high tolerancing power of nature, the disturbances upto a certain level can be absorbed. But with the rise in agricultural production, urbanization, modern technology, high standards of living, human interference with environment clubbed with rapid rise in human population, a stage has reached where nature is unable to sustain this ever growing pressure through its capacity for self stabilisation. This indiscriminate neglect and disregard of environment may result in:

1. Large scale land degradation.
2. Deforestation
3. Extensive pollution of rivers and other water bodies.
4. Depletion of ground water level and deterioration of ground water quality.
5. Large scale atmospheric pollution in cities and industrial locations.
6. Extinction of great number of plants and animal species.
7. Rapid depletion of natural resources.
8. Climate change.

9. Acid rain.
10. Global warming.
11. Rise in sea water level.
12. Ozone depletion.
13. Smog.

Various *reasons of concern for degradation* of environment are given hereunder:

(*i*) The economic and population growth has put great pressure on consumption of water, energy, biological resources.

(*ii*) Ignoring the long term consequences of development activities, and not adopting the appropriate technologies for the purpose of sustainable development.

(*iii*) Failure to recognise that our economic system, our health and our life styles depend on the continued adequate functioning of ecological systems.

It should be understood clearly that the availability of natural goods and services is controlled by the hydrologic cycle, biogeochemical cycles, the climate system, and the maintenance of biological diversity and functioning of ecosystems.

19.2. ENVIRONMENTAL POLLUTION (DEGRADATION)

Environmental pollution can be defined in many ways, such as;

(*i*) Environmental pollution is a reduction in quality of environment by the introduction of impurities.

(*ii*) Environmental Pollution means the act of destroying the purity or sanctity or mating foul or filthy.

(*iii*) Environmental Pollution may also be defined as the introduction by man into the environment of substance or energy liable to cause hazards to human health, harm to living resources and ecological system, damage to structures or amenities or interference with legitimate uses of the environment.

(*iv*) Environmental pollution may be described as the unfavourable alterations of our surroundings because of the acts and omissions of the man. In our large and industrial towns, we are already, experiencing the effects of air, water, soil and noise pollution which has already crossed human tolerance.

Therefore, in environmental pollution, we should study the causes and acts of destroying the purity of air, water and land and their effect on ecology *i.e.*, plants, animals, and human beings, and measures to control their effects. Air pollution is mounted in cities due to higher growth of populations and industrialisation. Water pollution has increased with the growth, of population and industries. Disposal of industrial, commercial, and household waste is becoming a problem. Noise has also become acute due to tremendous increase in motor vehicles.

Pollution has reached such magnitude that it not only threatens the health and well being of the population in a particular area, but also produces effect on global scale. Man dumps tonnes of chemical and nuclear waste into the rivers and sea. The Food eaten by him is spread with lethal pesticides, which on long run can have serious side effects like paralysis, kidney trouble, cancer etc.

The environmental deterioration by man is attributable to three major causative factors:

1. Over-population.
2. Urbanisation and Higher living standards.
3. Industrialisation.

The increasing amount of wastes generated by these three phenomena, degrade the quality of land, water, air and food. Increase in population demands more foods, water and land—three major items that are limited on earth. To increase food production, either more efficient method should be adopted, exerting more pressure on the limited cultivable area, or more land should be devoted for agricultural purposes diverted from forest land. Thereby reducing carbon dioxide sinks which protect the world from a 'global-warming' effect. Man counts on the synthetic biocides to overcome crop losses due to pests, and relies on chemical fertilizers to boost the crop yield. Farmers who used to grow one crop a year, now have opted to cultivate as many as three. These practices have resulted in increased yields, soil degradation, nutrient depletion, water pollution by plant nutrients and contamination of the total environment with pesticides. Pesticidal chemicals are potential contaminants of air, water and food.

Every industry is associated with an emission of one type or other pollutants directly or indirectly. Industries are not only the responsible for the pollution of air but also the contamination of water. The untreated effluents by industrial facilities are primarily responsible for the water problem. Pollutants are chemical, physical or biological agents that exert undesirable effects on human health, and the environment. The air we breathe, the water we drink, the food we consume and the land we live on are all contaminated with one or the other from of pollutants, as mentioned hereunder:

Types of Environmental Pollutants

1. *Chemical Agents.* Gases and particulates, heavy metals, pesticides, petroleum, solid and liquid wastes.
2. *Physical Agents.* Heat, odour, noise, radiation.
3. *Biological Agents.* Micro-organisms, population.

Environmental Pollution and Health

One of the report of World Health Organisation (WHO) reveals that 24% of global diseases and 23% of all deaths are caused by environmental exposures many of which can be prevented by better environmental management. The four diseases most influenced by poor environment were diarrhoea, lower respiratory infections, unintentional injuries and malaria.

1. *The diseases caused by air pollution.* Some of the major air-borne diseases are: common cold, cold influenza, chicken pox, tuberculosis, silicosis, black lung cancer, bronchitis, asthma etc.,

2. *The diseases caused by water pollution.* Some of the common water-borne diseases are: cholera, typhoid, dysentery, hepatitis, intestinal disorders etc.,
3. *Food-borne diseases.* Contaminated food leads to various diseases, which includes: bolutism, viron food poisoning, fungal food poisoning, staphy lococcal poisoning.
4. *Deficiency diseases.* These may be: avitaminosis, anaemia, rickets and osteomalacia, teeth and gum diseases.

Between 60 to 90 percent of all cancers are directly or indirectly related to the environment that is related to factors; water, air, solar radiation, living and working environments and personal choices of diet and ways of life such as tobacco smoking and alcohol consumption.

19.3. AIR POLLUTION

Air pollution may be defined as the "presence in the outdoor atmosphere of one or more contaminants or combination thereof in such quantities and of such duration as may be, or may tend to be injurious to human, plant or animal life, or property, or which unreasonably interfere with the comfortable enjoyment of life or property, or the conduct of business."

Air pollution can also be defined as "the presence of any solid, liquid, or gaseous substance (including noise) present in the atmosphere in such concentrations that may or tend to be injurious to human beings, or other living creatures, or plants, or property, or enjoyment.

Atmosphere is a dynamic system, and absorbs various pollutants from natural as well as man-made sources, thus acting as a natural sink. Gases like CO, CO_2, H_2S, SO_2 and NO_2 as well as particulate matter such as sand and dust, are continuously released into the atmosphere through natural activities such as forest fires, decaying of vegetation, winds and sand or dust storms. Man-made pollutants *e.g.*, CO, NO_2, CO_2, hydrocarbons, particulates are also released into atmosphere. Now-a-days the problem of air-pollution has alarmingly increased due to population and other human activities. The pollutants are now-a-days enter the atmosphere at a faster rate than are absorbed by natural sink, thus accumulating in the air, and disturbing the equilibrium in the atmosphere. Green house effect on global warming is outcome of excessive pollution.

All the pollutants, which are emitted directly from the identifiable sources, either from the natural hazardous events like dust storms, volcanoes, etc., or from human activities like burning of wood, coal, oil in homes or industries or automobiles etc., are called the *primary pollutants*. The following five **primary pollutants** contribute to about 90% of the global air pollution:

1. **Oxides of sulphur,** like sulphur dioxide (SO_2).
2. **Oxides of carbon,** like carbon monoxide (CO) and carbon dioxide (CO_2).
3. **Oxides of nitrogen,** like NO, NO_2, NO_3.
4. **Volatile organic compounds,** mostly hydrocarbon.
5. **Suspended particulate matter** (SPM).

These primary pollutants often react with one another or with water vapour, aided and abetted by the sunlight, to form entirely a new set of pollutants, called the ***Secondary***

Pollutants. These new pollutants are often more harmful than the original primary pollutants that produce them.

The important **secondary pollutants** are:

1. Sulphuric acid (H_2SO_4),
2. Ozone (O_3);
3. Formaldehydes; and
4. Peroxy-acyl-nitrate (PAN) etc.,

H_2SO_4 is formed by the simple chemical reaction between SO_2 and H_2O vapour, and is a much more toxic pollutant than SO_2, having far reaching effects on environment, since it causes *acid rains*.

Other secondary pollutants like ozone, formaldehyde, PAN, etc., are formed by the *photochemical reactions,* caused by sun light between two primary pollutants. Say for example, O_3 is formed due to the photochemical reaction between hydrocarbons (HC) and nitrogen oxide (NO). Similarly, aldehydes may be formed by photochemical oxidation of hydrocarbons in the atmosphere.

The major air pollutants are now discussed below in details, mentioning their hazardous effects on human body, alongwith their sources of origin.

1. Oxides of Sulphur

Sulphur dioxide (SO_2) is the basic air pollutant amongest all the oxides of sulphur. It is an irritant gas, and when inhaled, affects our mucous membranes. *It increase the breathing rate and cause oxygen deficits in the body, leading to bronchial-spasms in the affected persons.* Patients of a asthma are very badly affected by this pollutant.

Some quantity of atmospheric sulphur dioxide (SO_2) may dioxide to from sulphur trioxide (SO_3), which when inhaled, may dissolve in the body fluids to from sulphuric acid (H_2SO_4), which is a very strong corrosive acid. SO_3, thus, causes high and worse irritation even at lower concentrations, leading to *severe broncho-spasm*.

SO_2 is also responsible for causing acidity in fogs, smokes, and in rains, and hence is the major source of corrosion of buildings and metallic objects.

SO_2 mainly originates in the atmospheric air from the : *refineries* and *chemical plants, smelting operations, and burning of fossil fuels*. Thermal power plants may emit SO_2 in large quantities, depending upon the quality of coal burnt by them. *Open burning of garbage* as well as *municipal incineration plants* may also emit sulphur dioxide in the air.

2. Carbon Monoxide (CO)

Carbon monoxide is a colourless, odourless, and toxic gas, produced when organic materials like natural gas, coal, or wood are incompletely burnt. Vehicular exhausts are the single largest source of carbon monoxide, as the number of vehicles have increased and is increasing by leaps and bound over the world: Poorly maintained vehicles and those having inefficient engines, without being fitted with adequate pollution control devices, release greater amounts of carbon monoxide.

Carbon monoxide possesses about 200 times affinity for blood-haemoglobin, (Hb) than oxygen. Eventually, when inhaled, CO replaces O_2 from the haemoglobin, and forms what is known as carboxy-haemoglobin ($CO.H_b$). This carboxy-haemoglobin is of no use for respiratory purpose; and hence, CO inhalation impairs normed oxygen transport carried out by the blood.

Low levels of CO inhalations produce symptoms like headache, dizziness, reduction in reaction time, etc., Higher levels of CO inhalations may cause nausea, heart palpitations, difficulty in breathing etc., In extreme cases, when about half of the haemoglobin of the blood is used up informing carboxy-haemoglobin, death becomes a certainty.

In cities having higher numbers of automobiles and industries, CO is found in as high concentrations as upto 60 mg/m^3 (54 ppm) with still higher concentrations in tunnels garages, and near the road intersections or the running automobiles, as against the maximum permissible concentration of about 10 mg/m^3 (*i.e.* 10,000 μg/m^3).

The national ambient air quality standards in India prescribe the maximum permissible concentration of CO, on hourly weighted average basis, as equal to 4000 μg/m^3 (*i.e.* 4 mg/m^3) for residential areas.

3. Oxides of Nitrogen (NO_X).

Atmospheric nitrogen may combine with oxygen at high temperatures, as generated during fuel combustion, to form nitrogen oxide (NO). *The nitrogen oxide at low levels is relatively harmless, but at high concentrations may cause asphyxiation and respiratory discomfort, because it combines about 3 lakh times faster with haemoglobin than oxygen.*

The nitrogen oxide further readily combines with the atmospheric oxygen, to form nitrogen dioxide (NO_2), which again is a very harmful compound for human health, as it is responsible for causing *eye and nasal irritations, respiratory discomforts,* and even deaths, as caused by NO. Prolonged exposures to somewhat higher concentrations of NO_2 exceeding the permissible concentration, has also been attributed to having caused serious diseases like *emphysema* (swelling due to air in body tissues), increased susceptibility to pneumonia, and even lung cancer.

Under the national ambient air quality standards prescribed in India the maximum average annual concentrations of oxides of nitrogen as NO_2 for residential areas is 60 μg/m^3. Very high concentrations of oxides of nitrogen, such as about 40 to 50 mg/m_3 of NO_2, may cause great respiratory discomfort, leading to quick deaths.

The oxides of nitrogen originates into the atmosphere from automobile exhausts, incineration plants, furnace smokes, etc., all involving high temperatures due to combustion of fuel.

Oxides of nitrogen become more significant, as they are involved in the formation of secondary air pollutants, such as ozone.

4. Hydrocarbons

Hydrocarbons are the group of compounds consisting of carbon and hydrogen atoms. They are either evaporated into the atmosphere from the petroleum fuel supplies (such as petrol, diesel, etc.,) or are emitted out in the automobile exhausts as the remnants of petroleum fuel that did not burn completely. They may also be contained in the fumes of oil refineries.

Hydrocarbons are washed out of the air when it rains, and run into surface water. Their presence in water causes an oily film on the water surface. *In low concentrations and when washed away in quick succession by rains, they do not generally cause much harm; but at high concentrations, they may cause irritation of eyes and respiratory tract, and responsible for causing body cancers.*

The presence of hydrocarbons in the environment may prove to be quite hazardous, and hence needs to be properly monitored.

5. Suspended Particulate Matter (SPM)

The particulate matter in air may occur in air largely in solid form as particles of dust, smoke, fume, etc.,; and also in liquid form as mist and fog. The particles larger than a molecule but small enough to remain suspended in air are called **aerosols.**

The suspended particulate matter in the atmosphere is a variable component, and is introduced either through natural phenomena like winds, volcanic eruptions, pollens and spores, decomposing particles of organic matter, etc.,, or through human activities like mining, burning of fossil fuels, industrial processes, etc.,

The suspended particulate matter in air may prove to be harmful to human health, inspite of the fact that the human respiratory system has a number of mechanisms for protecting the lungs from the entry of particles from air alongwith respiration. Infact, the bigger particles (> 10 μ) can be trapped by the hairs and sticky mucus in the lining of the nose. Smaller suspended particles of upto 10 microns (μ) can, although *reachtracheo-bronchial system* (*i.e.,* the tubes leading to lungs), but get trapped therein the mucus. They are sent back to the throat by beating of hair like cilia, from where they can be removed by spitting or swallowing. However, very small suspended particles may still leach the *pulmonry* (*i.e.,* lung regions) and damage the lung tissues, causing diseases like asthma, bronchitis, and even lung cancer, when such particles bring with them toxic and carcinogenic pollutants attached to the surfaces of the particles.

The recent advancements have shown that the particles which are smaller than 10 micron in size, prove to be highly dangerous to human health. Such smaller particles of sizes upto 2.5 micron, called ***PM 2.5,*** *are hence being also monitored.*

Various microscopic particulate substances, called **aero-allergens,** may also be present in air, and may cause allergic reactions in sensitive human bodies. The reactions to such substances occur in our body mainly in the skin and the respiratory tract. Sneezing is one symptom of allergy, followed by skin *troubles andlor bronchitis, asthma,* etc.,

Such organic allergens have been largely found to originate from living things like plants, animals, etc.,, although, however, finely powdered industrial materials may also sometimes cause allergic reactions in sensitive persons. Pollen grains and fungal spores from local plants are one of the worst allergens.

6. Lead

Lead is another major air pollutant that is emitted by vehicles. High lead levels have been reported in the ambient air in our metro cities. Leaded petrol is the primary source of air-borne lead emissions in cities. The compulsory use of unleaded petrol in cities like Delhi, has, however, reduced the lead menace from the atmospheric air in the city.

7. Photochemical Smog and Production of PAN and Ozone

Smog is a mixture of smoke and fog. It can be of two types; *i.e.* photochemical and coal induced. The atmospheric conditions favourable for formation of *photochemical smog* are :

(*i*) The presence of high concentrations of hydrocarbons and nitrogen oxides in the atmosphere :

(*ii*) Air stagnation, as not to cause disposal of air pollutants; and

(*iii*) Abundant sunlight.

The formation of photochemical smog is restricted to highly motorised cities.

Smog is caused by the interaction of some hydrocarbons and oxidants like SO_2 or NO_x, under the influence of sunlight, giving rise to the formation of dangerous *peroxy-acetyl-nitrate* (PAN). The main constituents are nitrogen oxide, **PAN**, hydrocarbons, carbon monoxide, and large quantity of **ozone. While PAN in itself is an extremely harmful gas, the ozone gas is also quite harmful to the respiratory system**, as it may enter deeper into the lungs (lower lobes) than even the oxides of sulphur.

19.4. SOURCES (CAUSES) OF AIR POLLUTION

Air pollution is mainly **caused** by the industrial activity. Most industries consume lots of coal or petroleum products (the so called fossil fuels), and their burning produces various gaseous pollutants. SO_2, CO and CO_2 are the common pollutants produced by this activity. Other gaseous pollutants are also produced by different industries that use various organic chemicals. Gases like xylene, methane, toluene, etc., come under this category.

Cement producing factories cause immense air pollution, emitting huge quantities of dust. Ash is generated by coal-burning power stations and furnaces. Mining and ore processing also cause serious particulate air pollution. Pesticides manufacturing industries often have to use poisonous gases, and these gases often cause toxic air pollution. Accidental release of such gases from a pesticide manufacturing plant can cause a major disaster, as was experienced in the *Bhopal Gas Tragedy* which occurred in India on 03-12-1984. Chlorine and ammonia gases, which likewise, are routinely used in industries, also have a large potential to cause pollution.

Air pollution is also largely being caused by **automobiles.** In highly urbanised areas, the mixture of air pollutants of air pollutants produces secondary air pollutants, giving rise to *photochemical smog*. The term *photochemical* implies that the reactions are helped by the sun-light. Various nitrogen oxides are also the products of petrol combustion.

Various particulate pollutants are also produced by the industrial activity. Stone crushing, glass manufacturing, cement production, etc., generate lots of dust particles. All coal burning machines and thermal power plants produce lots of unburnt carbon. Motor cars, heavy transport vehicles, rail engines, generators, and other equipments using diesel oil, cause significant air pollution by generating particulate matter, especially carbon.

Air conditioners, refrigerators, and some sprays use chlorofluorocarbon (CFC) gas. This has proved to be one of the most harmful gaseous pollutant, as the compund is found to be destructive of the atmospheric ozone layer, which protects us from the harmful ultra violet radiations from the sun rays, causing ozone depletion, having global impacts. Jet aeroplanes also cause air pollution; and, so is the case with satellite launching vehicles (rockets).

The air pollution, can, thus, be caused by numerous activities, and humanity will have to be conscious of the pollutants that are being released by human activities into our atmosphere.

Major sources of air pollution are listed hereunder :

1. Burning of different kinds of fuels, including combustion-fired power plants.
2. Controlled burn practices used in agriculture and forestry management.
3. Motor vehicles generating air pollution emission,
4. Marine vessels.
5. Wood, coal, fuel oil or natural gas burning.
6. Oil refining, power plant operation and industrial activities.
7. Chemicals, dust, and crop waste burning in farming.
8. Fumes from paints, varnish, aerosol sprays and other solvents.
9. Waste deposition in landfills which generate methane.
10. Construction activities.
11. Use of Air conditioners, refrigerates etc.,
12. Jet aeroplanes.
13. Military uses, as nuclear weapons, toxic gases, germ warfare and rocketry.
14. Dust from natural sources usually large areas of land with little or no vegetation.
15. Pine trees, which emit volatile organic compounds.
16. Smoke and carbon monoxide from wildfires.
17. Volcanic activities, which produce sulphur, chlorine, and ash particulate.

Various sources of **air pollution can be grouped into the following categories:**

A. Fuel burning.
B. Industrial processes.
C. Automobile exhaust.

Each of these sources are describe here:

(A) Fuel Burning

These operations are performed for heat and power generation. The fuel burnt may be coal, gases, oil, refuse, wood etc., These are responsible for generating air pollution of the following types:

(*i*) **Smoke:** This is a product of incomplete combustion of coal, carbon monoxide, carbon dioxide etc.,

(*ii*) **Dust:** It is finely divided powder and remains in atmosphere in floating condition. These are generated in large quantity in mining, quarrying, construction, cement industries etc.,

(*iii*) **Fumes:** Solid particles generated by condensation from gaseous state, and often accompanied by a chemical reaction such as oxidation, are known as fumes.

(*iv*) **Gases:** These are formless fluids.

(*v*) **Smog:** This is a formation of heavy, cloudy, hazy, floating layer formed by the mixture of smoke, dust, fog, mist etc.

(B) Industrial Processes

As mentioned earlier industries like cement, chemical, textile, thermal power station, steel plants, oil refineries, fertilizers etc.,- are main contributors to air pollution. These are responsible for emitting harmful gases, fly ash, dust, soot.

With the proliferation of lead, mercury, oxides of nitrogen, ozone, cyanides, sulphur compounds etc., are responsible for health hazards and are increasing regularly. The Workers working in mining, construction and cement industries have to inhale lot of dust and suffer from Asthema, Tuberculosis, Silicosia, Byssionosis etc.,

(C) Automobile Exhaust

Most of the vehicles and construction equipment are operated by diesel engines and their exhaust is responsible for major air-pollution. The construction equipment may be either stationary equipment without prime mover like air compressors, diesel generating sets, concrete mixers, dewatering pumps or may be with prime movers *e.g.*, dumpers, transit mixers, compactors, shovels, excavators, graders, loaders, pavers etc.,

Automobile exhausts has following main constituents :

1. **Gases:** Carbon monoxide, nitric oxides, unburnt hydrocarbons, partial oxidation products.
2. **Particulates:** These include lead compounds, carbon and various other inorganic or organic materials.

Furthermore, additional hydrocarbons and fuel additives are contributed by the evaporation of gasoline from automobile tanks.

19.5. EFFECTS OF AIR POLLUTION

One of the most obvious local effects of particles in the atmosphere is a *reduction in visibility.* Soot and smoke is usually quite apparent in highly industrialised and motorised cities. The reduction in visibility often causes slowdown of air-traffic, needing costly instrument-guided landing systems.

The presence of gases in the atmosphere has been causing the most prominent effect of producing *objectionable odour, taste,* or *obvious corrosive or chemical effects on objects.* Due to their chemical and corrosive effects, the air pollutants have been damaging the exteriors of buildings and cars by breaking down the paint emulsions, causing heavy economical losses. The air pollutants have been quickly spoiling the fabric and leather dresses worn by human beings, alongwith causing skin damages on their exposed body parts. Several monuments and old beautiful buildings made in exposed stone and marbles, have been badly spoiled and damaged by the acidic fumes.

Air pollution has many adverse effects. It damages the building structures, materials, vegetations, causes health hazards, and obstructs visions. The effect of air pollution can be categorised as:

(*a*) Effects on human health,
(*b*) Effects on animals,
(*c*) Effects on plants,
(*d*) Effects on materials,
(*e*) Effects on atmosphere,
(*f*) Effective on aquatic life,
(*g*) Effect in Global and Regional environment.

(A) Effects on Human Health. Air pollution seriously affects the respiratory system. Carbon monoxide when absorbed into the lungs and reduces the haemoglobin available to carry oxygen to the body. This may cause injury to vital organs, sulphur dioxide also causes cough, shortens breadth and can cause irritation to the eyes. It when converted to sulphuric acid damage the lungs. It also helps in severe infections and allergic response. Even very low concentration of ozone can cause damage of the lungs. Hydrogen sulphide and other organic oxides can create odour problems. Hydrocarbons are believed to cause cancer in lung, whereas lead emitted by automobile exhausts is a cumulative poison and can cause brain damages.

Some of the chronic diseases due to exposure to polluted air are lung cancer, bronchitis, emphysema and asthma.

(B) Effects on Animals. The effects of air pollution due to its inhaling on animals are similar to those on human. Air pollutants also enter the animal body through ingestion of contaminated plant materials. Among the metallic contaminants, arsenic, lead and molybdenum are important. Air contaminated with ozone, flouride effects the forage crops, which when eated continuously by the animals affects the animals adversely.

(C) Effect on Plants. As we know that leaf veins function as the transport system for water, minerals and food. The leaf tissues form cells in between them. Air pollution damages these leaf veins and tissues. The damage can be in the form of Necrosis (refers to dead areas on a leaf structure), Chlorosis (resulting in yellowing of the leaf), Abscission dropping of leaves!. When the absorption of sulphur dioxide exceeds a particular level, the cells become inactive and are killed, resulting in tissue collapse and drying of plants adversely. Chlorine, ammonia, hydrogen sulphide, and ethylene also affects the vegetations adversely. Smog also causes injuries to the plants, and certain species are injured even with very small concentration of smog.

(D) Effects on Materials. Air pollution also adversely affects building structures and materials. Smoke, dust and oxides of sulphur are most destructive pollutants. Sulphur dioxide changes with moisture and air to sulphuric acid which is corrosive. Moisture, temperature, sunlight and air movement decide the rate of attack of pollutants. Aluminium, copper, iron and steel are corroded when exposed to contaminated air. Building structures are damaged/ disfigured due to polluted air, sooty deposits, normal weathering.

(E) Effects on Atmosphere. Increased combustion of fossil fuel and oil, enchances the, carbon dioxide concentration of the atmosphere. Carbon dioxide absorb heat and the radiative cooling effect of the earth is thus decreased. CO_2 is the single culprit responsible for half of global warming. This is known as "greenhouse effect" and is expected to raise the planet's temperature by 1.5—4.5°C in about 60 years.

'Greenhouse' is a shield that keeps our planet warm enough-to sustain life, but is now becoming a heat trap threatening to disrupt the global environment. A thin layer of certain gases, about 25 km like a green house glass letting heat through but stopping enough radiation back into space to warm out world. But now "Green house" layer is becoming' denser, keeping in more heat.

(F) Effects on Aquatic Life. Air pollutants mixing up rain can cause acidity in fresh water lakes, affecting aquatic life, especially the fish life.

(G) Effects on Global and Regional Environment. Air pollutants may cause acid rains, ozone depletion, and global warming, thereby adversely effecting the environment at regional on global levels.

19.6. ACID RAIN

Acid rain is commonly defined as "a condition in which natural precipitation becomes acidic after reacting with pollutants in the air". Acid rain is formed when industrial emissions of gases including sulphur dioxide, nitrogen oxides, and hydrogen chloride combines with water droplets in the atmosphere. This chemical reaction forms weak sulphuric acid, and nitric acid that cause acid rain.

The ill-effects of acid rain can be found on vegetation, soil, marine resources, moments as well as on man. The ill-effluents of acid rain have no national limits as nature knows no geographical boundaries.

The main causes of acid rain are the gases emitted by industries, petroleum refineries and motor vehicles which pollutes the atmosphere. In other words, acid rain simply means fall-out of acids caused by sulphur dioxide and nitrogen oxide with rainfall and thus increase in the amount of rainwater.

19.7. AIR QUALITY INDEX (AQI)

The air quality index tells us how clean or polluted our air is, and what associated health effects might be of concern for us. The AQI made its debut in 1968 in USA. In India the National Air Quality Index (AQI) was launched in New Delhi on September 17, 2014 under Swach Bharat Abhiyan.

The United States Environmental Protection Agency (EPA) has developed an Air Quality Index that is used to report air quality. This AQI is divided into six categories indicating increasing levels of health concern.

Table - 1

Air Quality Index AQI- Values	Levels of Health Concern	Colours
0 - 50	Good	Green
51 - 100	Moderate	Yellow
101 - 150	Unhealthy for Sensitive Groups	Orange
151 - 200	Unhealthy	Red
201 - 300	Very Unhealthy	Purple
301 - 500	Hazardous	Maroon

The Environmental Protection Agency (EPA), USA has established National Ambient Air Quality Standards (NAAQS) for each of the pollutants in order to protect public health. Table-2 shows the breakpoints for various pollutants.

19.7.1. Computing the AQI

According to Wikipedia (2012), the mathematical formula for computing Air Quality Index is given as follows.

$$\tau = \frac{I_{\text{Hight}} - I_{\text{Low}}}{C_{\text{High}} - C_{\text{Low}}}(C - C_{\text{Low}}) + I_{\text{Low}}$$

where : τ = The Air Quality Index

C = Pollutant Concentration

C_{Low} = The concentration breakpoint that is ≤C

Table - 2

O_3 (ppp)	O_3 (ppp)	$PM_{2.5}$ (mg/m³)	PM_{10} (mg/m³)	CO (ppp)	SO_2 (ppp)	NO_2 (ppp)	AQI	AQI
C_{Low} – C_{High} (avg.)	C_{Low} – C_{High} (avg.)	C_{Low} – C_{High} (avg.)	C_{Low} – C_{High} (avg.)	C_{Low} – C_{High} (avg.)	C_{Low} – C_{High} (avg.)	C_{Low} – C_{High} (avg.)	Low – High	Category
0 - 54 (8 - Hr)	–	0.0 - 12.0 (24 - Hr)	0 - 54 (24 - Hr)	0.0 - 4.4 (8 - Hr)	0 - 35 (1 - Hr)	0 - 53 (1 - Hr)	0 - 50	Good (Green)
55 - 70 (8 - Hr)	–	12.1 - 35.4 (24 - Hr)	55 - 154 (24 - Hr)	4.5 - 9.4 (8 - Hr)	36 - 75 (1 - Hr)	54 - 100 (1 - Hr)	51 - 100	Moderate (Yellow)

71 - 85 (8 - Hr)	125 - 164 (1 - Hr)	35.5 - 55.4 (24 - Hr)	155 - 254 (24 - Hr)	9.5 - 12.4 (8 - Hr)	76 - 185 (1 - Hr)	101 - 360 (1 - Hr)	101 - 150	Unhealthy for Sensitive Groups (Orange)
86 - 105 (8 - Hr)	165 - 204 (1 - Hr)	55.5 - 150.4 (24 - Hr)	255 - 354 (24 - Hr)	12.5 - 15.4 (8 - Hr)	186 - 304 (1 - Hr)	361 - 649 (1 - Hr)	151 - 200	Unhealthy (Red)
106 - 200 (8 - Hr)	205 - 404 (1 - Hr)	150.5 - 250.4 (24 - Hr)	355 - 424 (24 - Hr)	15.5 - 30.4 (8 - Hr)	305 - 604 (1 - Hr)	650 - 1249 (1 - Hr)	201 - 300	Very Unhealthy (Purple)
–	405 - 504 (1 - Hr)	250.5 - 350.4 (24 - Hr)	425 - 504 (24 - Hr)	30.5 - 40.4 (8 - Hr)	605 - 804 (1 - Hr)	1250 - 1649 (1 - Hr)	301 - 400	Hazardous (Maroon)
–	505 - 604 (1 - Hr)	350.5 - 500.4 (24 - Hr)	505 - 604 (24 - Hr)	40.5 - 50.4 (8 - Hr)	805 - 1004 (1 - Hr)	1650 - 2049 (1 - Hr)	401 - 500	

The concentration breakpoint that is $\leq C$

The index breakpoint corresponding to C_{Low}

The index breakpoint corresponding to C_{High}

Example 1. Suppose a monitor records concentration of then find AQI.

$$\tau = \frac{100-51}{35.4-12.1}(20-12.1)+51$$

$$= \frac{49}{23.3}\times 7.9+51$$

$$= 67.61 \text{ which is of Moderate Category}$$

Example 2. Suppose a monitor records CO of 14.0 ppm, then find the AQI.

$$\tau = \frac{200-151}{15.4-12.5}\times(14-12.5)+151$$

$$= \frac{49}{2.9}\times 1.5+151$$

$$= 176.3 \text{ which falls into Unhealthy Category}$$

If multiple pollutants are measured at a monitoring site, then the largest or **dominant** AQI value is reported for the location.

The Ozone AQI between 100 and 300 is computed by selecting the larger of the AQI calculated with a 1 - hour ozone value and the AQI computed with the 8 - hour ozone value.

Different countries have their own air quality indices, corresponding to different national air quality standards. In India the National Air Quality Index (AQI) was launched in 2014 under the **Swach Bharat Abiyan**. The latest index measures eight parameters. The pollutants considered are PM_{10}, $PM_{2.5}$, NO_2, SO_2, CO, O_3, NH_3 and Pb. for which National Ambient Quality Standards are prescribed.

QUESTIONS

1. Define and explain 'Environmental pollution'.
2. Define and explain 'air pollution.
3. Define air pollution, and differentiate between 'natural' and 'man-made' air pollution.
4. List out the pollutants causing air pollution and their sources of origin.
5. Explain the reasons for the occurrence of various air pollution disasters.
6. State the harmful effects of SO_2, CO, P_b and photochemical oxidants on human health, vegetation, and materials.
7. Classify and outline the characteristics of particulate pollutants.
8. Briefly pinpoint various causes of air-pollution.
9. Differentiate between primary and secondary air pollutants. Enumerate various such pollutants.
10. What is photochemical smog and how it is formed?
11. How does air pollution affects human health?
12. Enúmerate the major air pollutants, their characteristics, sources and health effects on human beings.
13. Enumerate the important local, regional, and global impacts of air pollution.
14. What is Air Quality Index? How is it useful?

Air Pollution Control

20.1. INTRODUCTION

The atmosphere, does possess self-cleansing properties, which continuously clean and remove the pollutants from the atmosphere under natural processes. So, long as the pollutants discharged by man into the environment, is lower than the natural cleansing capacity of the environment, we live happily without any air pollution problem. But as and when the discharged pollutants exceed the natural cleansing capacity, our environment becomes polluted, giving us diseases, spoiling our clothes, plants, eatables, buildings, etc., Attempts are then made to reduce the emission of pollutants from the automobiles or factories by adopting mechanical means, or by using high rise chimneys for better dispersion and dilution of pollutants over a longer range of environment. The natural self-cleansing process of the environment, and the engineering measures adopted to artificially clean the industrial and vehicular gases, before they are into the environment, are discussed in this chapter.

Preventing Air Pollution

In order to prevent air pollution following steps must be taken:

1. Everyone of us must feel responsibility and take initiative immediately to prevent air pollution.
2. We should start planting trees, as they purify the air, and thus keep the sky clean. Trees breathe carbon dioxide and release oxygen. Forests act as vast sinks for carbon dioxide and tend to reduce global warming.
3. Curtail vehicular emission by using them to minimum possible, using fuel efficient engines, keeping vehicles in maintained condition. Conserve fuel, encourage mass transport, use cars in pools, monitor the fuel emission.
4. Polythene, plastic containers, polyurethane foams and P.V.C. are eco-non-friendly items, therefore their use should be avoided, disposal of these plastics is not an easy task, as burning produces soot, corrosive gases and highly offensive odour.
5. Maximise use of solar energy—use of solar heaters, solar energy, electric plants be encouraged.

6. Air pollution monitoring should be carried out regularly.
7. While preparing a project, environment management plan should be prepared for formation, implementation and monitoring of environmental protection measures during and after commissioning of projects.
8. Emission levels of pollutants from chimneys should conform to the pollution control standards prescribed by the central or state pollution control boards, for which necessary arrangements must be made. Generally used air cleaning devices are:
   (*a*) Electrostatic precipitators
   (*b*) Cyclone separators
   (*c*) Wet collectors
   (*d*) Gravity settlers
   (*e*) Bag filters
   (*f*) Cyclone filters
   (*g*) Wet scrubbers
   (*h*) Absorbers or Adsorbers.
9. For indoor pollution, use proper ventilation so that used air or vitiated air gets removed by air.
10. Air pollution may be minimised by :
    - Selecting proper raw materials.
    - Making suitable changes in manufacturing process.
    - Modifying the equipment used.
    - Recovering or recycling.
11. For installing an industry, select a site which will have minimum adverse effects on people, animals, agriculture etc.
12. In order to control emission from motor vehicles proper maintenance including engine tuning, use of catalytic reactors etc., are necessary.
13. Equipment with latest technology, fuel efficient engines, plants with dust collectors etc., should be used so as to have better control over air pollution.

20.2. AIR AND ITS COMPOSITION

Air composition in atmosphere contains life-giving gases, like oxygen for man and animal, and carbon dioxide for plants. Air is an essential item for men, animals and for all living beings. Fortunately, air is nature's gift and no efforts are required for getting the air from the atmosphere. Fresh air is required for maintaining a good health. If supply of fresh air from the is inadequate, it will result in headache, suffocation, sleepiness, loss of appetite and overall in ability to concentrate in work. Such a condition may occur in a closed non-ventilated room, stuffy theatre halls, and other overcrowd public places. Therefore, the buildings are required to be well-ventilated and the stagnation of air at any place is prevented.

The atmosphere acts like a blanket for the earth and regulates the heat balance. It also protects us from the harmful ultra violet radiation of the sun. The atmosphere also serves as a storehouse for water vapour, which leads to precipitation. The moisture content in the atmosphere is responsible for the creation of several forms of condensation and precipitation (clouds, fogs, frost, rainfall, ice, hailstorm, snowfall etc.).

Advantages of the Air (Atmosphere)

1. The atmosphere is an insulating blanket of air around the earth. Without it the temperature at the equator would rise to about 80°C during the day and drop to as low as –140°C at night.
2. It is a reservoir of oxygen needed by the men and animals, and of carbon dioxide essential for plants.
3. It serves as a distribution system for moisture.
4. Without the atmosphere there would be no sound and no light.
5. The atmosphere shields the earth from lethal concentration of ultra violet rays.

Composition of Air

The composition of the air (atmosphere) is taken as the composition of pure dry air, which constitutes mainly of nitrogen (78%) and oxygen (21%) that together make up 99% of the air by volume. Their proportion remains almost constant within lower layers of the atmosphere close to the earth. Remaining 1% is accounted for by gases, like argon (0.93%), carbon dioxide (0.03%), hydrogen, helium and ozone. Besides water vapour, dust particles, smoke and impurities are also present in varying quantities.

The vapour content in the air ranges between zero to 5 percent by volume. The vapour is received through the evaporation of moisture and water from the water bodies, vegetation, soil cover, seas and oceans.

In the purer regions, like above the sea or mountain the air may contain 21.99% of oxygen, whereas in the overcrowded halls it may be as low as 20.65%. Carbon dioxide is obtained by combustion, fermentation and respiratory processes of animals. The quantity of carbon dioxide is decreased and a balance is maintained by vegetation, rain and high winds,. If CO_2 percentage increases upto 1.5, it will produce nausea, depression and headache etc.,

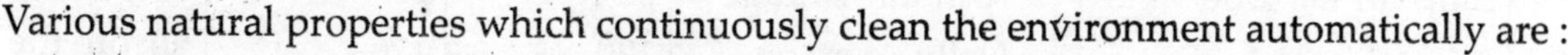

20.3. NATURAL SELF-CLEANSING PROPERTIES OF AIR

Various natural properties which continuously clean the environment automatically are :

(i) *Dispersion;*

(ii) *Gravitational settling* with or without *flocculation* of particles;

(iii) *Absorption* including *washout* and *scavenging;* and

(iv) *Adsorption.*

All these processes, naturally occur periodically in the ambient air, thereby removing or reducing the various pollutants entering the air. These processes are briefly discussed below :

1. Dispersion

Dispersion of pollutants by winds reduces the concentration of air pollutants at one place. This mechanism is, a *diluting* mechanism only. The dispersion through long chimneys helps in diluting the air pollutants near their sources only, and does not reduce their long-term undesirable effects on the community as a whole.

2. Gravitational Settling

Gravitational settling is a natural mechanism, under which large heavy particles from the ambient air settle down on buildings, trees and other objects.

This process also helps in removing **flocculated particles** formed by uniting of smaller particles over larger particles, till a floc particle, large and heavy enough to settle out under gravity is formed.

3. Absorption

In the natural **absorption process,** the *gaseous* as well as particulate *pollutants* from the air get collected in the rain or mist, and may settle down with that moisture. This phenomenon takes place when falling raindrops absorb pollutants, and is also known as **washout** or **scavenging.**

The gaseous pollutants are removed in dissolved state with moisture, either with or without chemical changes.

4. Adsorption

Adsorption is the phenomenon in which the gaseous, liquid, or solid pollutants present in the ambient air are kept *attracted, generally electrostatically,* by a surface, where they are concentrated and retained. Natural surfaces, such as soils, rocks, leaves, blades of grass, buildings, and other objects can absorb and retain pollutants. The particles may come in contact with such surfaces by *inertial impaction,* under which the pollutants are carried to such surfaces by winds.

20.4. AMBIENT AIR QUALITY STANDARDS

Ambient air quality standards prescribed in India are given in the following table.

S.No.	Pollutant	Time weighted average	Concentration in Ambient Air in μg/m³		
			Industrial Areas	Residential Areas	Sensitive Areas
1.	Sulphur Dioxide (SO_2)	– Annual Average – Daily Average	80 120	60 120	15 30
2.	Oxides of Nitrogen (NO_2)	– Annual Average – Daily Average	80 120	60 80	15 30
3(A).	Total suspended Particulate Matter (TSPM)	– Annual Average – Daily Average	360 500	140 200	70 100
3(B).	Respirable Particulate Matter (Size ≤ 10 μm) (RSPM)	– Annual Average – Daily Average	120 150	60 100	50 75

4.	Lead (Q_b)	– Annual Average – Daily Average	1.0 1.5	0.75 1.00	0.50 0.75
5.	Carbon Monoxide (CO)	– 8 hour Average – 1 hour Average	5,000 10,000	2,000 4,000	1,000 2,000

20.5. VENTILATION FOR CONTROLLING INDOOR AIR POLLUTION

The **indoor air pollution,** which results in polluting the air inside the buildings and in our houses is more dangerous than the outdoor air pollution, because we generally remain inside our houses for much longer a period than in the outside environment. The women folk in particular work for several hours in their kitchens, where a lot of smoke and air pollutants may exist, causing severe harm to their health.

The **most important source** of indoor air pollution in developing countries like India is the combustion of domestic **biomass fuel,** such as cow dung cakes, *wood, charcoal,* and *crop residues,* which are being used for cooking on a very large scale, particularly by the poor rural and tribal people. The incomplete combustion of such unprocessed solid biomass fuels increases the levels of **indoor air pollutants** such as *particulate matter,* and noxious gases like *carbon monoxide*, Author major by product of burning of such biomass fuel is the *fine soot particles* that can penetrate deep into the lungs causing inflammation of airways and lungs resulting in acute respiratory diseases and lung resulting in acute respiratory diseases and lung cancer. Exposure to these emissions is more likely to lead to lung cancer that even tobacco smoking.

The indoor air pollution has, infact, been found to be much worse than the outdoor air pollution, since a pollutant released indoor is thousand timer more likely to reach the lungs than a pollutant released outdoors. It is generally the women and young children and infants who face the maximum adverse effects of indoor air pollution. In poor household, there is no separate kitchen, and people usually stay in the same room where they cook, or burn fuel to heat during winters. Moreover, women who work on cooking, and their young children particularly the infants, always necessarily stay at the place of cooking or burning of traditional fuel. This exposes them to continuous indoor air pollution over long hours. This proves wrost for the children (upto the age of about 5 years), whose lugs are in the developing stages.

Promoting the use of cleaner fuel, such as biogas, LPG, solar cookers, improved stoves, and better ventilation of homes and kitchens should, therefore, be given top priority to reduce indoor air pollution. Moreover, children and pregnant women, who are most susceptible to the ill effects of indoor air pollutions, need to be protected on top priority from such indoor air pollutions.

Problems Due to Air Contamination

Thus, common problems felt by people due to air contamination are :

1. Dryness and irritation of the eyes, nose, throat and skin.
2. Headache.
3. Fatigue.
4. Shortness of breath.

5. Hypersensitivity and allergies.
6. Sinus congestion.
7. Coughing and sneezing.
8. Dizziness, and/or
9. nausea.

Air is considered unfit to breathe if it contains the following:

S. No.	Name of the gas	% volume in atmosphere
1.	Oxygen.	Less than 19.5
2.	Carbon dioxide.	More than 0.5
3.	Carbon monoxide.	More than 0.005
4.	Hydrogen sulphide	More than 0.001
5.	Nitrous oxide	More than 0.002
6.	Oxides of hydrogen	More than 0.002
7.	Methane	More than 0.5
8.	Flammable gases	More than 1.5

Ventilation

The air change in a building is essential to remove any smoke, odour, dust etc., to a reasonable level. Ventilation is defined as a process of removing or supplying air by natural or mechanical means.

Ventilation is necessary for the supply of fresh air for breathing and for maintaining a satisfactory thermal environment. Ventilation is generally defined as the supply of outside air to the interior for air motion and replacement of vitiated (state) or polluted air.

1. **Natural Ventilation.** The rate of ventilation by natural means through doors and windows and other openings depend on direction and velocity of wind (doors, windows and ventilators).
2. **Mechanical Ventilation.** Ceiling fans are mostly used for inducing air motion for comfort. Exhaust fans are also used at places like kitchen, bath rooms, stores etc., to replace the vitiated air with fresh air outdoor air.

Purpose of Ventilation

As stated above, ventilation is meant for supply of fresh air, and to replace the old hot used up (exhausted) air. The ventilation ensures the removal of bad effects of occupancy of an enclosed space:

(*i*) By providing necessary oxygen to remove oxygen deficit caused by respiration;

(*ii*) By removing and diluting CO_2 in the air;

(*iii*) By lowering down the temperature by removing hot used up air and replacing it by colder fresh air,

(*iv*) By reducing humidity, and

(*v*) By reducing body odours.

20.6. CONTROL OF AIR POLLUTION IN CONSTRUCTION ACTIVITIES.

For the purpose of air pollution control in construction activities, the pollution can be divided into following categories:

1. Particle control in chimneys.
2. Particulate control in atmosphere.
3. Dust control.

1. Particle control in chimneys

Plants used in construction, cement industries, thermal power plants etc., emit particulate through chimneys. These emit dusts of coal, stone, cement, lime, sand, etc., and combustion gases. The persons engaged on these plants and in nearby areas are badly affected by this type of pollution. These also adversely affect the nearby residential areas, and cultivated farm etc., Dust collectors are required to be installed with such plants to remove suspended particles from the chimney emission. The Dust Collectors are of several types, and their suitability is required to be decided depending upon the characteristics of the particulate.

2. Particulate Control in Atmosphere

In construction, cement, and mining industries, thermal plant areas, coal, stone aggregates, ore, sand etc., are stored at site or plant areas. These should be maintained in such a condition so as to prevent dust from becoming wind borne while in storage or being handled, loaded or transported. Following steps may help in this respect:

(*i*) Wetting of material to prevent dust becoming wind borne.

(*ii*) Location of stock piles in an area, protected from prevailing winds.

(*iii*) Installation of wind breaks *i.e.*, trees, wall etc.,, to protect stock piles from direct exposure to the prevailing winds.

(*iv*) Storage of materials in bins, silos or any other type of enclosures.

(*v*) Plant and plant areas should be cleaned regularly of accumulated dust, dirt etc.,

3. Dust Control

Dust emitted due to haulage, working of heavy earth moving equipment etc., are needed to be controlled by adopting following measures:

(*i*) Maintaining haul roads, moving the vehicles at reasonable speed, sprinkling water on unpaved haul roads.

(*ii*) Heavy earth moving epuipments should be provided with cabins so as to protect the operators.

20.7. CONTROLLING AIR POLLUTION IN INDUSTRIES

As discussed earlier, the atmosphere possesses self-cleaning properties, which continuously remove the pollutants from the atmosphere under natural process. So long as the pollution is lower than the natural cleansing capacity of the environment, there is no problem of air pollution. But, when the pollutants exceed the natural cleansing capacity, our environment is said to be polluted, and gives us the diseases, spoiling our clothes, plants, eatables, buildings etc.,

The pollutants emission from automobiles and industries can be reduced by using high rise chimneys for better dispersion and dilution or by adopting mechanical means for collecting the particulates.

Different methods for controlling air pollution in industries are:

1. Dilution method.
2. By installing Engineering devices.
   (*i*) Devices for reducing particulate matter.
   (*ii*) Devices for reducing gaseous pollutants.

These methods are being discussing in following paragraphs in detail.

20.7.1. Dilution Method (Through High Rise Chimneys)

Tall chimneys are used to be disperse the exhaust pollutants over a great area and thereby reduce the concentration of the pollutants, such as particulate matter, carbon monoxide, nitrogen oxide, and sulphur oxides to the levels required by the government environmental policy and regulations.

The high rise chimneys spread the pollutants over larger distances, and, thus reducing the pollution near the emission source. The main drawback in this method is that, it only reduces the concentration of pollution at particular place (area), rather than reducing or removing the pollution load from the total environment, as a whole.

20.8. PARTICULATE EMISSION CONTROL DEVICES

The commonly used devices, which control particulate matter are:

(*i*) Gravitational settling chambers.
(*ii*) Centrifugal collectors including Cyclone collectors.
(*iii*) Wet Scrubbers (collectors) including Spray towers, Wet cyclone scrubbers, and Venturi-scrubbers.
(*iv*) Electrostatic precipitators; and
(*v*) Fabric filters.

These mechanical devices are explained in brief as under:

1. Gravitational Settling Chambers

These are suitable for removing particles of size more than 50 μm at about 40 to 50%

efficiency. The emitted smokes, when made to pass through a settling chamber (Fig. 20.1), drop some of their larger sized particles in the chamber. From air pollution point of view smaller sized particles (< 50 μm) are most troubulesome, hence this device is used normally as pre-cleaners to remove large and possible abrasive particles. In this device large volume of the chamber allows the air to flow at a low velocity, giving time for the particle to settle out.

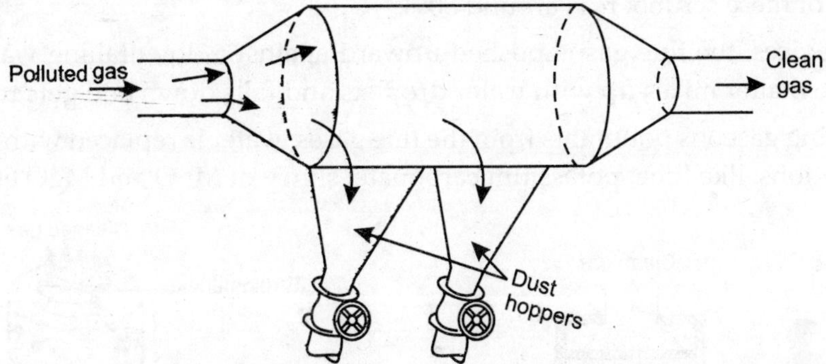

Fig. 20.1. Gravitational settling chamber.

2. Centrifugal Collectors

These are used to remove particles of size between 5 and 25 μm with 50 to 90% efficiency. These are suitable for handling large volume of gases and are widely used in industries producing large quantities of gas, like cement and fertiliser plants, petroleum refineries, asphalt mixing plants etc.

A cyclone collector is a specially designed closed chamber, in which the velocity of the inlet gas transformed into spinning vortex, and the particles from the gas thrown out under the centrifugal force. The particles thrown out on the walls of the chamber, slide down to the hopper, and are thus removed. The efficiency of the device depends on the generated centrifugal force, which in turn depends on mass of the particles, inlet gas velocity, and radius of cyclone.

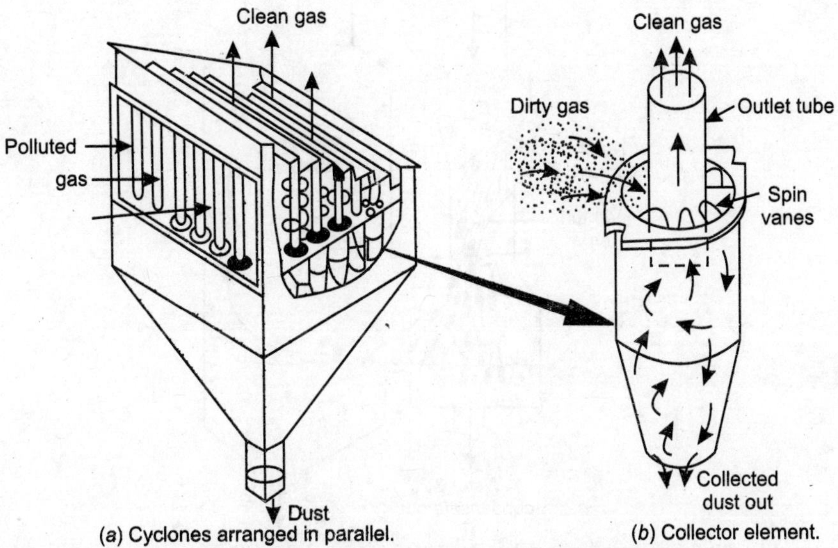

(*a*) Cyclones arranged in parallel. (*b*) Collector element.

Fig. 20.2. Cyclone Collector (Cyclone Separator).

3. Wet Scrubbers (Collectors).

These include spray tower, wet cyclonic scrubber, and venturi-scrubber. The spray tower is suitable for removing particles of more than 10 μm size and can be made to remove gaseous pollutants also, alongwith removing particulate matter. Wet cyclonic scrubbers and venturi-scrubbers are suitable for removing particles of sizes more than 2.5 μm and more than 0.5 μm. The efficiency of these scrubbers is around 80%.

In these devices, the flue gas is pushed upward against a downfalling water (or liquid). The particulate matter mixes up with water droplets and falls down and gets removed.

For removing gaseous pollutants from the flue gases, water is replaced with other aqueous chemicals solutions, like lime, potassium carbonate, slurry of MnO and MgO etc.,

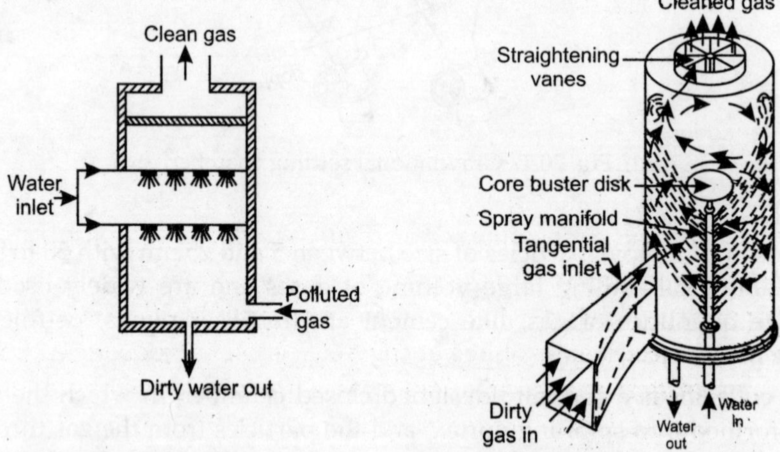

Fig. 20.3. Spray Tower. **Fig. 20.4.** Wet-Cyclone Scrubber.

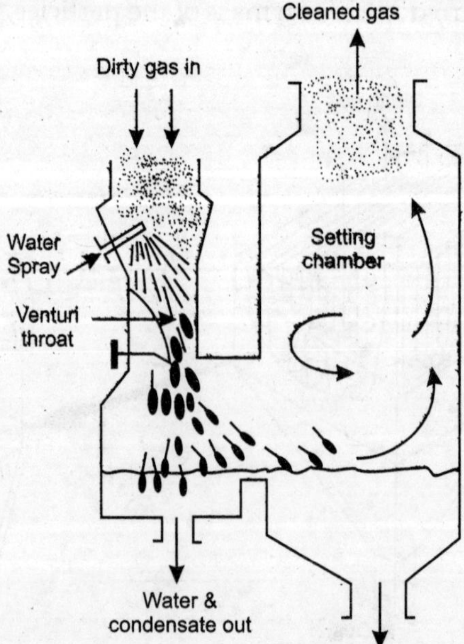

Fig. 20.5. Venturi-scrubber (also for removing gaseous pollutions).

4. Electrostatic Precipitator (ESP).

This is used to remove the particles of more than 1 μm size. These devices have efficiency of about 95 to 99%, and can remove even small particles. In these, the flue gas is made to pass through a highly ionised zone, where the particles get electrically charged and are separated out from the gas with the help of electrostatic forces in the powerful electric field.

These are widely used in thermal power plants, pulp and paper industries, mining and metallurgical industries, iron and steel plants, chemical industries.

Electrostatic precipitator has two sets of electrodes, insulated from each other, that maintain an electrostatic field between them at high voltage. The flue gases are made to pass between these two sets of electrodes. The electric field ionises the dust particles that pass through it attracting them to the electrode of opposite charge. The other electrode is maintained at a negative potential of 30,000 to 60,000 volts. The dust particles are removed from the collecting electrode by tapping the electrode periodically.

The ESP is costly but has low maintenance cost and is frequently employed with pulverised coal fired power plants for its effectiveness on very fine ash particles and is superior to that of any other type. It is commonly employed when the volumetric throughput of gas is high.

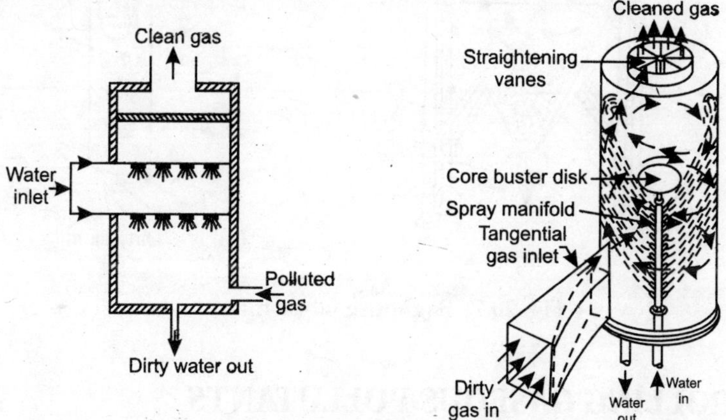

Fig. 20.6. High Voltage Electrostatic Precipitator (Tube type).

5. Fabric Filters

This device is suitable for removing particles of size even lesser than 1 μm and efficiency of more than 99%, provided the flue gas is dry. Otherwise, there is a risk of condensation inside the filter, which may cause clogging.

In this system, the flue gas is allowed to pass through a woven or felted fabric, which filters out the particulate matter and allows the gas to pass. The small particles are retained on the fabric initially through interception, and later on, when a dust mat is formed the fabric starts collecting particles more efficiently.

A baghouse filter unit, in which a large number of filter bags are arranged so that continuous removal of collected material is possible. These bags consist of porus structure material, fabric or fibrous matter has the properly to retain the particulate matter as the dirty gas passes through the voids or layers of filter. In a bag filter, air enters from the bottom and flows out along its sides, leaving the dust to form a cake inside the filter.

The common filter materials used in filter bags are cotton fibre, wool, nylon, orlon fibre glass etc., Filter efficiency depends upon the density and depth of filter packing and the ratio of gas flow volume to filter cloth surface area.

A baghouse filter unit contains several vertically hanging fabric cylindrical bags. The upper ends of the bags are closed, and lower ends are attached to a hopper.

The framework, housing the hanging bags, is provided with an automobile shaking device for cleaning the bags of the collected dust.

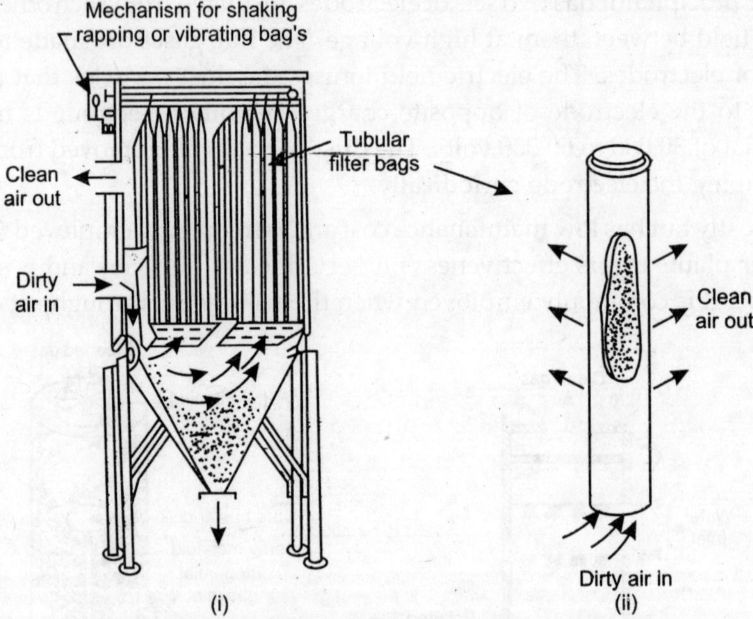

Fig. 20.7. Baghouse fabric filter unit.

20.9. CONTROLLING GASEOUS POLLUTANTS

In order to reduce the pollution load entering the environment from industries, several measures are adopted. Some of them are:

1. **Replacement of burning fuel** by electric or solar energy, as it eliminate the very production of pollution in the combustion process.
2. **Use of better quality of fuel,** such as LPG (liquid petroleum gas) and LNG (Liquefied natural gas) in place of coal, as they will produce less polluted gas in the emissions.
3. **Absorption Units:** Absorption by liquids is a most common method of gas purification. Absorption involves the transfer of a pollution from the gaseous phase to the liquid phase. The process is accelerated by high interfacial surface areas, turbulence, and large, mass diffusion coefficient.

   Absorption units include spray towers and venturi-scrubbers which can simultaneously be used for removing particulate pollution alongwith gaseous pollutants. These units have already been discussed earlier.

4. Adsorption Units

An alternative to and sorption by liquids is adsorption of pollutants on solids. Adsorption is the capture and retention of gas molecules from the gas phase by the surface of the solid adsorbent. The most suited solid for this purpose are : activated carbon, silica gel, activated alumina etc., In these units, the dirty gases are made to pass through the beds of such adsorbent materials, wherein the pollutants are effectively caught and removed. Activated carbon beds can very effectively catch hydrocarbons, H_2S, and SO_2.

5. Combustion or Incineration Devices

The combustion or Incineration equipment may be used to purify polluted gases, when the pollutants in the gas streams are oxidisable to an inert gas. Pollutants, like hydrocarbons, and carbon monoixde (CO) can be easily burnt, oxidised, and removed in such equipments.

Ultimate goal of any combusion operation is the oxidation of the contaminants to CO_2, H_2O vapour, SO_2 and NO_2. The CO_2 and H_2O vapour are harmless, while other two remain a problem unless they can be easily separated from the gas stream.

Both, *Direct flame combustion* by **afterburners** and *catalytic combustion* (20.8), have been used in commercial applications. **Catalytic incineration** can be used when combustible materials in the waste polluted gas, are too low to make direct-flame incineration feasible. This generally happens when the energy of the polluted gas is lesser than 3.7 MJ/m^3. At all such energy contents, the gas burning becomes self supporting after initial ignition, thereby making the use of direct flame incinerators, possible. Secondary supporting fuel may also sometimes be used to make such a method feasible.

It has also to be ensured that the final combustion product from the incinerator is easily disposable and less toxic than the original gas pollutants being removed.

When the combustible fuel value of the polluted gases are lower, some *catalytic materials* have been found to accelerate the rate of oxidation, without themselves undergoing a chemical change, thus reducing the incineration time, or making the incineration feasible even with low-energy polluted gases.

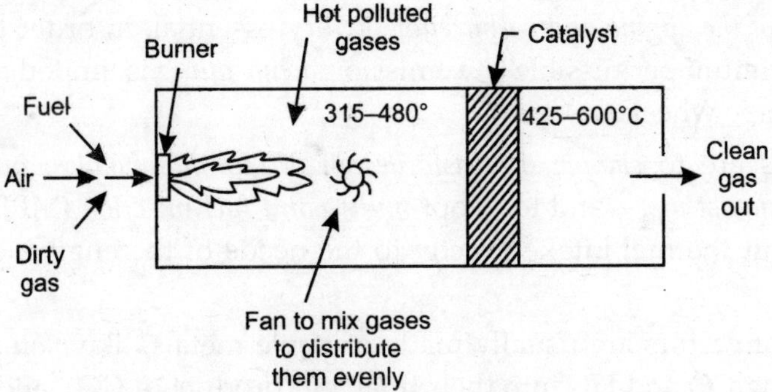

Fig. 20.8. Catalytic Incineration.

Catalytic incinerators have often been successfully used to control SO_2, NO_x, CO, hydrocarbons, etc., Besides being costly, their major drawback is their susceptibility to poisoning by sulphur and lead compounds, even in trace amounts.

20.10. CONTROLLING AIR POLLUTION FROM AUTOMOBILES

Automobile exhaust is a major source of air pollution in urban areas. Pollutants like load, carbon monoxide, sulphur dioxide, oxides of nitrogen particulate, hydrocarbons, etc., are emitted by automobile exhaust. The pollution due to auto-exhaust can be reduced and controlled by following measures:

1. By designing efficient engines, which will be more fuel efficient and emits better quality exhaust gases.
2. By efficient planning of city roads, which will permit faster movement of vehicles.
3. Attachment of exhaust filters to the vehicles will purify the auto-exhaust.

The contents of the pollutants in the smoke emitted by the automobiles depend upon the type and condition of the engine of the automobile.

An ideal internal combustion engine (I.C. engine), when gets *mixture of Air* and Fuel (14.7: 1 ratio) burns them up completely to CO_2, water, and nitrogen, without producing and CO. But in actual practice, two and three wheeler *petrol engines receive high fuel content mixture* (12 : 1 ratio), during cold start, idling, acceleration, and high speed cruising; and similarly *diesel engines receive low fuel content mixture* (16 : 1 ratio) when running at a medium speed. In practice, the combustion of fuel in automobiles, is never complete, the exhaust gases from such engines, invariable contain products of incomplete combustion; *i.e.* CO, HC, H_2, as well as small fractions of nitrogen oxides. These pollutants are highly dangerous to the overall environment, and to the life in general.

The emissions from petrol engines of two, three and four wheelers (including cars, having 4 stroke engines), contain heavier concentrations of HC and CO; whereas, the four stroke diesel engines of diesel vehicles, contain heavier concentrations of NO alongwith thick smoke and particles. The auto emissions also contain gaseous pollutants, like SO_2, and *lead compounds*, especially when **lead containing fuel** is used.

Efforts are onto reduce lead and sulphur content to as low a level as possible in the gasoline. Besides this, the up keep and maintenance of the auto engines, is also of utmost importance. The proper *tuning of the engine and carburettor* is very essential, in order to control pollutant emissions. The maximum permissible CO emission, *while idling* is limited to 3.0% for cars, and 4.5% for two and three wheelers.

The other steps are to *change the basic design of our automobiles, incorporating catalytic convertors in their exhaust pipes,* and to adopt *multi-point fuel injection* (MPFi) *system* with dual intake valves to limit the fuel intake exactly to the needs of the engine, with no excess fuel usage at any point.

The **catalytic convertors** are usually made of noble metals, like *platinum, palladium,* etc., and help in oxidising CO and HC into their final and product of CO_2, and also to reduce NO into nitrogen. These noble metal catalysts are highly active, and resist sulphur poisoning.

The Hon. Supreme Court of India, has already ordered on 29.4.1999 that all the new cars (Petrol as well as diesel driven) to be sold in Delhi and its surrounding towns included in the National Capital Region (NCR) must confirm to the stricter international pollution control norms.

The solution of the problem of increasing vehicular air pollution also lies in using cleaner fuel like CNG (Compressed Natural Gas). That is why, the Hon. Supreme Court of India, vide its order of July 1998, directed for the replacement of all pre-1990 autos (Three wheelers) and taxis in the NCR with new vehicles running on CNG or other clean fuels (LPG). The introduction of CNG run buses and three wheelers in Delhi has certainly helped in controlling the vehicular air pollution of Delhi, but their increasing numbers is not allowing the air pollution to come under full control.

20.11. EFFECTS OF ENCLOSED SPACE ON POLLUTANTS

When an enclosed space is inhabited by human and/or animals, etc.,, the following effects are produced :

1. Reduction in Oxygen Content

Fresh air usually contains 21% of oxygen (by volume)*; whereas, the exhaled air contains only about 17% oxygen. The continuous respiration will, therefore, continuously reduce the available oxygen in an enclosed space. More is the number of occupants respirating in the given space, more deficiency of oxygen will be created. Such oxygen deficits cause headache, nausea, anoxenia, vertigo, etc., In order to avoid such oxygen deficits in a given building, ensure supply of sufficient quantity of fresh air through ventilation.

2. Increase of Carbon Dioxide

Fresh air normally contains about 0.04% of carbon dioxide; whereas the exhaled air contains as high as 50% of carbon dioxide. On an average, about 17 litres per hour of CO_2 is exhaled by an adult human being through respiration. CO_2 is not toxic, and its smaller concentrations upto about 0.04% rather stimulates respiration; but the higher concentrations of about 1.5% cause nausia, depression, and headache. When CO_2 concentration rises to about 2.5% or so, it may extinguish a burning candle. Still higher concentrations of the order of 5% may cause fainting and prove fatal.*To avoid such CO_2 excesses in a building, provide sufficient entry of fresh air, through proper ventilation.*

3. Increase in Temperature

A person, on an average, products about 69 kC of heat per hour. A large part of this heat is lost by leakage through the rooms. Inspite of such leakage, the room temperature may considerably rise, if it is occupied by many persons.

The heat so produced by occupants of the room is either to leak out to the atmosphere, or is to be ultimately absorbed by their bodies. And since a body temperature is to be maintained at 37°C (98.4° F), this extra heat has to be continuously dissipated by the occupants themselves. Higher outside temperatures retard this dissipation, reduces leakage, and thereby causing discomfort to the occupants. If things become severe, it may lead to heat exhaustion, heat cramp, and finally heat stroke. *Continuous supply of fresh air through proper ventilation is, therefore, very necessary to avoid heat effects.*

4. Increase in Humidity

Moisture, in the form of *water vapour*, is released by inhabitants of a building @ about 1.08 kg per day (45 gm per hour). This increases the humidity of the room/building, thereby decreasing evaporation and cooling of the bodies of the inhabitants, and thus causing them discomfort. This moisture, supported by the dust particles, also acts as carrier of germs and pathogenic bacteria from one person to another, causing waterborne diseases, like infuenza, flu, viral fever, tuberculosis etc.,

5. Increase in Organic Matter and Odours

Organic matter and odours are released by human beings from their skins, clothings, and mouths. The increase of these is an enclosed space, occupied by several persons, may cause nausia, headache, and may even aggravate ones existing illness. The *dust particles* present in the room, also help the water vapour in carting organic matter, odours, and diseases from one person to another.

Even serious diseases like scarlet fever; whooping cough, measles, and lung diseases, etc., get spread through crowded buildings and shops. In order to avoid such bad effects of occupancy of a building, it is absolutely necessary to design it in a manner to ensure proper and sufficient ventilation.

20.12. ODOUR AND ODOUR CONTROL

Taste and smell are the two chemical senses. In the case of smell (odour), some individuals have the ability of detect very minute quantities of substances. Unfortunately, the human nose is the only good measuring device known, but it is non-dependable. Since, people have mixed reactions to a given order. What is offensive to one is acceptable to another, as demonstrated by people's varied responses of the different perfumes worn by men and women. There is marked disagreement as to the offensive of selected odours.

Unfamiliar odour is more easily detected and is more likely to cause complaints than a familiar one. Time and place are important psychological factors in a person's reaction to given odours. A tolerable odour in a person's working environment can be intolerable in his home. The odour of freshly baked bread can be delightful, but people who live near a bakery and smell it 24 hours a day may find it less pleasing. The effects of disagreeable odours are primarily nuisance effects.

Some odours at high intensity may lead to nausea. At night, odours can interfere with sleep, Thus affecting a person's well being. Offensive odours from poorly operated dumps, chemical plants, meat packing plants or fish market may cause the value of nearby property to fall.

Types of Odours.

Odours can be classified into following groups :

1. Spicy *e.g.* cloves.
2. Flowery *e.g.* rose.
3. Fruity *e.g.* orange
4. Resinous *e.g.* turpentine or eucalyptus oil.
5. Brunt *e.g.* tar.

6. Foul *e.g.* sulphuretted hydrogen.

From the above, we can define the odour as the "preception of smell". Whether pleasant or unpleasant, odours are induced by in haling air-borne volatile organics or inorganics.

Effects of Odour

Odour affects human beings in a number of ways. Strong unpleasant or offensive smells can interfere with a person's enjoyment of life especially if they are frequent and/or persistant. Major factors relevant to perceived odour nuisance are :

- Offensiveness
- Duration of exposure to odour.
- Frequency of odour occurrence.
- Tolerance and expectation of the receptor.

Though foul odour may not cause direct damage to health, toxic stimulants of odour may cause ill health or respiratory symptoms. Secondary effects in some may be nausea, insomnia and discomfort. Very strong odour can result in nasal irritation, breathing problems or asthma.

Sources of Odour

Most commonly reported odour-producing compounds are hydrogen sulphide (rotten egg odour), and ammonia (sharp pungent odour), carbon disulfide, products of decomposition of proteins (especially of animal origin), phenols, and some petroleum hydrocarbons are common odourants.

Most offensive odours are created by the anaerobic decay of wet organic matter such as flesh, manure, feed or silage. Warm temperatures enhances decay and foul odour production. Various sources of odour can be classified as :

1. Point Sources.
2. Area Sources.
3. Building Sources.
4. Fugitive Sources, such as emissions from soil bed.

Most of the man-made odour sources are : Garbage which lies uncollected for days together, increased sewage production and improper sewage treatment practices produce unpleasant odour. Large livestock operations, poultry farms, tanneries, slaughter houses, food and meet processing industries, and bone mills are major contributors to odour pollution.

Odour Measurement and Monitoring

1. **Odour Intensity.** It is the strength of the perceived odour sensation. Widely used scale for odour intensity is :

0	—	No odour
1	—	Threshold level

2 — Definite odour
3 — Strong odour
4 — Overpowering odour.

For odour intensity measurement, generally a panel of 6 to 12 persons of normal health are employed. The members of the panel sniff the air at a given location at the same time and report individually the nature of intensity of the odour. By averaging the valves recorded be assigned to the odour intensity at a location.

Odour Control Methods

Odour can be controlled by following two approaches:

1. To reduce concentration so that the smell is less intense and therefore less objectionable. In this approach, odour control is achieved:
   - By reducing the source of the odour:
   - Diluting the odour by greater dispersal:
   - Removing the odour from a gas stream by absorption, absorption, or oxidation: or
   - Chemically converting the odour product to one less odorous.
2. To change or mask the quality of the odourant so that the smell becomes more pleasant and acceptable to the populace. In this approach attempts are made to:
   - Hide the original odour by introducing a more powerful odour having pleasing scent, or
   - By modifying the original odour.

20.13. THE AIR (PREVENTION AND CONTROL OF POLLUTION) ACT, 1981

The Origin of the Act lies in the United Nations Conference on the Human Environment held Stockholm in 1972. In this conference decisions were taken to take appropriate steps for the preservation of the natural resources of earth which, among the other things include the preservation of the quality of air control of air pollution.

The **main objectives of this Act** are:

1. To provide for the prevention control, and abatement of air pollution.
2. To provide for the establishment of Central and State Boards with a view to implement the Act.
3. The confer the powers to implement the provisions of the Act on the Boards and assign to them functions relating to pollution.

The **scope of this Act,** is to preserve the quality of air and control of air pollution. Where 'air pollution' have been defined as the presence of air pollutant in the atmosphere.

Other Important Terms as Defined in the Act.

1. **"Air Pollutant"** means any solid, liquid, or gaseous substance (including noise) present in the atmosphere in such concentration as may be or tend to be injurious to human beings or other living creatures or plants or property or environment.
2. **"Central Board"** means the Central Pollution Control Board, Constituted under section 3 of the Water (Prevention and Control of Pollution) Act, 1974.
3. **"Emission"** means any solid or liquid or gaseous substance coming out of any chimney, duct of flue or any other outlet.
4. **"State Board"** means in relation to a State in which the Water (Prevention and Control of Pollution) Act, 1974 is in force and the State Government has constituted for that State.

The **main provisions under this** Act are:

Chapter II (sections 3 to 15) deals with the provisions related to Central and State Boards for the prevention and control of air pollution.

Chapter III (Sections 16 to 18) deals with the Powers and Functions of the Boards.

Functions of Central Pollution Control Board (CPCB)

The main function of the Central Board is to implement the legislations that has been created to improve the quality of air, and prevent and control air pollution in the country. Other functions mentioned in the Act, includes

1. Advice the Central Government on matters concerning the improvement of air.
2. Coordinate the activities of the State Board and resolve disputes.
3. Provide technical assistance and guidance of the State Boards.
4. Plan and organise the training of persons engaged to be engaged.
5. Lay down standards for the quality of air.
6. Collect and disseminate information in respect of matters relating to air pollution.
7. Perform such other functions as may be prescribed in the Act.

Functions of the State Pollution Control Boards (SPCB)

Its important functions are:

1. In advise the State Government on matters concerning the prevention, control or abatement of air pollution.
2. To inspect any control equipment, industrial plant or manufacturing process and to give directions to such persons as it may consider necessary.
3. To inspect air pollution control areas.
4. To lay down standards for emission of air pollutants into the atmosphere from industrial plants and automobiles.
5. To perform such other functions as may be prescribed or entrusted to it by the Central Board or the State Government.

Section 37 to 43 (In Chapter VI) of the Act deals with the penalties and procedures for different offences by companies and Government Departments.

QUESTIONS

1. Enumerate and describe the various natural properties which clean the ambient air (environment) over time by removing the various air pollutants.
2. Enumerate the engineering measures that are adopted to reduce emissions of air pollutants from
   (*i*) Factories, and
   (*ii*) Automobile, giving brief description of such measures.
3. Enumerate and describe briefly (with neat sketches) the various engineering devices that are used to control the emissions of gaseous air pollutants from industries.
4. Write a note on air pollution caused by automobiles and its control.
5. Write notes on the following:
   (*i*) Electrostatic precipitators.
   (*ii*) Spray towers.
   (*iii*) Venturi scrubbers.
   (*iv*) Fabric filters.
   (*v*) Catalytic incineration.
   (*vi*) Cyclone collectors.
   (*vii*) Gravitational settling chambers.
   (*viii*) Dynamic precipitators.
6. Briefly explain the salient features and provisions of the Air (Prevention and Control of Pollution) Act, 1981.
7. State the merits and demerits of bag filter over other devices.
8. How does air pollution affects human health?
9. What is air and its composition, discuss in detail.
10. What is natural self-cleansing properties of air? Discuss in detail.
11. What is ventilation? Discuss its role in controlling the indoor pollution.
12. How and which of the air pollutants affect the plant and animal life? Discuss.
13. Discuss various control measures for air pollution in construction activities.
14. Enumerate the engineering measures that are adopted to reduce emission of air pollution from industries.
15. Discuss the controlling measures for air pollution from automobiles.
16. Mention the national ambient air quality standards of important air pollutants prescribed in India.
17. Explain the principles of working of a cyclone.
18. Discuss bag filter. State its merits and demerits over other devices.
19. Write short note on:
    (*i*) Electric static precipitator,
    (*ii*) Incinetor for the combustion of air pollutants,
    (*iii*) Venturi-scrubber.
20. List the factors that are considered while selecting a particular device for removing particulate matter.

Noise Pollution and Control

21.1. INTRODUCTION

Sound in the environment is caused by vibrations in the air that reach human ears and stimulate a sensation of hearing. When the sound becomes loud, or unwanted, it becomes **noise.** Since the unwanted sound produces several undesirable effects on our body health, it can be termed as an environment pollutant.

Sound is caused by the vibration of molecules and is the periodic mechanical disturbance in fluids and solids. Noise radiates from vibrating surfaces and vibrating gases. Sound travel in pressure waves and is described in two measures :

(*i*) The frequency of the waves that determines the pitch of the sound, and

(*ii*) The amplitude (or magnitude) which is a measure of loudness. The frequency of sound is expressed in terms of cycles per second, and the unit of amplitude is decibel (dB).

Noise can also be defined as that *unwanted sound pollutant, which produces undesirable psychological and psychological effects in an individual, by interfering with one's social activities like work, rest, recreation, sleep, etc.,* The noise, as in air pollutant, infact differs from other pollutants in the sense that it is *transient* in nature, and is not a continuing or persisting phenomenon.

Noise may be defined as unwanted sound at a wrong time and at wrong place. Generally prolonged and loud sound is considered as noise. Meaningful sound is generally meant for inviting attention or excepting a consequent response, such as the cry of a baby. Excessive noise can result in loss of hearing. Today, noise pollution has become a serious health hazard in the big cities, as it disturbs the sleep, increases heart beat, effects respiratory system, dizziness, loss of physical control and other changes due to stress, nausea and vomiting may be caused, deterioration in concentration and even mental disorientation have been noticed temporarily.

Loud continuous noise reduces the working efficiency, interferes with communication increase the frequency of errors in working which may even cause accidents.

21.2. EFFECTS OF NOISE

Effects of Sound at different levels are

135 dB – Painful.

110 dB – Uncomfortable.

88 dB – Hearing impairment on prolonged exposure.

65 dB – Intrusive.

Effects of noise are of following two types :

(*a*) **Physiological Effects.** Immediate permanent hearing impairment may be caused at high levels of about 150 dB. Sound levels in the sewage of range of 120–150 dB, effects on respirating system, dizziness, less of physical control, and other changes due to stress, nausea, and vomiting may be caused. Even sounds of the level of 70 dB may have measurable physiological effects.

(*b*) **Psychological and Other Effects.** Detrioration in concentration and even mental disorientation have been noticed at high noise levels through temporarily. Loud continuous noise reduces the working efficiency, interferes with communication, increases the frequency of errors in working which may even cause accidents.

The various effects of noise can be divided into the following categories :

1. Noise induced annoyance.
2. Noise induced diseases.
3. Sleeplessness.
4. Communication interference.
5. Noise induced hearing loss.
6. Effect of noise on wild life.

1. Noise Induced Annoyance

One of the most important effect of noise in humans is annoyance and irritation due to disturbance. A noise can be said to be annoying when a person does not like same, and hence wants the sound to be put off, or wish to leave the noisy area at the earliest.

2. Noise Induced Diseases

Noise may produce several undesirable physiological and psychological diseases in human beings. The diseases caused may include: *anxiety, tenseness, nervousness, headaches, fatigue, nausea, high blood pressure, high pulse rate, greater perspiration* etc.,

3. Sleeplessness

The noise may induce sleep disturbances including *shorter sleep durations, more frequent awakenings*, etc.,

4. Communication Interference

Noise can badly disturb communication when a person is speaking on telephone; or when individuals are talking face to face.

5. Noise Induced Hearing Loss.

Exposure of human and animal ears to intense noise for a long duration may cause damage to the inner ear, thereby decreasing ones ability to hear. On lower scale, the damage may cause the individuals to complain of tinnitus in the ears, while longer exposure may lead to deafness/permanent loss of hearing.

With the increasing industrialisation and commercialisation of our society, and with consequential increase in noise producing auto-mobiles and machines, noise is becoming a slow poison, and is becoming more and more prominent as a pollutant.

21.3. NOISE MEASUREMENT

Sound is produced in the environment by alternating pressure changes in the air and is caused by the vibration of solid objects. These vibrations cause the surrounding air to undergo compression, then rarefaction, again compression, then rarefaction and so on. Such alternating compression, and rarefaction of the surrounding air produces sound waves which propagate in the form of sinusoidal path.

The time between the successive peaks or troughs of oscillation is called the **period (P)**, and its inverse, which represents the number of times a peak arrives in one second is called the **frequency (*f*)**. The distance between successive peaks or troughs is called the wavelength (λ), which is related to frequency (*f*). Here, frequency of the waves determine the pitch of the sound, and the amplitude (or magnitude) is a measure of the loudness. The frequency of the sound is expressed in terms of cycles per second, and the unit of amplitude is decibel (dB).

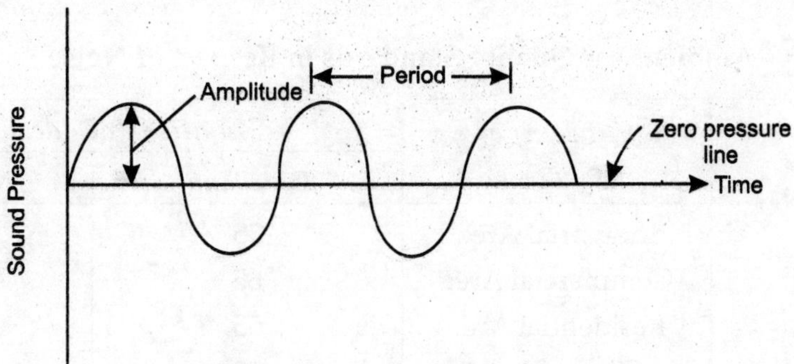

Fig. 21.1. Sinusoidal Sound Wave.

The unit of loudness expressed by the mathematical exponent of number 10 is called bel. One-teeth of a bel (decibel) is taken for quantitative measurements. The decibel scale is logarithmic; each 10-decibel increase represents a tenfold increase in noise intensity. Human perception of loudness also conforms to a logarithmic scale, a 10-decibel increases is preceived as roughly a doubling of loudness.

Thus, 30 decibels is 10 times more intense than 20 decibels and sounds twice as loud; 40 decibels is 100 times more intense than 20 and sounds 4 time as loud; 80 decibels is 1 million times more intense than 20 and sounds 64 times as loud.

The **kind of sound**–whether noisy, musical, abrasive, soothing, displeasing or pleasant depends on a combination of characteristics in addition to pitch, amplitude, and loudness. A vibrating object produces many frequencies even-though, as with a musical instrument.

Speed of sound. Sound travels at different speeds, depending on the density and elasticity of the medium through which it travels. The denser and less elastic the substance, the slower the speed of the sound waves. Sound travels at a velocity of 344 m/sec (*i.e.* 20640 m/min or 1238 km/hr) in air when at a temperature of 20°C. At a lesser temperature, the air is denser and sound waves travel at a slightly slower speed. The velocity increases about 0.6 m/sec for each centigrade rise in temperature.

Note 1. Sound intensity varies as square of the sound pressure.

Note 2. An increase of 10 dB in sound pressure level will correspond to the sound pressure or loudness of sound, increasing by 10 items.

Note 3. One dB is the faintest sound which can be perceived by human ear, and the maximum sound which can be tolerated by human ear is about 180 dB.

21.4. NOISE LEVEL STANDARDS IN INDIA

In order to legally enforce control on noise levels, the Government of India has notified the **Noise Pollution (Regulation and Control) Rules 2000,** under the *Environment (Protection)* Act, 1986. Under these Rules, Ambient Air Quality Standards in respect of Noise, have been specified in respect of different types of areas uses (such as industrial area, commercial area, residential area and silence zones), as shown below:

Ambient Air Quality Standards in Respect of Noise

Area Code	*Category of Area/Zone*	*Limits in dB (decibels).*	
		Day Time	*Night Time*
(A)	Industrial Area	75	70
(B)	Commercial Area	65	55
(C)	Residential Area	55	45
(D)	Silence Zone	50	40

Notes : 1. Day time shall mean from 6.00 a.m. to 10.00 p.m.

2. Night time shall mean from 10.00 p.m. to 6.00 a.m.

3. Silence zone is an area comprising not less than 100 metres around hospitals, educational institutions, courts, religious places or any other area which is declared as such by the component authority.

21.5. SOURCES OF NOISE AND THEIR NOISE LEVEL

Outdoor noise is usually produced by the following sources :

1. *Traffic* including air traffic, and rail and road traffic,

2. *Industries;* and
3. *Other miscellaneous activities.*

All these sources of noise are discussed hereunder:

1. Traffic Noise

Traffic usually produces a lot of noise, particularly in big cities and is responsible for causing a lot of nuisance to the society. The amount and type of noise produced by traffic is largely dependent upon the type of traffic. Say for example, the noise produced will depend upon the type of aircraft or automobile used for the traffic. A jet aircraft produces more noise than a propeller type; a diesel jeap or a truck produces more noise than a petrol car; badly maintained automobiles produces more noise than new and good conditioned vehicles; diesel locomotives produce more than the electric trains; badly designed and maintained rail tracks produce much more noise than those in good condition. Noise levels produced by different types of traffic are given below :

Noise Levels of Different Sources of Traffic

S. No.	*Source of Noise*	*Noise level in dB*
1.	Air traffic	
	(*i*) *Jet aircraft* at take off stage at about 300 m	100—110
	(*ii*) *Propeller type aircraft* at take off stage at about 300 m	90—100
2.	Rail traffic (at about 30 m)	90—110
3.	Heavy road traffic	80—90
4.	Medium road traffic	70—80
5.	Light road traffic	60—70

In order to ensure that the vehicles on roads donot produce excessive noise, the Union Environment Ministry has prescribed noise standards for different types of vehicles, as shown below :

Noise Standards for Different Types of Vehicles

S. No. (1)	*Type of vehicle* (2)	*Noise level in dB* (3)
1.	Two wheelers	80
2.	Cars	82
3.	Passenger or Commercial vehicles	—
	(*i*) upto 4 MT (Tonnes)	85
	(*ii*) between 4 MT to 12 MT	89
	(*iii*) More than 12 MT	91

2. Industrial Noise

Noise is the essential by-product of industry; its intensity and nature being dependent upon the type of work carried out in the industry. Industrial noise are usually produced by rotating, reciprocating or any other types of machinery, or by high pressure high velocity gases, liquids or vapour involved in the industrial processes. The usual noise level of the industries is of the order of 60 to 95 decibels (dB).

3. Noise Produced by Other Sources

Several other human activities, such as blaring of loudspeakers and sirens, shouting of hawkers, playing of children, general life and activity, ringing of temple and church bells, etc., do produce noises of different levels, and tones.

Acceptable Noise Levels in India

S.No.	*Location/Areas*	*Noise Levels in dB*
1.	Rustlling of leaves	10
2.	Rural Areas	25 – 35
3.	Sub urban Areas	30 – 40
4.	Urban Residential Areas	35 – 45
5.	Resiential & Business Urban Areas	40 – 45
6.	City Areas	45 – 55
7.	Hospitals and Libraries	35 – 40
8.	Apartments, hotels, homes	35 – 40
9.	Normal conversation	50 – 60
10.	Restaurants	50 – 55
11.	Schools	45
12.	Commercial areas	65
13.	Siren	100

Noise Pollution in Construction Industry

High level noises are created due to collision or friction of metal against metal, aggregate against aggregate, and metal against aggregate; movement of compressed air into the atmosphere, and the combustion of fuel in the burner. Areas of high level noises are: crushers, burners, blowers, exhaust fans, pug mills, jack hammers, etc., Operators in most plants are therefore housed in cabins, where exposure level to noise is reduced. If an employee is required to work near plant area outside cabin, it should be for short durations, as prolonged exposure in excessive noise is irritating, produce mental fatigue and reduce the efficiency of persons around the plants, especially asphalt plants and crushing plants where noise level is too high.

21.6. TYPES OF SOUND

Sound is generally divided into following three types :

(*i*) Continuous sound.

(*ii*) Intermittent sound.

(*iii*) Impulsive sound.

These are explained hereunder:

1. **Continuous Noise.** This is an uninterrupted sound level that varies less than 5 dB during the entire period of observation. A running fan is an example of such a sound.
2. **Intermittent Noise.** This is a noise which continues for more than 1 second and is then interrupted for more than 1 second. A drilling machine used by a dentist produces such types of sound.
3. **Impulse Noise.** This is characterised by a change of sound pressure of atleast 40 dB within 0.5 second with a duration of less than 1 second. The noise produced from firing of a weapon would fall in this category.

Noise Rating System

As discussed above, a noise may consist of different types of sounds with different pressure levels operating for different time intervals. The frequencies of these sounds may also vary. Combined resultant noise will, in fact, be responsible for determining the human response and degree of annoyance caused by it. The combined impact of different sound pressures lasting for different period is, therefore, worked out, and is usually represented by a term called ***equivalent noise level*** (Leq), which is also called the '***equivalent continuous equal energy level.***

21.7. NOISE CONTROL MEASURES

Following are some of the common measures used for noise control and prevention

1. Replacing noise producing machinery with quiet alternatives.
2. Interrupting the path of sound with insulating materials.
3. Use of baffles, sound proof materials.
4. Mounting the noise and vibration source on flexible supports.
5. Providing conveyors to avoid fall of material on metals from height.
6. Doing proper maintenance and lubrication.
7. Protecting the workers by providing earplug, earmuff, helmets, control cabins etc., Thus, shielding the receiver.
8. Modify the present practices and procedures so as to minimise noise level.
9. Tress are supposed to be the best for noise control, therefore, more and more trees should be planted along the roads, rail tracks and around the aerodromes. These should be planted in plenty in colonies, hospitals, school campuses etc., to protect the people from noise pollution from outside.
10. Plan traffic movement for smooth flow of vehicles.
11. Rigidly fix the machines on concrete blocks and place absorbent material to reduce industrial noise due to impact and vibrations. Steel springs, felt, rubber and cork reduce transmission of mechanical vibrations.

12. Area or rooms producing noise may be segregated so as to create a buffer zone. Sound proof office rooms are desirable.
13. Ensure smooth working of machines for proper repair and maintenance.

21.8. NOISE ABATEMENT AND CONTROL

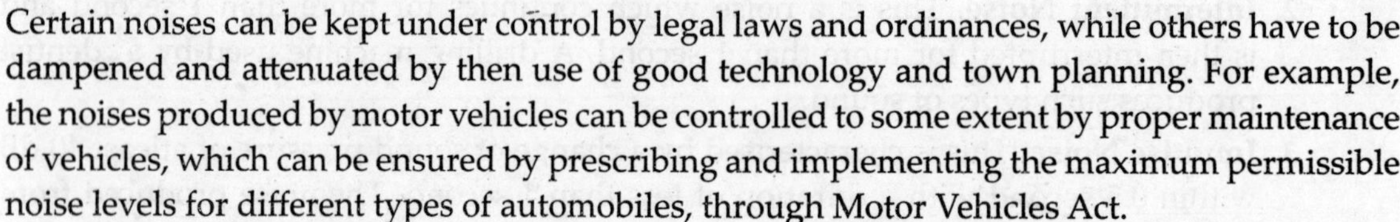

Certain noises can be kept under control by legal laws and ordinances, while others have to be dampened and attenuated by then use of good technology and town planning. For example, the noises produced by motor vehicles can be controlled to some extent by proper maintenance of vehicles, which can be ensured by prescribing and implementing the maximum permissible noise levels for different types of automobiles, through Motor Vehicles Act.

Similarly, industrial noises can be brought under the control of Factories Act by specifying maximum permissible noise levels and other checks. Public blaring of loudspeakers and playing of television sets at loud levels can be prevented by general legal laws of public nuisance or laws specifically made for noise pollution. Other noises are to be reduced by better design technology to be used in the modern day gadgets like fans, air-conditioners, washing machines, refrigerators, mixers and grinders, etc.,

Another method for abating noise effects is to use **proper town planning techniques, for** construction of houses and offices away from the major sources of noise. Proper segregation, **zoning** and separation of residential complexes from the commercial and industrial ones, by means of physical barriers, roads, railway lines, parks or green belts, do constitute good town-planning.

Raising the thick and high vegetation and growing of trees along the sides of roads and railway lines, offer cheaper barriers to cause some noise reduction.

Locating the noisy sources on the downwind side of the residences, may be another important consideration in good town planning, because noise will then travel farther in the downwind direction away from the residences.

From the discussions made above, it is clearly seen that the control of noise can be achieved by adopting following **approaches :**

1. **Reducing the sound produced.** Noise producing sources should be rectified so that intensity of noise is reduced at the source of itself.
2. **Interrupting the path of the sound.** Some barrier is made or the path of the sound is channelised so that desired location can be segregated from the effect of the noise.
3. **Protecting the recipient.** The application of personal protection by using earmuffed or cotton plug, the adverse effect of noise can be controlled.

   Thus, when we consider the noise in the

   (*i*) Industry,

   (*ii*) Community,

   (*iii*) Home, we should keep in our mind that all the problems can be solved by these three possible solutions.

21.9. NOISE AND VIBRATION HAZARDS

As discussed above, vibration is closely associated with noise, and is a potential occupational health hazard, especially in constructional and industrial fields. Workers come to contact with intense, quite forceful physical vibrating tools and machinery.

Employers should provide protection for workers from the harmful effects of noise and vibration by measures including :

(*a*) Replacing hazardous machines and providing less hazardous ones.

(*b*) Reducing the exposure of workers (duration). *e.g.*, in the use of jackhammers, vibratory tools, pneumatic drills, rivetting tools, presses etc.

(*c*) Providing personal hearing protection.

Vibrations in machines or hand operated tools also create noise. The vibration transmitted to the body may injure muscles and joints, and affect blood circulation. While using these tools the workers should wear gloves, which help to cushion the vibrations.

Common causes of vibrations in machines are imbalance, misalignment, bearing wear or its damage, mechanical looseness, coupling problem, gear problems, loosened nuts, bolts and rivets etc., The solution to these problems is regular inspection and removal of defect before it is increased.

Too much noise and vibrations also produce mental fatigue and reduce the efficiency of the worker. Noise for a running machine can be reduced by enclosing the source of noise, use of baffles and sound proof materials etc., Noise and vibrations can also be controlled to some extent by proper maintenance, checking, and proper lubrication etc., Sometimes too much noise also adversely affects the hearing capacity of the workers.

Following are the examples of noise reduction in equipment and workshops :

1. Best position of a sound source is to keep the sound source away from all reflecting surfaces.
2. The motors, pumps etc., should be mounted on heavy bases to avoid noise transmission.
3. All structure borne noise can be significantly reduced by mounting vibration source on flexible supports (springs, cork, foam, rubbers etc.,).
4. As the greater the mass and fall height, the louder the noise, therefore, the convey or should be mounted so that the material falls on edge of the hopper so that the free height is minimised.
5. Exhaust outlets should be fitted with silencers or mufflers.
6. Sound insulating screens must be provided to reduce noise from stationary plant.
7. If the level of noise is high, then appropriate earmuffs or ear plugs should be provided for the persons working near the source of such noise.

QUESTIONS

1. Indicate suitable methods to prevent and reduce noise pollution.
2. Define 'noise' and explain as to how and why it should be regarded as environmental pollutant.
3. Explain in brief the major factors and actions that may help in noise abatement in a modern society.
4. Differentiate between continuous noise, intermittent noise and impuse noise.
5. Discuss various sources of noise, and their typical noise levels, in a modern society.
6. Write a brief note on noise abatement.
7. Write a detailed note on 'noise pollution, its causes, effects, and remedies.
8. Explain in detail about "Ambient Air Quality Standards in respect of noise in India".

PART – 4

Environment Impact Assessment and Some Important Topics

Environment Impact Assessment (EIA)

22.1. ENVIRONMENT IMPACT ASSESSMENT (EIA)

Environment Impact Assessment or EIA can be defined as the study to predict the effect of a proposed activity (project) on the Environment. It aims at establishing quantitative values for selected parameters, which indicate the quality of environment before, during and after the proposed activity. EIA is the study of probable changes in various socio-economic and bio-physical characteristics of the environment, which may result from a proposed or impending action.

The EIA is considered to be a major step in environment protection. To-day it is a formal process practiced all over the world. In 1989 the World Bank adopted EIA for major development projects, in which the borrower country had to undertake EIA under Bank's supervision.

In 1994, the Union Ministry of Environment (Protection) Act 1986, promulgated EIA notification making Environmental Clearance (EC) mandatory for expansion or modernisation of any activity or setting up new projects.

An efficient and honest EIA, besides environmental protection can also help in optimum utilization of resources and saving in time and cost of the project.

22.2. MAIN FEATURES OF EIA

Briefly the main features of EIA include the evaluation of present status and possible changes in environmental components and to find ways to avoid negative impacts of any proposed action. It also envisages to find ways to enhance positive effects of the project.

It is worth noting that even though all major, medium and even minor projects would have both Negative and Positive effects; the word **impact** in the context of EIA has developed almost exclusively a negative connotation.

22.3. EIA PROCESS

The EIA process is an interdisciplinary and multistep procedure to ensure that environmental considerations are included in decisions regarding projects that may impact environment.

The stages of an EIA-Process will depend upon the requirements of the country or donor. However, most EIA processes have a common structure. Normally, the Environment Impact Assessment consists of eight steps with each step equally important in determining the overall performance of the project. These eight steps are briefly described below :

(1) **Screening :** In this first stage of EIA it is determined whether the proposed activity/ project requires an EIA and if it requires an EIA, then the level of assessment required.

(2) **Scoping :** This stage identifies the key issues and impacts that should be further investigated. This stage also defines the boundary and time limit of the study.

(3) **Impact Analysis :** Here EIA identifies and predicts likely environmental and social impacts of proposed activity / project and evaluate their significance.

(4) **Mitigation :** This step in EIA recommends the actions to reduce and avoid the potential adverse environmental consequences of developmental activities.

(5) **Reporting :** This stage represents the result of EIA in a form of a report to the decision making body and other stakeholders and interested parties.

(6) **Review of EIA :** It examines the adequacy and effectiveness of the EIA report and provides additional information necessary for the decision-making.

(7) **Decision Making :** It decides whether the project proposal has to be rejected, approved on need some improvement.

(8) **Post-Project Monitoring (PPM) :** This stage comes into play once the project is commissioned and starts functioning. It checks whether the negative impacts of the project do not exceed the legal standards and implementation of the mitigation measures are in a manner as described in the EIA-report.

22.4 METHODOLOGIES FOR ENVIRONMENT IMPACT ASSESSMENT OF ANY PROPOSED PROJECT

Some important methods currently used for EIA-evaluation are summarised below. These methods might vary from simple to complex *i.e.* easily visible to higher order impacts.

22.4.1. Ad hoc Methods

Ad hoc methods indicate broad areas of possible impacts by listing composite environmental parameters like-flora and fauna, lakes and forest likely to be affected by the proposed activity. It does not define any specific parameters within the impact area.

22.4.2. Checklist Method

It is the most common formal method used for impact identification. It indicates and explains the environmental features that need to be considered in any project or activity.

The checklist is either 'simple' or 'structured'. In the structured checklist impacts are weighted. In weighted checklist importance is attached according to severity of each parameter.

Below is given (Table -1) examples of a simple and structured checklist for a land Development Project.

Table. 22.1

S.N.	Item	Impacts	Weightage
1.	Local Economy	• Public Fiscal Balance • Employment	A or 4
2.	Natural Environment	Air, Water Quality, Flora, Wild Life, Natural Disasters	B or 3
3.	Services Public / Private	Water Supply, Medical care, Education, Electricity, Transport, Housing, Shopping, Security, Entertainment	B or 3
4.	Aesthetics	Attractiveness, Landmarks	C or 2
5.	Social Impacts	Displacement, Social Interaction, Friendliness, Overall Contentment	C or 2

Both these lists are subject to change and can be improved to suit local conditions as experience with its use is gained. For weightage also scale A to D or 4 to 1 can be chosen. Other weightage like 1 to 10 etc., can be chosen by the analyst.

22.4.3 Matrix Method

The matrix methodology incorporates both a list of project activities and a checklist of potentially impacted environmental characteristics. The two lists are then related in a matrix which identifies cause and effect relationship between specific activities and impacts.

Matrix methodology may either specify which actions impact which environmental characteristics, or it may simply list the range of possible actions and characteristics in an open matrix to be completed by the analyst. Matrix method is versatile, as it can be used for small as well as large-scale projects alike. They require information about both the environment components and project activities. The **cells** of matrix are filled in using experts' subjective judgement, or by using extensive data base.

Example of a Simple Matrix

This is simply a list of environmental aspects listed along the vertical axis, against which we determine whether an activity would have an adverse, (negative) effect, no effect or beneficial (positive) effect. A simple mark (✓) is made under the appropriate column. (Table -2).

Table. 22.2

S.N.	Activity	Adverse Effect	No Effect	Beneficial Effect
1.	Land transformation		✓	
2.	Land Use			✓
3.	Water Resources	✓		
4.	Air Quality	✓		
5.	Noise & Vibration	✓		
6.	Transport System			✓

Alternately, the activities can be listed along the horizontal axis with the environmental aspects in the vertical axis and a mark (✓) is given to the pairs that interact between themselves. Here is an example of a matrix prepared for a Pulp and Paper industry (Table 3). This is a table of simple interaction matrix. An improved version of this is the Significance or Importance Rated Matrix.

In Importance Rated Matrix the cells are divided into two parts, the upper part showing the magnitude of the effect on environment and lower portion showing importance of the effect.

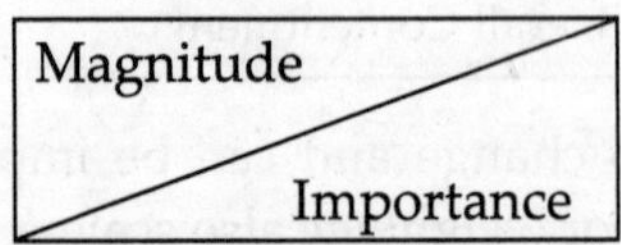

Table. 22.3

S.N.	Environmental components ⇩	Project Activities								
		Plant constriction	Farming of Kenaf	Use of Pesticides & Fertilizer	Transport of Raw Material	Water Intake	Solid Waste	Effluent Discharge	Emission	Employment
1.	Surface Water Quality			✓			✓	✓		✓
2.	Surface Water Hydrology					✓				
3.	Air Quality				✓				✓	
4.	Fisheries			✓				✓		
5.	Terrestrial Wildlife Habitat	✓								

6.	Terrestrial Wildlife	✓								
7.	Land Use Pattern		✓							
8.	Highways / Railways				✓					
9.	Water Supply			✓				✓		
10.	Agriculture		✓							
11.	Housing									✓
12.	Health						✓	✓	✓	
13.	Socio-Economic									✓

The magnitude and Importance are evaluated on a chosen scale and assigned value like A or 4, B or 3, C or 2 and D or 1 in decreasing order. The analyst can chose their own range of scale.

Referring to Table 3, (row 1) if we want to assign the magnitude and importance of impacts, it may look like as follows,

Project Activities

S.N.	Environmental Components	Plant constriction	Farming of Kenaf	Use of Pesticides & Fertilizer	Transport of Raw Material	Water Intake	Solid Waste	Effluent Discharge	Emission	Employment
1.	Surface Water Quality			A / B			C / B	B / B		

22.5. THE LEOPOLD MATRIX

This is one of the very famous EIA matrix. Leopold Matrix (1971) analyses 100 specific actions and 88 environmental components. Out of these relevant components and actions are chosen for a particular project. Both, the magnitude and significance, are rated on a scale of 1 to 10. This is based on analyst's decision and experts' opinion. Below is an example of water reservoir project Table (4). Showing magnitude and significance of predicted impacts on scale 1 to 10. The upper left of the cell shows magnitude of interaction and lower right the magnitude of importance.

22.6. SOME OTHER TYPES OF MATRICES

(1) **Modified Graded Matrix (Lohani & Than 1980) :** In this each activity is assigned a relative value based on its priority level in the developmental activity. This method is particularly helpful when it comes to identifying activities that have a higher potential for causing damage in the developmental site and specify areas that need most attention.

(2) **Impact Graded Matrix :** This type of matrix can identify potential impact areas, specify mitigation measures and even list out the agencies that can aid in the mitigation of these impacts.

(3) **Loran Matrix :** Unlike the Leopold Matrix, which has 100 activities and 88 impacts, this matrix has 234 developmental activities and only 27 impacts to be assessed. Rest of the procedure is similar to Leopold Matrix.

Table. 22.4

Proposed Action / Resources	Immigration of Labour	Dam Construction	Transmission Line	Reservoir filling	Heavy Metal Discharge	Growth of Aquatic Weeds	Relocation of Inhabitants	Leopold Method Total
Health	5 / 8	4 / 6		5 / 8	4 / 7	6 / 6		24 / 35
Biodiversity		3 / 4		3 / 6	3 / 7	5 / 5		14 / 22
Archaeological Artifacts	4 / 6			8 / 8				12 / 14
Tourism			7 / 6	7 / 6				14 / 12
Downstream Water Pollution		7 / 7		7 / 8	2 / 4			16 / 19
Social & Economic Aspects							8 / 7	8 / 7
Forestry		4 / 2						4 / 2
Fishery		2 / 5			2 / 5			4 / 10

Navigation				6/5				6/5
Aquatic Plants				6/6				6/6
Leopold Method Total	9/14	20/24	7/6	42/47	11/23	11/11	8/7	

22.7. MATHEMATICAL MODELLING

Mathematical models are modern tools for prediction work. However, this method can make accurate prediction, only when the input data is reliable and adequate. Here it is to be realized that model output is not necessarily an end in itself but may be an input for assessing changes in economic, social and ecological terms.

This method requires expert handling, computer knowledge and scientific understanding.

22.8. EXPERTS' ADVICE

Expert advice should be sought for predictions which are inherently non-numeric and particularly suitable for estimating social and cultural impacts. In this local experience provides invaluable insight. Expert opinion is also likely to be needed to assess the implications of any modeling predictions.

22.9. ECONOMIC STUDY/PROJECT APPRAISAL

It is important to realize that environmentally sound development bring long-term economic benefits. Unfortunately, short-term gains are often given priority.

The most common method of project appraised followed is the Cost-Benefit or Benefit-Cost (B.C) or Cost-Effectiveness Analysis. However, it has not been found easy to incorporate environmental impacts into the traditional cost-benefit analysis, primarily because of the difficulty in quantifying and valuing environmental effects and emotional consequences.

A better EIA can provide some information on such effects and quantify, to some extent, their importance. This information can be used in BC analysis. Cost effective analysis can be used to determine which is the most efficient and least-cost method of meeting environmental objectives.

Valuing the environment raises complex and controversial issues. Environmental health effects present similar problem. Some effort has been done to translate these issues in monetary terms.

This would require to classify the benefits and losses into tangible and intangible groups *i.e.* which can be expressed in monetary terms or otherwise.

22.10. FINAL REPORT

The final report of an EIA is often referred as an Environmental Impact Statement (EIS). The EIS "is the detailed analysis that serves to ensure that the policies and goals defined by National Environmental Policy (NEP) are infused into ongoing programmes and actions of various agencies including the governments."

EIS_s are prepared for projects that the proposing agency views as having significant prospective environmental impacts.

EIS should provide a discussion on significant environmental impacts and reasonable alternatives (including a non-action alternative) which would either avoid or minimize adverse impacts or enhance the quality of human environment.

Experts preparing an EIA report must be able to appreciate that the final report will be available to a wide range of people, hence EIS should be concise, clear, to the point, and shall be supported by evidence that the agency has made the necessary environmental analysis.

The World Bank (WB)-1991 has prescribed following points to be covered in the main text of an EIS.

1. A description of the programme.
2. A summary of the EIA methodology.
3. The policy, legal and administrative framework.
4. A summary of Baseline data
5. A description of Government and Non-Government Organizations (NGO_s) involved.
6. Environmental Impacts.
7. The Environmental Action Plan including Mitigation.
8. Recommendations and guidelines to decision makers.
9. A statement of provision of auditing.
10. Appendix.

    The appendix should contain,

    (a) Glossary of terms/units

    (b) List of team members involved.

    (c) Records of public meetings and consultations

    (d) Sources of information

    (e) Photographs.

22.11. ENVIRONMENTAL MITIGATION

It describes the programmers to offset known negative impacts to any existing natural resource such as a stream, wetland, endangered species, archaeological site or historical structure/ monument.

To mitigate means to make less damaging. Environmental mitigation is typically a part of an environmental crediting system. It involves allocating debits and credits. Debit occurs when a natural resource is damaged or destroyed, and credit occurs when it is improved or preserved.

Obviously, the success of mitigation depends upon the correctness of EIA.

22.11.1. Mitigation Measures

Mitigation measures are means to prevent, reduce or control adverse environmental effects of a project, and include restitution for any damage to the environment caused by those effects though replacement, restoration, compensation or by any other means.

For different project phases, special mitigation measures have to be adopted. For example during construction there may be excess dust emission for which water spray is an effective measure. On the close of construction operation use of low sulphur fuel may help in controlling the air pollution. During the operation of the project dust collection system and emission control devices have to be brought in. Similarly, for noise control standard distance between noise source and receptor to be maintained.

22.12. ENVIRONMENTAL AUDIT

In simple terms environmental audit is an assessment of the extent to which an organization is observing practices which minimize harm to the environment. The ISO14000 has defined Environmental Audit as, "A systematic, documented verification process of objectively obtaining and evaluating audit evidence to determine whether specified environmental activities, events, conditions, management systems, or information about these matters conform with audit criteria and communating the results of this process".

22.12.1. Objectives of Environmental Audit

The overall objective of environmental auditing is to help safeguard the environment and minimise the risk to human health. The key objects, therefore, are:

(a) Determine how well the environmental management systems and equipments are performing.

(b) Verify compliance with relevant laws, and

(c) Minimize human exposure to risks.

22.12.2. Scope of Audit

Audits can address one topic, or a whole range of issues. The greater the scope, more elaborate should be the organization, the time spent as well as the depth of investigation.

22.12.3. The Typical Audit Process

A typical audit process contains three steps, as described below, (*i*) **Pre-Audit Steps (Activities):** It includes assigning the priorities and their selection criteria; selection of Audit Team, their availability, their travel and lodging arrangement and assign their responsibilities; making of audit plan which should define scope, prepare questionnaire; staff and resources requires.

(*ii*) **On site steps (Activities at the site):** In this steps following activities are performed:- Review of background information; review of audit plan; assess management strength; collect data; ensure protocol steps are completed; ensure reliability of findings; evaluate audit findings and prepare report; discussion with stakeholders on findings and gather audit evidence.

(*iii*) **Post-Audit steps (Activities):** Here the main job is to issue draft report and collect feedbacks; improvement in the draft report and highlight the requirements for action plan and fix the deadline and finally-follow up action plan.

22.12.4. Benefits of Environmental Audit

Benefits of any Environmental Auditing depends upon the objectives and scope of the audit. A properly implemented environmental audit plan provides a range of benefits for an organisation. These are summarized below:

(1) It provides a framework measuring environmental performance of the project.
(2) It fixes accountability on environment managers.
(3) It brings accuracy to the outcome.
(4) It provides valuable information for future planning.
(5) It reduces uncertainties in predictions.
(6) It helps in mitigation efforts.
(7) It brings saving of time and money.

In short, chief among the benefits from environmental auditing are the detection of compliance problems before these problems pose serious liabilities, cost-saving through operating efficiency, and reduced risks from environmental hazards.

It may be noted that non-compliance with environmental laws and regulations can impose fines, penalties, legal liabilities and even criminal sanctions for responsible officials.

22.12.5. Some disadvantages of Environmental Auditing

The main risk in audit programme is chances of making wrong conclusion. Some of the limitations are,

(*a*) Extra cost: It is additional burden on the organization.
(*b*) Periodic and frequent changes in rules and regulations.
(*c*) Administrative and legal problems involved in enforcing remedial measures.

QUESTIONS

1. Define Environment Impact Assessment. What are its main features ?
2. Describe EIA Process.
3. Differentiate between a simple and a structured checklist.
4. Explain Matrix method of EIA.
5. Write a note on Leopold Matrix.
6. How Project Appraisal is part of EIA ?
7. According to World Bank. What points should be covered in the Final Report ?
8. What is meant by Environmental Mitigation ?
9. What are the objectives of Environmental Audit ?
10. What are the advantages of Environmental Audit ?

Some Important Topics

23.1. LIGHT POLLUTION

The definition of **light pollution,** also known as **photo pollution** or **luminous pollution**, is the excessive, misdirected or invasive use of artificial outdoor lighting.

Luminous pollution is caused by using outdoor lights when and where they are not necessary. Poorly designed residential, commercial, and industrial outdoor lights also contribute significantly to light pollution. It is estimated that nearly 30% of outdoor lighting is wasted due to the poor design.

Light pollution has dire effects on our environment and energy resources as well as wildlife ecology and astronomical research. It also affects the quality of life and safety of humans. Some of the effects are described below:

Effect on Environment: The International Dark-Sky Association (IDA) estimates that excessive night time lighting releases more than 12 million tons of carbon dioxide each year. According to IDA it would take nearly 702 million trees to absorb the CO_2 Produced by wasted light.

Wildlife: Luminous pollution affects the feeding, sleeping, mating and migration cycles of all wildlife.

Mammals: Mammals such as bats, raccoons, coyotes, deer, and moose can experience difficulty in foraging for food at night due to over illumination. There is increased risk due predators. A decline in reproduction is also noticed.

Birds: Birds like owls nighthawks use moonlight and starlight to hunt and migrate at night. Artificial light sources can overwhelm natural light sources which results in deviation in their migration route.

Amphibians, Reptiles and Insects also get affected in different ways. However, **astronomers** is the group of people most affected by light pollution. Astronomy is one of the oldest sciences in human history, and has made countless contributions to technology with applications like Personal Computers (PCs), communication satellites, mobile phones, Global Positioning Systems (GPS), Solar Panels and Magnetic Resonance Imaging (MRI) Scanners. Today, astronomy helps us determine the Sun's effect on climate change, and identify any potential threats to Earth from space. In order to conduct observation and research, astronomers require dark skies.

Light pollution affects humans causing various health problems, including sleep disorders, anxiety, depression, diabetes, some types of cancer, cardiovascular diseases, immunological disorders, and obesity.

Hormones regulates the sleep and wake cycle, is acutely disturbed by light pollution. Hormone is activated by darkness and repressed by light.

Specific Categories of Light Pollution are:

- Over Illumination.
- Glare – Occurs due to unshielded lighting.
- Sky glow – Refers to the combination of all reflected light and upward light escaping into the sky.
- Light Trespass – It occurs when unwanted light enters one's property.
- Other sources include – Vehicular light, welding spark, lightening etc.,

23.2. VISUAL POLLUTION

Visual pollution is the term given to unattractive and man-made visual elements, a landscape, or any other thing that a person does not feel comfortable looking at. It, therefore, is an aesthetic issue and it impairs one's ability to enjoy a vista or view. It creates harmful changes in the natural environment. An overcrowded area causes visual pollution.

The main sources of visual pollution are-excessive and overcrowded advertisements, telecom and electric wires, poles, towers, signboards, billboards, posters and hoardings, unproductive deforested land, badly designed buildings and huge constructions, smoke spewing chimneys, graffitti etc.,

In India roadside butchering of animals and birds are a major cause of visual pollution. Stray cows and stray dogs and their easing out on the street is also a problem related to visual pollution.

23.3. NUCLEAR/RADIOACTIVE POLLUTION

> *"Radioactive pollution can be defined as the release of radio-active substances or high-energy particles into the air, water or earth as a result of human activity, either by accident or by design."*

Radioactivity : It is defined as the 'spontaneous emission of radiation in the form of particles or high energy photons resulting from a nuclear reaction.'

Radio-active Elements : A radio-active element is an element whose nucleus breaks down and it gives off energy. Examples of such elements are Radium (Ra), Uranium (U) and Radon (Rn).

Radio-active atoms are unstable; that is they have too much energy. When the nucleus is disturbed they release extra energy- here they are said to decay. After releasing all of their energy, the atom becomes stable and are no more radio-active.

Nuclear Fallout : Radio-active pollution that is spread throughout the earth's atmosphere is termed as **fallout**. It is also known as **Black Rain**. It is the residual radioactive material propelled into the upper atmosphere following a nuclear blast or a nuclear reaction conducted in an unshielded manner.

The two best known examples of fallout are the bombing of Japanese cities Hiroshima and Nagasaki by America in 1945 and the Chernobyl Nuclear Power Station disaster in 1986. This created havoc killing lacs of people and promoting dangerous diseases.

Nuclear Radiation : These are the particles and photons emitted during reactions that involve the nucleus of the atom. It may also be defined as the emission or transmission of energy in the form of waves or particles through space or through material medium. The rays or particles are named as rays, x-rays and neutron radiation.

Sources of Radiation : Since the beginning of time, all living things have been, and are still being, exposed to radiation. However, most of us are unaware of all the natural and man-made sources of radiation in our environment.

The chart below shows the public's exposure to ionizing radiation which is about 620 millirem. Of this total natural sources of radiation accounts for about 50%, while man-made sources account for the remaining 50%.

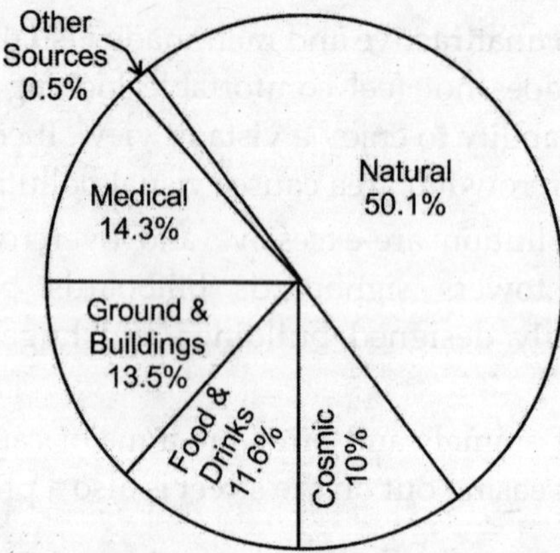

Source : National Physical Laboratory.

Fig.23.1.

Natural Sources of Radiation : The Earth itself is a source of terrestrial radiation. Radioactive materials like Uranium, Thorium and Radium exist naturally in soil and rocks. Essentially all air contain Radon, which is responsible for major dose we receive. It comes from decay of natural Uranium. Our body also contains material called radio-nuclides.

Man-made or Human Sources of Radiation : Ever since radiation was discovered, people have benefitted from its use in medicine and industry. Today man-made sources of radiation accounts for 21% in developing countries and rest in industrialized countries.

Radiation used in medical applications is the largest man-made radiation. Majority of this exposure is from x-rays. In the field of nuclear medicine radiopharmaceuticals are used for diagnoses.

Other sources of man-made radiation include products such as : Building and road construction materials, combustible fuels, x-ray security systems, televisions, smoke detectors, fluorescent lamp starters, luminous watches, tobacco, ophthalmic glass, ceramics and of course the mobile phones and transmission equipments. Testing of Nuclear weapons has exposed many areas to nuclear radiation.

Biological Effects of Radiation : Whether the source of radiation is natural or man-made, whether it is a small dose of radiation or a large dose, there will be some biological effects.

All biological effects begin with the consequence of radian interactions with the atoms forming the cells. Biological symptoms of radiation on human health are divided as follows:

(i) The acute symptoms due to short term intense exposure : These symptoms occur within few days or weeks. When radiation levels are near 100 rems, it can cause blood cancer (leukemia). These are the result of accident or catastrophe like atomic bomb explosion or leakage.

(ii) Effect of long term, low level radiation exposu.. : These symptoms occur a long period exposure and are easily distinguished from common diseases. An example is that of Marie Curie and her daughter Irene Joliot Curie who died of cancer caused due to constant exposure to radio-active elements during research.

Radiation may also cause genetic damage which can even be passed to the future generations.

23.3.1. Direct and Indirect Effects of Radiation

Ultimately radiation affects the **cells**. These effects may be Direct or Indirect.

Direct Effects : When radiation interacts with atoms of DNA (Deoxiribo-Nucleic-Acid) molecules or some other cellular components critical to survival of cell, it is said to be direct effect. Such an interaction may affect the ability of the cell to reproduce and thus, survive.

Indirect Effects : Cells are made up largely of water. Interacting with radiation water may break into Hydrogen (H) and Hydroxyls (OH). These may recombine to form Hydrogen Peroxide (H_2O_2) which destroys the cells.

23.3.2. Control of Radioactive Pollution

The realization of the bad effects of radiation hazards in recent years made people to think for protection against the radiation hazards. Global radiation level which was 45 mrem per year has been raised to 155 mrem per year. Though there is no control on natural sources of radiation but definitely earnest efforts can be made to decrease the level of pollution from anthropogenic sources.

Preventive measures recommended are listed below,

(1) Nuclear devices should never be exploded in air. If unavoidable they should be exploded undeground in desert areas.

(2) Production of radio-isotopes should be minimized.

(3) Containment may be employed to decrease the radioactive emission. It can be done by using tightly sealed boxes for storing.

(4) In nuclear and chemical industries, the use of radio-isotopes may be carried under a jet of water (or soil) instead of powder or gaseous forms.

(5) Nuclear medicine therapy should only be applied whenever absolutely necessary.

(6) The dose to which an individual is exposed should not exceed the recommended value.

*(**Note** : rem : roentgen equivalent man. It is unit of radiation dosage. 1 rem = 100 erg/g. It is observed that 1 rem carries 0.055% of cancer risk)*

23.4. SMELL / ODOUR POLLUTION

Smell pollution or **odour** pollution is also an indicator of changes in environment which adversely affect human beings in particular.

Odour can be defined as the "perception of smell" or in scientific terms as "a sensation resulting from the reception of stimulus by the olfactory sensory system." whether pleasant or unpleasant, odours are induced by inhaling air-borne volatile organic or inorganic substances. Odour is subjective, what is offensive to one person may be acceptable to another. However, the smell has the connotation of bad odour which causes discomfort and distress.

Sources of Odour : Most commonly reported odour-producing compounds are hydrogen sulphide (having rotten egg odour) and ammonia (with sharp pungent smell). Carbon disulphide, mercaptans (pungent smelling gas), product of decomposition of proteins (specially of animal origin), phenols and some petroleum hydrocarbons are other common odorants.

More offensive odour is created by the anaerobic decay of wet organic matter such as flesh, manure, fodder or silage (green wet crop). Odour can arise from many sources. Most of these are man-made *e.g.* Garbage, Improper dumping garbage on vacant lands is a common phenomenon. It leads to foul smell due to putrefaction of dumped garbage. Unscientific design of landfill, increased sewage production and improper sewage treatment practices produce unpleasant odour.

Large livestock operations, poultry farms, tanneries, slaughter houses, fish markets, food and meat processing industries and bone mills are among major contributors to the odour pollution. Agricultural activities like decaying of vegetation, production and application of compost etc., also contribute to the menace.

Effects of Odour on Human Health : Odour may affect human beings in a number of ways. Strong unpleasant, or offensive smells can interfere with a person's enjoyment of life especially if they are frequent and / or persistent.

Toxic stimulants of odour may cause ill-health and damage the respiratory symptoms. Secondary affects, in some, may cause nausea, insomnia and discomfort. Very strong odour can result in nasal irritation; trigger symptoms in individuals with breathing problems or asthma.

23.4.1. Measurement and Monitoring of Odour

Odours are measured adopting **Olfactometric** testing methods, which are psycho-physical methods. The odour measurement by alfactometer can be divided into two parts :

(a) Determination of the Threshold Concentration of Odoriferous Gases: For determining threshold concentration by olfactrometry testing procedure, a diluted odorous mixture and an odour free gas (as a reference) are presented separately to eight panelists in succession.

In comparing the gases emitted from each port, the panelists are asked to report the presence of odour. Using penalist responses threshold concentration can be estimated.

(b) Determination of the Type and Intensity of Odour: For odour intensity measurement, generally a panel of 6 to 12 persons of normal health is employed. The panel members smell the air at a given location at the sametime and report individually the nature and intensity of the odour. By averaging the values of penalists a single value is assigned to the odour intensity at a given location. Generally odour intensity increases with odourant concentration. Odour intensity (*P*) and odorant concentration (*S*) are related as follows,

$$P = \text{k } 10 \text{ g } S. \text{ where } K \text{ is a constant}$$

Odour Unit : The point at which 50% of the penalists cannot smell the odour and 50% can, is called the **Perception Threshold** and is equal to 1 odour unit per cubic meter.

23.5. SMOG (COMBINATION OF SMOKE AND FOG)

Photochemical SMOG is a classical example of a secondary air pollution. Smog reduces visibility and effect plants, animals and human life. Three types of smog has been identified.

(1) Classical or Sulphurous Smog (London Smog): It consists carbon base soot and sulphur dioxide. In humid air, water condenses on carbon particles and forms an irritating air mass termed as smog.

(**Note:** A major disaster occurred in London in 1952 killing 12,000 people due to sulphurous smog)

(2) Photochemical Smog (Los Angels Smog): It is made up of unburnt gaseous hydrocarbons and oxides of nitrogen, both of which are present in the emission of internal combustion engines (motor vehicles). These chemicals play an important role in the formation of ozone (O_3) and Peroxyacytyl Nitrate (PAN). Los Angeles, California (USA) first experienced a serious air pollution in the year 1944. Many people died and thousands got sick.

(3) Winter Smog (Delhi Smog): This is caused by collection of particulate matter (very fine type of dust and toxic gases) in the air due to stagnant movement of air during winters. According to an estimate about 10,000 people lose their lives every year in Delhi, one of the most polluted cities.

The cause of winter smog is **thermal inversion.** (**Note:** Normally the colder air sinks below and warmer air rises up, carrying low-lying pollutants higher into the troposphere. Sometimes, however, a layer of dense, cool air beneath can be trapped below layer of less dense warm air, causing **temperature inversion**. Consequently, pollutants are not dispersed and enhance harmful effects of air pollution.

23.6. ACID RAIN

Acid rain formation is another classical example of a secondary air pollutant. When coal, petrol or natural gas is burnt in power plants, factories and automobiles, sculpture dioxide and nitrogen oxide are produced. A large part of these oxides entering the atmosphere react with water and other chemicals in the air to form sulphuric acid (H_2SO_4) and Nitric Acid (HNO_3) alongwith some other harmful pollutants like sulphates and nitrates. The process of formation of acids is hastened due to presence of catalysts like metal ions of Mn_2, Fe_3, Ni_2, Cu_2 in water droplets. Reactions in simple form are mentioned below,

$$2SO_2 + O_2 \rightarrow 2SO_3$$

$$SO_3 + H_2O \rightarrow H_2SO_4$$

and

$$NO_2 + NO_3 \rightarrow N_2O_5$$

$$N_2O_5 + H_2O \rightarrow 2HNO_3$$

Traces of Hydrochloric Acid (HCl) are also found in acid rain.

Normal rainwater is slightly acidic with pH = approximately 6. It may be 4 in some cases and minimum reported had been down to 2.4.

The Ill-effects of Acid Rain : Acid rain may cause extensive damage to materials and terrestrial ecosystems such as water, fish, vegetation, soils, buildings, monuments etc., Five negative effects of acid rain are described below.

(1) **Effect on Human Beings:** Acidification effect human nervous system, respiratory system and digestive system. The heavy metals released by acid rain may cause potential threat to human health.

(2) **Effect on Aquatic Biota:** Due to acidification of water bodies, many bacteria and blue green algae are killed disrupting the whole ecological balance. Freshwater lakes are fairly alkaline, due to acidification phytoplankton and zooplankton are affected. Snails, oysters etc., have their shells of calcium carbonate can die due to acidification.

(3) **Effect on Terrestrial Ecosystem:** Acidification of soil adversely affects soil fauna and lead to reduced forest productivity. Demineralization of soil occurs as calcium, magnesium, sodium, potassium cations are replaced by acid cations like aluminium etc., Also nitrogen fixation ability of bacteria diminishes rapidly.

(4) **Injury to Vegetation:** Exposure to low level of acidic water causes diffuse **Chlorosis** in leaves. Acidic water containing sulphur dioxide (SO_2) affects the functional cells near the stomata.

(5) **Effect on Buildings:** Acid rain causes extensive damage to buildings and structural material of marble, limestone, slate etc., The attack of acid rain on marble is termed as **stone leprosy**. Acid rain corrodes houses, monuments, statues, bridges, fences, railing etc., It may also damage oil based paints and automobile coatings.

Prevention of Acid Rain Pollution: Some small steps can go long way in reducing the damage caused by acid rain, briefly described here.

Energy Conservation: We should restrict our energy consumption by using cooling and heating equipment only whenever and wherever needed.

Transportation: Automobiles are the major contributors to acid rain pollution. We should find alternative modes of transportation in order to decrease the use of fossil fuels.

Alternative fuels: An excellent way to prevent acid rain is to stop using non-renewable fuels and use renewable sources of energy like solar, hydro and wind energy.

23.7. GLOBAL OZONE PROBLEM

Ozone is a tri-atomic form of oxygen (O_3) which occurs naturally throughout the atmosphere. It is blue in colour and has a Pungent Smell.

Only a trace of O_3 exists near the Earth's Surface. Its maximum concentration (about 10 ppm) is in the stratosphere (15 to 50 km above earth's surface).

Ozone is formed in the stratosphere by a photochemical reaction occuring under the influence of solar radiation as given below.

$$O_2 + h_v \rightarrow O + O \text{ (photolyses)}$$

(from UV radiation)

$$O + O_2 \rightarrow O_3$$

Ozone is also formed and released at the ground level in a very small quantity of about 0.05 ppm by motor vehicles and factories, during the burning of fossil fuel.

Close to the ground it is a harmful pollutant for plants, animals and human beings. It can cause chest pain and irritate the respiratory tract causing lung infections.

In stratosphere, ozone strongly absorbs ultraviolet radiation coming from the sun and is broken into O_2 and O

$$O_3 + h_v \text{ (from UV radiation)} \rightarrow O_2 + O \text{ (photo dissociation)}$$

These O_2 and O so formed quickly recombine to form O_3.

By this mechanism the harmful energy of UV-radiation is dissipated as heat.

$$O_2 + O \rightarrow O_3 + \text{heat}$$

23.7.1. The Stratospheric Ozone

The ozone layer has been estimated as about 0.29 cm or 290 DU (1 Dobson Unit : 0.01 mm) thick above equator and upto 0.40 cm above the poles towards the end of winter season.

It changes with change in seasons and latitudes being thickest during spring and thin during rainy season.

Ozone Layer Depletion (Ozone Hole) : Ozone Layer thickness above 200 DU is considered normal. If it falls below 200 DU, it is termed as **ozone hole**. The overall reduction in global - average of ozone layer is now estimated at about 8%.

23.7.2. Effect of Ozone Layer Depletion

The ozone layer strongly blocks (absorbs) the harmful shortwave UV-radiation from the sun and thereby protects life on earth from radiation damage. Thinning/depletion of ozone layer will mean arrival of more of UV-radiation causing serious damage to different life forms in the bio-sphere.

Based on its effect on living organisms, ultraviolet radiation from the Sun has been classified as UV-A, UV-B and UV-C radiation.

- UV-A is low-energy, long wave length radiation and is not harmful.
- UV-B is higher energy short wave length radiation and harmful to both animals, humans and plants.
- UV-C has very high energy and can quickly damage life forms on earth. However, it is completely absorbed by the protective ozone layer.

23.7.3. Ozone Depleting Agents

Chloro-fluoro-carbons (CFCs) have been found to be main culprit causing most of ozone-layer depletion. Though these are harmless chemicals in the troposphere, CFCs rise up slowly into stratosphere where they produce active chlorine radicals Cl and ClO, under the influence of high energy UV-rays. These radicals act as catalyst and destroy ozone through its conversion into O_2 which is unable to absorb U-V radiation.

$$Cl_2 + h_v \rightarrow Cl + Cl$$

$$Cl_2 + O_3 \rightarrow ClO + O_2$$

Other gases which help in destruction of ozone layer are methane (CH_4) and Nitrous Oxide (N_2O) which act on ozone with a series of complicated reactions.

23.7.4. World's Concern Over Ozone Layer Depletion

Ozone depletion and the ozone hole have generated worldwide concern over increased cancer risks and other negative effects. The Montreal protocol on ozone layer depletion was designed to protect the fragile ozone layer. The provisions of the protocol include the requirement that signatories to the protocol base their future programmes on the current scientific, environmental, technical and economic information provided by worldwide experts. Now 197 countries have signed the protocol. They have also ratified all the amendments in the original document.

23.8. FLY ASH AND ITS MANAGEMENT

Fly Ash: It is the finest of coal ash particles. Fly ash is the fine powder formed from the mineral matter in coal, consisting of non-combustible matter in coal, and a small amount of carbon that remains from incomplete combustion. Fly ash is generally light **tan** in colour and consists mainly of silt sized and clay sized spherical particles.

Properties of fly ash vary significantly with coal composition and plant operating conditions. Fly ash is referred to as either **Cementations** or **pozzolanic**. A cementations material is one that hardens when mixed with water. A pozzolanic material will also harden with water but only after activation with an alkaline substance such as **lime**. These two properties make some fly ashes useful for cement replacement in concrete and many other building applications.

Fly Ash in India: The global energy demand is ever increasing. Much of this growth will continue to be concentrated in the developing world, primarily in India and China, as industrialization, population growth and the unprecedented expansion of the middle class will propel the need for energy in general and coal in particular.

The Indian coal is generally of low grade having high ash content of the order of 30-45% producing large quantity of fly ash at coal/lignite based thermal power stations in the country. This large quantity of ash, which not only requires large areas of precious land for its disposal but also one of the sources of pollution of both air and water.

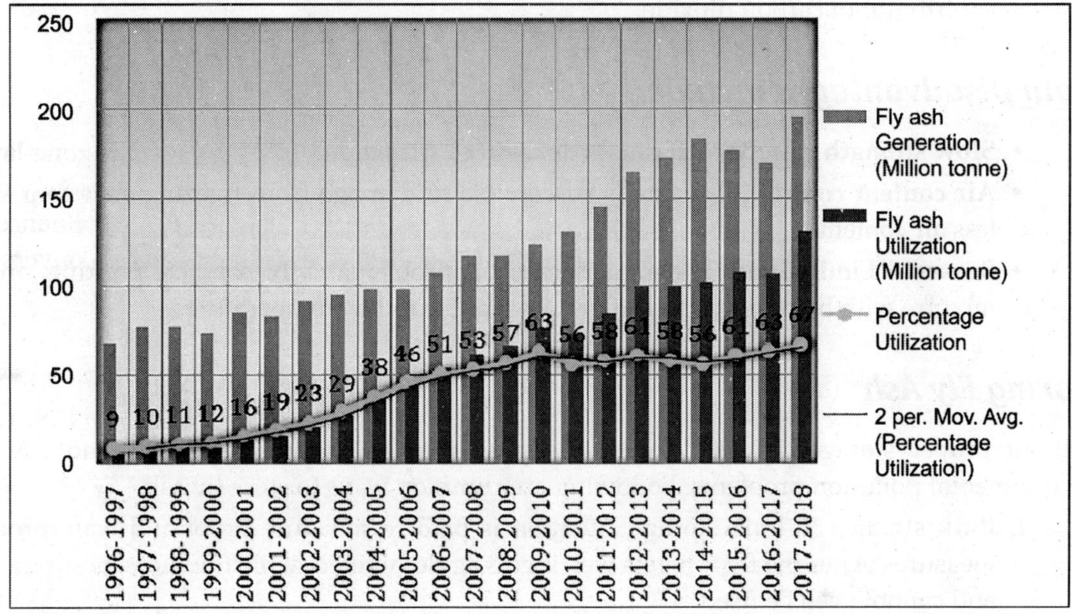

Fig.23.2. Fly ash – yearwise generation.

Source: Journal of Engineering Design & Analysis Vol. 2, Issue 1-2019.

India is the top most country in production of fly ash producing almost 200 million tonnes of fly ash per year. However, India has achieved tremendous increase in its utilization which now stands at 67% of the total produced.

23.8.1. Applications and Benefits of Fly Ash

Fly ash can be a used as prime material in many cement-based products, such as poured concrete, concrete blocks, and bricks. One of the most common uses of fly ash is in **Portland Cement** concrete pavement or PCC pavement.

Fly ash can be a cost-effective substitute for Portland cement. It is also recognized as an environmentally friendly material because it is a byproduct and has low **embodied energy,** the measure of how much energy is consumed in producing and shipping a building material. By contrast Portland cement has a very high embodied energy.

23.8.2. Advantages and Disadvantages of Using fly ash

Following are the main advantages of fly ash utilization :

- Prevents contamination of water both surface and groundwater.

- **Prevents soil Erosion:** Helps restrict use of topsoil for manufacturing of bricks.
- **Useful in variety of construction works:** Fly ash is a proven resource material for many applications of construction industry. It is used in manufacturing of Portland cement, bricks and tiles which show a better strength.
- **Used in agriculture:** As an agent for acidic soil and as a soil conditioner.
- **Environmental benefit:** It reduces the use of Portland cement which is a major contributor of carbon dioxide.

Main Disadvantages include:

- **Slow strength gain:** Which causes delay in construction.
- **Air content control:** Concrete is susceptible to damage from freezing as it contain less air content.
- **Seasonal Limitations:** The winter season is problematic for concrete pouring, and mixture with high in fly ash is more susceptible to low temperatures.

Storing Fly Ash

Without proper storage, fly ash can be easily affected by external elements and cause environmental pollution problems. Following are common ways to store fly ash.

1. **Bulk storage**: In bulk storage attention is paid to moisture proof and waterproof measures. After the fly ash gets wet, it gets agglomerated, which reduces its strength and cannot even be used.
2. **Bag storage**: Moisture proof is also essential in bag storage as well. The bagged fly ash must be stored in dry places. Bag storage is not suitable for long term storage of fly ash as its quality drops with time.
3. **Silo storage**: This a safe method of storing fly ash. Welded steel plate silos are widely used. They are very strong and air and moisture tight silos of large capacity can be constructed.

23.9. GREENHOUSE EFFECT

Greenhouse : A greenhouse is a framed structure covered with transparent material (glass) and large enough to grow crops. Greenhouse is the body that allows the short wave length incoming solar radiation to come in, but does not allow the returning long wavelength outgoing infrared radiation to escape. Hence, inside of the greenhouse is kept warm.

In 1827, French Scientist J.B.J. Fourier imagined Earth as a giant greenhouse. He found that the atmosphere is largely transparent to incoming shortwave solar radiation, while being absorbing to outgoing long-wave thermal radiation. This helps in keeping the planet (Earth) warm giving life to every plant and animals.

Naturally occurring blanket of gases around the earth which help in keeping the earth warm to a comfortable temperature are called Green House Gases (GHGs). These primarily include Carbon dioxide, Methane, Water Vapour, Nitrous oxide and Chloro-fluoro-carbons. Ozone also contributes to the effect. These gases comprise 1 to 2% of the atmosphere. Without these GHGs the average temperature of earth would have fallen by –20°C.

The past about 100 years of human activities have changed the composition of the atmosphere to the disadvantage of life on earth. Burning of fossil fuels, destruction of forests and emission from industrial and agricultural operations result in large quantities of GHGs into the atmosphere. This increased amount of these gases absorb more and more long wave length radiation and thus raise the temperature of the earth globally which may result is great disasters. The phenomenon of rise in temperature of the earth's atmosphere is termed as **Global Warming**.

23.9.1. Relative Contribution of GHGs

Carbon dioxide seems to be having maximum contribution to the Greenhouse effect. Following table shows the relative contribution in percent of various GHGs.

Gas	% Contribution
Carbon dioxide	50-60%
Methane	19-20%
CFCs	14-17%
Ozone	8%
Nitrous oxide	4-6%
Water vapour	2%

Concentration of GHGs is ever increasing. The trend is shown below. (Approximate)

Gas	Concentration ppm	% annual increase
Carbon dioxide	356-368	0.4
Methane	1.7	1.0
Nitrous oxide	0.3	0.3
CFCs	0.0005	5

23.10. GLOBAL WARMING POTENTIAL (GWP)

It is the potential of a green house gas to cause global warming. It is expressed as a multiple of the warming caused by the same mass of Carbon dioxide (CO_2). Carbon dioxide has been assigned the value of 1.

For example, if methane (CH_4) has a global warming potential of 21, it means that 1 kg of methane has the same impact on climate change as 21 kg of carbon dioxide and thus one kg of methane would count as 21 kg of CO_2 equivalent.

23.10.1. Global Warming Potential (GWP) of Greenhouse Gases

Gas	GWP
Carbon dioxide (CO_2)	1
Methane (CH_4)	21
Nitrous oxide (NO_2)	296
CFC	5700-11900

Source **:** Green Facts.

On the global basis developed countries or industrially advanced countries are responsible for 85% of GHGs. Following is the list of countries and their GHGs percent emission.

Country	% GHGs Emission
China	22.7
USA	15.6
India	5.7
Russia	5.4
Japan	2.9
Brazil	2.6
28 EU countries	10.9

Total GHGs emission is expected to hit all-time high in 2019. Global greenhouse gas emissions is increasing at the rate of 2% per year, reaching 51.8 gigatonnes in CO_2 equivalent (Gt CO_2 eqt.)

*[**Note :** All the data provided in above pages are from different sources. There may be considerable variation in data provided by different agencies.]*

23.10.2. Effect of Increasing GHGs

Predicting the consequences of increasing GHGs concentration is a difficult task for researchers. An increase in concentration of greenhouse gases produces a warming effect. According to NASA, the average global temperature on Earth has increased by a little more than $1^{O}C$ ($2^{O}F$) since 1880. Two thirds of the warming has occurred since 1975, at the rate of roughly 0.15-$0.20^{O}C$ per decade.

The main effect of increased greenhouse gas emissions is **Global Warming**. The increase in the average temperature of Earth is changing living conditions on the planet. The **main** consequences of the phenomenon are briefly discussed here,

- **Thawing of Glacial Mass :** Among the most dramatic evidence that Earth's climate is warming is the dwindling and disappearance of mountain glaciers around the world.
- **Desertification of Fertile Areas :** Global warming is having a profound impact on

the processes of soil degradation and is contributing to the desertification of the most arid areas on the planet.

- **Increase in Intensity of Hurricanes :** Though intensification of Greenhouse effect does not cause these extreme climatic events, but it does increase their intensity.
- **Migration of Species :** Many animal species will be forced to migrate in order to survive the changes in the climate pattern altered by the progressive increase in temperature.
- **Impact on Agriculture and Livestock :** Global Warming has already altered the length of the growing seasons in large parts of the earth. It has also caused proliferation of insects, invasive weeds and crop diseases. Climate change affect important species in multiple ways : reproduction, metabolism and new diseases.
- **Sea Level Rise :** Rising sea level is one of the effects of climate change. It is estimated that sea level has swelled over 23cm since 1880 with about 7.5 cm gained in last 25 years. Scientists feel that melting of polar ice could lead to sea level rise resulting in flooding of coastal areas. It is feared that if all the polar ice melts due to rise in temperature the sea level would rise by more than 60 m, submerging large coastal area or even drowning some islands.

23.10.3. How Global Warming can be Checked

The goal is simple. Carbon dioxide is the climate's worst enemy. It is released when fossil fuels are burnt for energy. By using less of it, we can contribute to check the climate change and save money also. The best of the ways to check global warming are – awareness, use of renewable energy, efficient design of buildings which require less cooling and heating, use energy efficient appliances, reduce water use (less pumping), afforesting etc.,

QUESTIONS

1. What are the sources of light pollution ?
2. What are the negative effects of light pollution ?
3. Differentiate between Luminous Pollution and Visual Pollution.
4. What are the ill-effects of Acid Rain ?
5. How acid rain pollution can be prevented ?
6. 'Smog' is a classical example of secondary air pollution, explain ?
7. In what ways "Flyash" is useful ?
8. What is the theory behind Greenhouse effect ?
9. What is the relation between Carbon dioxide and Global warming ?
10. Define and explain Global Warming Potential (GWP).
11. How the increasing quantity of GHGs will affect the planet ?
12. How smell (odour) pollution affects human beings ?
13. What causes Nuclear Pollution ?
14. What measures should be adopted to control Nuclear pollution ?

APPENDIX-I: Question Bank

1. **Measuring Dilution:** In a stream AB, a treated wastewater enters between A and B. The Sodium Concentration in part A is 9.9% mg/L and the flow rate is 19.5% m^3/s The concentration of Sodium in the waste stream C is 300 mg/L and the flow rate is $2m^3/s$. Find the Concentration of Sodium in part B.

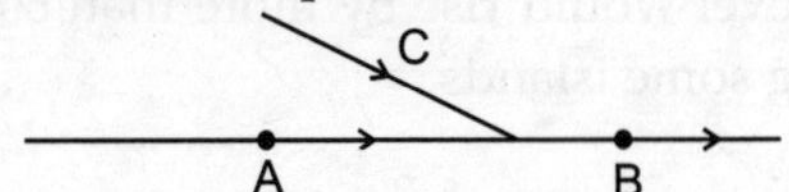

Solution: Writing the mass balance between A and B

$$C_AQ_A + C_WQ_w = C_BQ_B$$

$$\therefore C_B = \frac{C_AQ_A + C_WQ_w}{Q_B} = \frac{C_AQ_A + C_WQ_W}{Q_A + Q_W}$$

$$= \frac{9.9 \times 19.5 + 300 \times 2}{19.5 + 2} = \frac{793.05}{21.5}$$

$$= 36.88 \text{ mg/L}$$

2. Using the data given in the following table, find the dry mass of the solid waste sample

Component	Mass	Moisture content %	Dry Mass kg.
Food Waste	40 kg	71	11.60
Paper	5 kg	6	4.70
Card board	10 kg	5	9.50
Plastic	5 kg	2	4.90
Garden waste	13 kg	60	5.20
Wood	5 kg	20	4.0
Metals	2 kg	1	1.98

Total weight of sample = 80kg.

Dry mass of solid waste = 41.88kg.

Percent solid waste = $\frac{41.88}{80} \times 100 = 52.35\%$

3. A groundwater sample was found to contain 625 mg/L total dissolved solids (TDS). What is the percent of TDS in the sample?

Solution: Mass of I liter of water = 1 kg = 1000 g.

$625 \text{ mg} = 625 \times 10^{-3} \text{ g}$

$\therefore$ % TDS in sample $= \frac{625 \times 10^{-3}}{1000} \times 100 = 0.0625\%$

4. An amount of 25.9 mg Hydrochloric Acid (HCL) is added to distilled water and the total solution volume is made to one liter. The atomic weights of H and Cl are 1 and 35.5 respectively. Neglecting the Dissociation of water Find. The pH of the solution.

Solution: $HCl \rightarrow H^+ + Cl^-$

1 mole of HCl gives 1 mole of Hydrogen lons

(35.5+1) g of HCl gives 1g of Hydrogen ions

$\therefore$ 25.9 mg will give $\frac{1}{36.5} \times 25.9 = 0.709$ mg of $H^+ = 7.09 \times 10^{-4}$ moles of H^+

$$pH = -\log_{10}\left[H^+\right] = -\log_{10}\left[7.09 \times 10^{-4}\right] = -\log_{10} 7.09 + 4\log_{10} = -0.85 + 4 = 3.15$$

5. Find calorific value of coal with the help of following data

$C = 78\ \%$, $H_2 = 4\ \%$, $O_2 = 16\ \%$ and $S = 2\%$.

Solution: Following formula is used for calorific value Q (kJ/kg).

$Q = 337\, C + 1442\left(H - \frac{O}{8}\right) + 93 \times 2 = 29,356\ kJ/kg$

or 29.35 KJ / g

6. The residential area of a city has a population density of 15000 persons per km^2 and an area of 1,20,000 m^2. If the average water consumption is 150 litres per capita per day. Find the average and maximum sewage flow per day.

Solution: Population Density = 15000 per km

Area $= \frac{1,20,00}{1000 \times 1000} \text{km}^2 = 0.12 \text{ km}^2$

$\therefore$ Population of the residential area = $0.12 \times 15000 = 1800$ persons.

Average sewage flow $1800 \times 150 \times 80\% = 1800 \times 150 \times \frac{80}{100}$

$= 2,16,000 \text{ litres/day} = 216 \text{m}^3/\text{day}$

Maximum sewage flow = $\frac{18 + \sqrt{P}}{4 + \sqrt{P}} \times$ Average sewage Flow

$\frac{18 + \sqrt{1800}}{4 + \sqrt{1800}} \times 216 = 281 \text{m}^3/\text{day} = 3.25 \text{ litres/s}$

7. Calculate the quantity of air required per minute to remove 250 ppm of BOD from 7.5 million liters of sewage per day. Given that 30 m^3 of air is required per million liters for each ppm removed.

Solution: Quantity of air required $= 30 \times 7.5 \times 250$

$= 56,250 \text{ m}^3$ per day

$= 39 \text{ m}^3$ per minute.

8. Design an oxidation pond for treating sewage produced by 6250 persons. The sewage contributed is at the rate of 96 liters per capita per day and the average BOD is 300 mg/L

Solution: Quantity of sewage to be treating per day $= 625 \times 96 = 6,00,000$ litres

B.O.D content per day $= 300 \times 6,00,000$ mg $= 18 \times 10^7$ mg $= \dfrac{18 \times 10^7}{10^6} = 180$ kg

Let the organic loading in the pond = 300 kg/hectare/day

$\therefore$ Surface area required $= \dfrac{180}{300} \times 10,000 = 6000 \text{ m}^2$

$\therefore$ Let the Length : Breadth = 2:1

$\therefore \; L \times B = 6000$

$2B \times B = 6000$, or $B^2 = 3000$

$\therefore B = 54.77\text{m}$ (Say 55m)

$\therefore$ Length of the Pond = 110m.

Depth: Let the effective depth = 1.0m and total dept 2m.

$\therefore$ Capacity $= 110 \times 55 \times 1 = 6050 \text{ m}^3$

$\therefore$ Capacity = sewage per day $\times$ detention time in days.

$\therefore$ Detention time = capacity/sewage per day.

$= \dfrac{6050}{600} = 10.08$ day (say 10 days)

$\therefore$ Tank Dimensions are: 110 m $\times$ 55 m $\times$ 2 m with a detention period of 10 days.

9. Two samples of water A and B have pH values of 4 and 6 respectively. Then how many times sample A is more acidic than sample B.

Solution: $p(H) = -\log_{10}\left(H^+\right)$

$\left[H^+\right]_A = 10^{-4}$ and $\left[H^+\right]_B = 10^{-6}$

$\therefore \dfrac{\left[H^+\right]_A}{\left[H^+\right]_B} = \dfrac{10^{-4}}{10^{-6}} = 100$

10. A water treatment plant $1 \text{ m}^3/\text{s}$ has filter boxes of dimension $5 \text{ m} \times 9 \text{ m}$. Loading rate is $100 \text{ m}^3/\text{day}/\text{m}^2$ when three of the filters are out of service for back washing, what will be the loading rate?

Solution: Total daily filtered volume $= 1 \times 60 \times 60 \times 24 = 86,400 \text{ m}^3/\text{day}$

Total area required $= \dfrac{86,400 \text{m}^3/\text{day}}{100 \text{ m}^3/\text{day}/\text{m}^2} = 864 \text{ m}^2$

Area of single filter $= 5 \times 9 = 45\ m^2$

Total number of filters $= \frac{864}{45} = 19$

Working filters $= 19 - 3 = 16$

$\therefore$ Total surface are $45 \times 16 = 720\ m^2$

$\therefore$ New loading rate $= \frac{86,400\ m^3 / day}{720\ m^2} = \frac{120\ m^3 / day / m^2}{}$

11. The pH of water admitted into a treatment plant was 6.0 in the morning. In the next 24 hours it changed to 8.0 due to inflow of raw water. Assuming linear variation of hydrogen ion concentration with time, find the mean pH value over this 24 hour period.

Solution: $\left[H^+\right]_{t=0} = 10^{-6}, \left[H^+\right]_{t=24\ Hours} = 10^{-8}$

Mean $\left[H^+\right] = \frac{10^{-6} + 10^{-8}}{2} = \frac{10^{-6}[1 + 10^{-2}]}{2} = 0.505 \times 10^{-6}$

$\therefore$ Mean pH $= -\log_{10}(0.505 \times 10^{-6}) = -(-6.296) = 6.296$

12. An aerobic reactor receives wastewater at a flow rate of $500\ m^3$/day having a COD of 200 mg/L. The effluent COD is 400 mg/L. Assuming that waste water contains 80% biodegradable waste. Find the daily volume of methane produced by the reactor.

Solution: Organic matter stabilized per day

$= 500 \times (2000 - 400) \times 1000 \times .8$

$= 640 \times 10^6\ mg$

$= 640\ kg$

Assuming 1 kg organic matter produces $0.35 m^3$ methane

$\therefore$ 640 kg will produce $640 \times 0.35 = 224\ m^3$ of methane

13. Find the settling velocity of spherical discrete particles of 0.5 mm diameter and specific quantity of 2.65 in water at $20^\circ C$

Solution: Assume Laminar Flow

$\rho_w = 1000\ kg / m^3$

Dynamic viscosity of water $10^{-3}\ N - S / m^2$

Stokes' equation $v = \frac{g(\rho_s - \rho_w)d^2}{18\mu} = \frac{9.81\,(2.65 - 1) \times 1000 \times (0.5 \times 10^{-3})^2}{18 \times 10^{-3}} = 0.224$ m/s.

14. A domestic sewage contains 500 ppm of suspended matter. During primary sedimentation 70% of the suspended matter is removed, calculate the quantity of sludge, if it contains 7% solids.

Solution: Initial suspended solids = 500 ppm.

Suspended solids removed $= \frac{70}{100} \times 500 = 350$ ppm

$\therefore$ Quantity of sludge expected $= \frac{350}{.07} = 5000$ litres.

15. Design a trickling filter for treating 5 million liters of sewage per day. The BOD of sewage is 120 ppm.

Solution: 5 million at 120 ppm $= 120 \times 5 = 600$ kg per day

Assume rate of filter loading = 1200 kg of BOD per hectare-meter per day.

$$\therefore \text{Volume required } = \frac{600}{1200} = 0.5 \text{ ha} - \text{m}$$

Let the depth of filter = 1.8m

$$\therefore \text{Surface area of the filter } = \frac{0.5 \times 10{,}000}{1.8} = 2778 \text{ m}^2 = 2778 \text{ m}^2$$

Adopt 3 circular trickling filters.

$$\therefore \text{Area of each } \frac{2778}{3} = 926 \text{ m}^2$$

$\therefore$ Diameter of each = d

$$\therefore \frac{\pi}{4} d^2 = 926$$

$$d^2 = \frac{926 \times 4}{\pi}, \text{ or } d = 34.3 \text{ m} = 35 \text{ m (say).}$$

Provide 4 trickling filters (one as stand by)

Diameter = 35 m.

Depth = 1.8 m.

16. In a hydropower project the available head is 500 m and discharge is $6\text{m}^3/\text{s}$. Find the power developed (Potential), Considering efficiency of the plant as 0.84.

Solution: Power $= \rho g h \eta \theta = 1000 \times 9.81 \times 500 \times 6 = 29430000 \text{ W} = 29430 \text{ kW}$

In H.P. $= 29430 \times 1.341 = 39465.6$ hp.

17. A town of population 100 thousand is to be supplied water at the rate at 120 liters per had per day. Given that 80 % of water appears as sewage and peak flow is thrice the average. Design the rising main and sump well. Given: Length of rising main 85 m and mean velocity in the rising main as 1.2 m/s.

Solution: Average quantity of sewage $= \left[\dfrac{100{,}000 \times 120}{1000 \times 24 \times 60 \times 60}\right] \times 0.8 = 0.11\text{m}^3/\text{s}$

Assume peak flow as 3 times the average flow.

Maximum flow of sewage $= 0.333 \text{ m}^3/\text{s}$

Design

(*i*) Rising pipe, $Q = A \times V, \; A = \dfrac{Q}{V} = \dfrac{0.333}{1.2}$

$$\frac{\pi}{4} d^2 = \frac{0.333}{1.2} \qquad \text{or } d^2 = \frac{0.333 \times 4}{1.2 \times 3.14} = .3535$$

d = 0.59 m. adopt 600 mm dia pipe

(*ii*) **Sump Well:**

Let the detention period is $\frac{1}{4}$ Hour (15 minutes)

Maximum flow for 15 minutes $= 0.333 \times 15 \times 60 = 300 \text{ m}^3$

Capacity of rising main $= \frac{\pi}{4} \times (0.6)^2 \times 85 = 24 \text{ m}^3.$

$\therefore$ Total capacity of sump well $= 300 + 24 = 324 \text{ m}^3$.

Provide 4 sumps with one stand by.

$\therefore$ Capacity of each sump well $\frac{1}{3} \times 324 = 108 \text{ m}^3$.

Giving a depth of 3 m

Area of cross section of sump well $= \frac{108}{3} = 36\text{m}^2$

$\frac{\pi}{4} D^2 = 36, \quad D^2 = \frac{36 \times 4}{\pi}, \quad D = 6.77 \text{ m}.$

18. A generator is emitting a sound pressure of 1000 μ bar. What is the noise produced in dB?

$P_o = 20\mu\text{pa}$ (Reference Level)

$1 \mu \text{ bar} = 10^5 \mu\text{pa}$,

$P = 1000 \mu \text{ bar} = 1000 \times 10^5 \mu \text{ pa}$

$$SPL = 20 \log_{10}\left(\frac{P}{P_o}\right) = 20 \log_{10}\left(\frac{1000 \times 10^5}{20}\right)$$

$= 20 \times 6.698 = 134 \text{ dB}$

19. For noise measurement formula for sound pressure level (SPL) is 20 $\log_{10} \frac{P}{P_o}$, P_o is reference level. Find the resultant noise in dB, if $P = 0.0002 \mu$ bar

Solution $1 \mu \text{ bar} = 10^5 \mu \text{ pa}$

$\therefore P = 0.0003 \times 10^5 \mu \text{ pa}$

$P_o = 20 \mu \text{ pa}$.

$$SPL = 20 \log_{10}\left(P/P_0\right) = 20 \log_{10}\left(\frac{0.0002 \times 10^5}{20}\right)$$

$= 20 \times 0 = 0$ (Zero) dB.

20. The equivalent sound power levels in dB of the four sources with noise levels of 60 dB, 69 dB, 70 dB and 79 dB will be?

Solution Sound level $L = 20 \log_{10} \frac{P_1}{20 \mu \text{ Pa}}$

$$20 \log_{10}\left(\frac{P_1}{20 \mu \text{ Pa}}\right) = 60 \quad \text{or} \quad \log_{10} \frac{P_1}{20 \mu \text{ Pa}} = 3$$

$$\therefore \frac{P_1}{20 \mu \text{ Pa}} = \text{antilog } 3 = 1000$$

Sim City $\frac{P_2}{20 \mu \text{ Pa}} = \text{antilog} \frac{69}{20} = 2818$

$$\frac{P_3}{20 \mu \text{ Pa}} = \text{antilog} \frac{70}{20} = 3162$$

$$\frac{P_4}{20\ \mu\ Pa} = \text{antilog}\ \frac{79}{20} = 8912$$

$$\text{Equivalent}\ \frac{P_e}{20\ \mu\ Pa} = \sqrt{1000^2 + 2818^2 + 3162^2 + 8912^2} = 9918$$

$Leq = 20\ \log_{10}(9910) = 79.928\ dB.$

21. The Concentration of SO_2 in ambient atmosphere was measured as $30\ \mu g/m^3$. Under the same conditions, find the above $30\ \mu g/m^3$. Under the same conditions. Find the above SO_2 concentration in ppm.

Given, $\frac{P}{RT} = 41.6\ mol/m^3$, Molecular weight of $SO_2 = 64$

Solution: $1\ m^3$ of air has 30 μg SO_2

Volume of 30g $SO_2 = 10^6\ m^3$

$$\therefore 30g\ SO_2 = \frac{30}{64}\ \text{mole}\ SO_2,\ \text{From}\ PV = n\ RT$$

$$V = \frac{nRT}{P} = \frac{n}{P/RT} = \frac{30/64}{41} = 0.01127 m^3\ ppm$$

22. Water table drops by 2.5 m in an irrigable land of area 60 hectares. If porosity and specific retention are 0.35 and 0.11 respectively, what will be the change in storage?

Solution : Porosity = Specific yield (Sy) + specific retention (Sr)

$\therefore Sy = 0.35 - 0.11 = 0.24$

$$Sy = \frac{\text{Volume of water taken out}}{\text{Volume of aquifer}} = \frac{\text{Change in storage}}{\text{Volume of aqifer}}$$

$\therefore$ Change in storage = Volume of Aquiter $\times 0.24 = 60 \times 2.5 \times 0.24$ 36 ha – m

23. Results of a water sample analysis are as given below,

Cation	Concentration (mg/L)	Equivalent Weight
Na^+	40	23
Mg^{+2}	15	12.2
Ca^{+2}	50	20
K^+	5	39

(Milli equivalent weight of $CaCO_3 = 50mg/meq$)

Find the hardness of water in mg/L as $CaCO_3$.

Solution :

$$\text{Total hardness} = Mg^{2+} \times \frac{\text{Equivalurt weight of}\ CaCO_3}{\text{Equvalent weight of}\ Mg^{2+}} + \left[Ca^{2+}\right] \times \frac{\text{eq. wt of}\ CaCO_3}{\text{eq wt of}\ Ca^{2+}}$$

$$= 15 \times \frac{50}{12.2} + 50 \times \frac{50}{20} = 61.47 + 125 = 186.47 \text{ mg/L}$$

24. If the energy content in kJ/kg of various components is as follows,

Food waste	Paper	Card board	Plastics	Garden waste	Wood	Tin
4650	16750	16300	32600	6500	18600	700

Prepare a computational table and find energy content of the following sample.

Solution :

Component	Mass (kg)	Energy kJ/kg	Total Energy kJ
Food waste	43	4650	1,99,950
Paper	8	16750	13,4000
Card Board	13	16300	2,11,900
Plastic	8	32600	2,60,800
Garden waste	16	6,500	1,04,000
Wood	8	18,600	1,48,800
Metals	4	70	2,800
Total	**100**		**10,62,250**

$$\therefore \text{Energy content} = \frac{10,62,250}{100} = 10622 \text{ kJ/kg}.$$

25. A 200 ml sample of water has an initial pH of 10. If 30 milliliters of 0.02 N H_2SO_4 is required to titrate the sample to pH 4.5, find the total alkalinity of water.

Solution:

In titration each milligram will neutralize 1 mg of alkalinity. Therefore, the concentration of alkalinity expressed as milligrams per litre is

$$= \frac{30 \text{ mg}}{200 \text{ mL}} \times 1000 = 150 \text{mg/litre}.$$

26. A grit chamber is designed to remove particles with diameters of 0.2 mm, specific gravity 2.65, settling velocity for these particles is between 0.015 to 0.021 m/s. Flow velocity is 0.3m/s. Determine the channel dimensions for a waste water flow of 10,000 m^3/day

Solution:

(i) Assume a rectangular cross section with depth $= 1.5 \text{ m} \times \text{w}$

$$\therefore \text{Cross Sectional area} = \frac{Q}{v} = \frac{10000}{24 \times 3600 \times 0.3} = 0.385 \text{m}^2$$

Also, cross – sectional Area $= 1.5 \times w \times w = 1.5 \times w^2$

Now, $1.5 w^2 = 0.385 \text{m}^2$

Or, $w^2 = \frac{0.385}{1.5} = 0.2566$

Or, $w = 0.5$ m.

$\therefore$ Depth (D) $= 1.5 \times 0.5 = 0.75$ m.

(*ii*) Assume a settling velocity $(v_t) = 0.021$ m / s.

Detention time $(t_d) = \frac{D}{v_t} = \frac{0.75}{0.021} = 35.7$ S

(*iii*) Length of the channel $L = t_d \times v$

$= 35.7 \times 0.3 = 10.71$ m

$\therefore$ Dimensions are: Width = 0.5 m, Depth = 0.75 m, Length = 10.71 m.

27. Design a primary clarifier for a town having a population of 39000. The formation of sewage may be assumed to be 130 liters per capita per day.

$\therefore$ Daily flow of sewage $= 39000 \times 130 = 50{,}70{,}000$ litres $= 5070\ m^3 = 5100\ m^3$(say)

Assure detention period = 1.5 hours

$\therefore$ Tank capacity $\frac{5100}{24} \times 1.5 = 318.75\ m^3$

Assuming a depth of 2.5 m

Surface area (A) of tank $= \frac{318.75}{2.5} = 127.5\ m^2$

Let Length (L) $= 4 \times$ width (B).

$A = L \times B = 4B \times B = 4B^2 = 127.5\ m^2$

$\therefore B = 5.64$ m.

$L = 22.56$ m

Overflow rate $= \frac{5100}{5.64 \times 22.56} = 40.1 m^3 / m^2 day.$

(**Note:** Recommended overflow rate is between 40 to 50 $m^3 / m^2 day$)

28. In an unseeded sewage sample, following data is provided.

Initial DO = 10 mg/L

Final DO = 6 mg/L

Volume of the bottle = 300 ml.

Sample Volume = 7 ml.

Calculate 5- day BOD of the unseeded sample

Solution: Ratio of dilution $= \left[\frac{300}{7}\right] = 42.86$

5 – day BOD = loss of oxygen in ml/l × Ratio of dilution

$= (10 - 6) \times 42.86$

$= 171$ mg/L.

29. In a test of sewage of a town for total solids, following observations were made.

Weight of sample of sewage = 1500 g

Weight of solids after evaporation of liquid = 1.0 g

Weight of dry residue after ignition = 0.6g

Calculate: total solids, fixed solids and volatile solids.

Solution: Total solids in ppm $= \dfrac{1.0 \times 10^6}{1500} = 666.7$

Fixed solids in ppm $= \dfrac{0.6 \times 10^6}{1500} = 400$

Volatile solids in ppm = 666.7 – 400 = 266.7

30. A city has a population of 1,50,000, calculate the water to be provided for fire fighting per day.

Solution: Water required in kilolitres $= 100\sqrt{p} = 100\sqrt{150} = 1225$ kilolitres.

31. A town has a population of 200,000 at present (2020), what will be its population in 2050, if the rate of increase of population is 2%.

Solution: A basic equation is as follows,

$$N_t = Pe^{rt}$$

N_t = Number of people after t years

e = Base of natural logarithm (2.71828)

r = rate of increase

t = time in years

$N_t = 200000 \times 2.71828^{\frac{2}{100} \times 30} = 200000 \times 2.71828^{0.6} = 3,64,423$

32. If in New – Delhi Railway station 6 lakh people board the trains daily. What will be per day water requirement?

Solution: As per National Building code 2016

Total water required $= 6 \times 70 = 42$ lakh litres per day.

33. Predict the population for the years 2031, 2041 and 2051 with the help of the following data.

Year	Population in Lakh	Increment Lakh
1971	10	–
1981	12	2
1991	16	4
2001	21	5
2011	24	3
2021	29	5
		Average increase = 3.8 Lakh

Solution: Using Arithmetical Increase Method.

$P_n = P + nc$

Where P_n is population after n decades.

$C = \frac{dp}{dt}$ rate of change of population.

Population in the year $2031 = 29 + 3.8 \times 1 = 32.8$ Lakh

Population in the year $2041 = 29 + 3.8 \times 2 = 36.6$ Lakh

Population in the year $2051 = 29 + 3.8 \times 3 = 40.4$ Lakh.

34. Population of a neighbouring country over the decades is shown below, using Geometric Increase Method, Find its population in 2051

Year	Population (Millions)	Increment (millions)	Geometric Increase Growth rate
1991	112	–	–
2001	142	30	$\frac{30}{112} = 0.2678$
2011	179	37	$\frac{37}{142} = 0.26$
2021	220	41	$\frac{41}{179} = 0.23$

Solution: Using the formula, $P_n = P\left(1 + \frac{I_G}{100}\right)^n$

Where, I_G = Geometric Mean %, P = Present population, n = number of decades.

Geometric Mean = $(0.2678 \times 0.26 \times 0.23)^{\frac{1}{3}} = 0.25 = 25\%$

$P_{2051} = 220 \times (1 + 0.25)^3 = 430$ millions.

35. Considering the following population data, predict the population for the years 2021, 2031, 2041 using incremental increase method,

Year	Population Lakh	Increment (X) Lakh	Incremental Increase (Y)
1961	8.5	–	–
1971	10.1	1.6	–
1981	12.0	1.9	+ 0.3
1991	16.9	4.9	+ 3.0
2001	20.8	3.9	– 1.0
2011	25.8	5.0	+ 1.1
	Total	17.3	3.4
	Average	3.46	0.85

Solution: Population

Using the formula: $P_n = P + nx + \{n(n+1)/2\} Y$

Population in the year 2021,

$$P_{2021} = 25.8 + 1 \times 3.46 + \left\{\frac{1 \times 2}{2}\right\} 0.85 \;\; = 25.80 + 3.46 + 0.85 \;\; = 30.11 \text{ Lakh}$$

Population in the year 2031

$$P_{2031} = 25.8 + 2 \times 3.46 + \frac{2(2+1)}{2} \times 0.85 \;\; = 25.8 + 6.92 + 2.55 \;\; = 35.27 \text{ lakh}$$

Population in the year 2041

$$P_{2041} = 25.8 + 3 \times 3.46 + \frac{(3+1)}{2} \times 0.85 \;\; = 25.8 + 10.38 + 5.1 = 41.28 \text{ lakh.}$$

36. Compute the fire water demand for a city with population 5 lakh, by various methods.

Solution:

(i) Kuichling formula:

$$Q = 3182\sqrt{500} \;\; = 71{,}151 \text{ lit/min} \;\; = \frac{71{,}151 \times 60 \times 24}{10^6} = 102.45 \text{ MLD}$$

(ii) Freeman Formula: $Q = 1136\left[\frac{P}{5} + 10\right]$

$$Q = 1136\left[\frac{500}{5} + 10\right] = 1136 \times 110 \;\; = 124{,}960 \text{ lit/minute} = 179 \text{ MLD (approx).}$$

(iii) National Board Formula:

For congested city

$$Q = 4637\sqrt{p}\left[1 - 0.01\sqrt{p}\right]$$

$$Q = 4637\sqrt{500}\left[1 - 0.01\sqrt{600}\right]$$

$$= 4637 \times 22.36\,[1 - .2236] \; = 4637 \times 22.36 \; \times .7764 = 84{,}499 \text{ lit/min}$$

When $P \le 2$ lakh, $Q = 54600$ lit / min.

+

For residential city:

With small low buildings, Q = 2200 lit/min.

With large higher buildings, Q = 4500, lit/min.

37. A block has a population of 75,000. It has to be supplied water at the rate of 180 liters per head per day. Calculate the Horse Power of the Pump to fill a 50m high overhand tank. The length of the rising main is 250m and its diameter is 30 cm. The motor efficiency is 92% and that of pump is 62% Assume f = 0.035 and Peak demand as 50% more than the average demand.

Solution: Average demand $= 75000 \times 180 = 1{,}35{,}00000$ liters/day $= 13{,}500 \text{ m}^3/\text{day}$

$$= \frac{13{,}500}{24 \times 60 \times 60} = 0.15625 \text{ m}^3/\text{s}.$$

Peak hour demand $= 1.5 \times 0.15625 = 0.234 \text{ m}^3/\text{s}$

$$\text{Velocity through the main} = \frac{0.234}{\frac{\pi}{4} \times (.3)^2} = \frac{4 \times .234}{3.14 \times .09} = 3.3 \text{m/s}$$

$\therefore$ Head loss $= \frac{fLv^2}{2gd} = \frac{0.035 \times 250 \times 3.3^2}{2 \times 9.81 \times .3} = 16$ m

Total head = 50 + 16 = 66m.

Overall efficiency $= 0.92 \times 0.62 = 0.57$ p

Horse power $= \frac{\gamma QH}{75 \times \eta} = \frac{1000 \times 0.15625 \times 66}{75 \times 0.57} = 241$ hp.

Sludge Volume Index (SVI):

It reflects the physical state of sludge.

$$SVI = \frac{\text{Percentage of sludge by volume}}{\text{Percentage of suspended solid by weight}}$$

Also, $SVI = \frac{\text{Settled volume of sludge in 30 min (\%)}}{\text{Suspended solids in mixed liquor}}$

The desirable sludge volume index is 100 or less than 100 ml per gram.

38. Calculate sludge volume index (SVI) when 160 ml of sludge collected in 30 minutes on drying weighed 1 kg. Comment on the sludge.

Solution: $SVI = \frac{160}{1} = 160 > 100.$

This shows troublesome sludge

Sludge Density Index: (SDI):

This is an alternative to SVI and it is expressed as,

$$SDI = \frac{\text{\% of suspended solid by waste}}{\text{\% of sludge by volume}} \times 100$$

$$SDI = \frac{\text{Suspended solid in mixed liquor \%}}{\text{Settled volume of sludge in 30 min \%}} \times 100$$

The desirable value of SDI for a 'good' sludge is about 1 to 2. At the value 0.3 it is termed as poor sludge.

39. Calculate SDI when 100 ml of sludge is collected in 30 minutes on drying weight 800 mg.

Solution: $SDI = \frac{0.8}{100} \times 100 = 0.8 < 1$ Hence the sludge is problematic.

40. A primary clarifier treats 2 million liters of sewage per day containing 250 ppm of suspended matter. The clarifier removes 55% of suspended matter. If the moisture content formed at the bottom of clarifier is 96%. Calculate the quantity of sludge per day in bulk and weight. If the moisture content was 98% what would be quantity of sludge per day in bulk and weight. Take density of sludge as 100 kg/m^3.

Solution: Volume of suspended matter in sewage $= 250 \times 2 = 500$ lit.

$\therefore$ Volume of suspended matter removed $= 0.55 \times 500 = 275$ litres.

For 100 parts of sludge there are 4 parts of solids.

$\therefore$ Volume of sludge per day $= \frac{275 \times 100}{4} = 7150$ lit. $= 7.150 \text{ m}^3$

7.150 m^3

$\therefore$ Weight of the sludge $= 1000 \times 7.150$ kg $= 7.150$ tonnes.

In case moisture content was 98% volume of sludge per day $= \dfrac{275 \times 100}{2} = 14300$ lit $= 14.3$ tonnes.

41. **Design a septic tank for a residential colony with a population of 1200 persons. The rate of water supply is 120 liters per head per day. Also estimate the size of the soak well, if effluent from the septic tank is to be discharged in it?**

Solution: Flow of seepage/day $= 1200 \times 120 = 144000$ liters

Assuming detention period of 18 hours

Capacity of the tank $= \dfrac{1,44,000}{24} \times 18 = 1,08,000$ liters

If the tank is to be cleared annually, then the sludge storage capacity at the rate of 12 liters (assumed) per person per year $= 12 \times 1200 = 14,400$ litre.

$\therefore$ Total capacity of Tank = 1,08,000 + 14,400 = 1,22,400 liters

25% of this is provided for future expansion = 30,600 liters.

Final capacity of tank = 1,53,000 liters = 153 m^3.

Provide 3 units of septic tank with capacity $153/3 = 51\ m^3$ Capacity.

Taking the depth of the tank = 1.70 m. (Extra for Frce Board = 30 mm.)

Plan area of the tank $= \dfrac{51}{1.7} = 30\ m^2$

Let the ratio of length: width = 3.

$\therefore L = 3W$ and $A = 3W \times W = 30\ m^2$

Or $W^2 = 10$, W = 3.16m, L = 9.5m

Approximating,

Length = 10m, Width = 3m and Depth = 2.0 m (including free board).

42. **If pH of a juice is 4.1, what is the hydrogen-ion concentration in it?**

Solution: $pH = -\log_{10}(H^+)$

$\therefore -\log_{10}(H^+) = 4.1$

$H^+ = 10^{-4.1} = 10^{-5} \times 10^{0.9} = 7.94 \times 10^{-5}$ mol/L

43. **The alkalinity and hardness of a water sample are 250 mg/L and 350 mg/L $CaCO_3$. What is its carbonate hardness and non-carbonate hardness?**

Solution: Carbonate hardness is equal to total hardness of alkalinity, whichever is less,

$\therefore$ Carbonate hardness = 250mg/L < total hardness (350mg/L)

Non Carbonate Hardness = Total Hardness – alkalinity

$= 350 - 250 = 100$ mg/L. as $CaCO_3$.

44. **The amount of CO_2 generated (in kg) while completely oxidizing one kg of CH_{14} to the end products is?**

Solution:
$$CH_{14} + 2O_2 \rightarrow CO_2 + 2H_2O$$
16g 44g

16g CH_4 when completely oxidized gives 44 g CO_2

$\therefore$ 1 kg of CH_4 when completely oxidized leads to $\frac{44}{16} \times 1 = 2.75$ kg of CO_2.

45. A wastewater sample has initial BOD of 200 mg/L. The first order BOD decay co-efficient is 0.38/day. The BOD consumed (in mg/L) in 5 days is?

Solution: Initial BOD = Ultimate BOD

$$BOD_{10} = BOD_u\left[1 - e^{kt}\right]$$

$$= 200\left[1 - e^{-0.38\times5}\right] = 200\left[1 - 2.718^{-1.9}\right]$$

$$= 170 \text{ mg/L}.$$

46. The composition of certain MSW sample and specific weights of its components as below.

Component	% by weight	Specific weight kg/m³
Food	50	300
Dirt & Ash	30	500
Plastics	10	65
Wood & yard waste	10	125

The specific weight of the sample is?

Solution:

Let specific weight of MSW is $x \text{ kg/m}^3$ $\rho = \frac{\text{Total weight}}{\text{Total Volume}}$

Or $\frac{100}{x} = \frac{50}{300} + \frac{30}{500} + \frac{10}{65} + \frac{10}{125}$

Or $\frac{100}{x} = 0.4605$

Or $x = 217 \text{ kg/m}^3$

47. An existing 30 cm diameter sewer is laid at a slope of 1:300 to carry a discharge of $1800 \text{m}^3/\text{day}$. Use the following diagram to find the depth and velocity at the peak discharge. Use manings N = 0.015

Solution: $Q = A \times V = \left(\frac{\pi}{4} D^2\right) \times \frac{1}{n} R^{2/3} S^{1/2}$,

Hyd. radius, $R = \frac{A}{P} = \frac{\pi/4\, D^2}{\pi D} = \frac{D}{4}$

$$\therefore Q = \frac{\pi}{4} \times (0.3)^2 \times \frac{1}{.015} \times \left(\frac{0.3}{4}\right)^{2/3} \left(\frac{1}{300}\right)^{1/2}$$

$$= \frac{\pi}{4} \times .09 \times \frac{1}{.015} \times 0.1778 \times 0.0577$$

$$= 0.0483 \text{ m}^3/\text{s} = 4173 \text{ m}^3/\text{day}$$

$$V = \frac{Q}{A} = 0.69 \text{ m/s.}$$

$$\therefore \frac{q}{Q} = \frac{1800}{4173} = 0.431$$

From graph, $\frac{d}{D} = 0.45$

Also from graph, $\frac{v}{V} = 0.75$

$\therefore d = 0.45 \times 30 = 13.5$ cm

$v = 0.75 \times V = 0.75 \times 0.69$

$= 0.5175\text{m/s} \approx 0.5\text{m/s}$

48. A river has a flow of 1250 MLD, BOD of 5 mg/L and DO level of 8 mg/L before receiving a wastewater discharge at a particular location. The saturation DO level is 11.5 mg/L for the river. Wastewater discharge of 125 MLD with BOD of 250 mg/L and DO level of 2.5 mg/L falls at the location. Assuming complete mixing of river water and wastewater, what will be the immediate DO deficit in mg/L?

Solution: $DO_{mix} = \frac{DO_W Q_W + DO_R Q_R}{Q_W + Q_R}$

$$= \frac{2.5 \times 125 + 8 \times 1250}{125 + 1250} = 7.5 \text{ mg/L}$$

Deficit

$\therefore DO = DO_{sat} - DO_{mix} = 11.5 - 7.5 = 4$ mg/L.

49. Find out the pH of the mixture which will be formed by mixing the following two volumes.

	Volume	pH
Solution A	100 ml	6
Solution B	900 ml	5

Solution:

pH of A = 6 ie $[H^+]$ of A $= 10^{-6}$ mole/L

pH of B = 5 ie $[H^+]$ of B $= 10^{-5}$ mole/L

$\therefore$ Concentration of H^+ in A $= \frac{10^{-6} \times 100}{1000} = 10^{-7}$ mole/L

Concentration of H^+ in B $= \frac{10^{-5} \times 900}{1000} = 9 \times 10^{-6}$ mole/L

Concentration of $[H^+]$ in mixture $= 10^{-7} + 9 \times 10^{-6} = 10^{-7}[1 + 90]$

$= 91 \times 10^{-7}$ mole/Lit

$\therefore pH = -\log_{10}(91 \times 10^{-7}) = 5.04$

50. Solids analysis was conducted of wastewater from a factory. Samples were collected in 300 mL bottles as shown below,

Total Solids Test Data			
Dish No.	**1**	**2**	**3**
Weight of Emply Dish (g)	51.494	51.999	50.326
Weight of Dish + dry solids (g)	51.587	52.081	50.347
Weight of dish + Ignited solids (g)	51.541	52.042	50.383

(i) Determine the concentration of total solids in mg/L

(ii) Find the total volatile solids content

Solution:

(i)

	Sample (1)	Sample (2)	Sample (3)
Weight of Dish + Solids (g)	51.587	52.081	50.437
Weight of Dish (g)	51.494	51.999	50.326
Weight of solids (mg)	93	82	111

Volume used = 300 mL = 0.3L

$$\text{Average concentration of solids} = \frac{\frac{93}{3}+\frac{82}{3}+\frac{111}{3}}{3} = \frac{93+82+111}{3 \times 0.3} = 317.7 \text{ mg/L}.$$

(ii)

	Sample (I)	Sample (2)	Sample (3)
Weight of dish + solids (g)	51.587	52.081	50.437
Weight of Ignited solids (g)	51.541	52.042	50.383
Weight of volatile solids (mg)	46	39	54
Concentration of volatile solids	$46/0.3 = 153$	$\frac{39}{3} = 130$	$\frac{54}{0.3} = 180$

$$\text{Average concentration} = \frac{153+130+180}{3} = 154 \text{ mg/L}.$$

51. To determine the BOD of a wastewater sample, 5, 10 and 50 mL of the wastewater were diluted to 300 mL and incubated at 20°C in BOD bottles for 5 days. The results are given below :

S.No.	Initial volume of waste water (ml)	DO (mg/L)	DO after 5 days (mg/L)
1.	5	9.2	6.9
2.	10	9.1	4.4
3.	50	8.4	0.0

Find average BOD of the waste water.

Solution:

Sample I : $BOD = (DO_i - DO_f)$ Dilution Factor

$$= (9.2 - 6.9) \times \frac{300}{5} = 138 \text{ mg / L}$$

Sample II : $BOD = (9.1 - 4.4) \times \frac{300}{10} = 141 \text{ mg / L}$

$$\text{Average, BOD} = \frac{138 + 141}{2} = 139.5 \text{ mg / L}$$

(**Note :** Sample 3 data is excluded as $DO_f = 0$, Hence BOD cannot be determined).

52. A landfill is to be designed to serve a population of 2 lakh for a period of 25 years. The Solid Waste (SW) generated is 1 kg/person/day. The density of uncompacted Solid Waste is 110 kg/m^3 and a compaction ratio of 3.8 is suggested. The ratio of compacted fill (Solid Waste + Cover) to compacted solid waste is 1.5. Find the landfill volume required.

S.No.	Initial volume of waste water (ml)	DO (mg/L)	DO after 5 days (mg/L)
1.	5	9.2	6.9
2.	10	9.1	4.4
3.	50	8.4	0.0

Solution:

Solid waste generated per day : $= 1 \times 200{,}000 = 200{,}000$kg

Total solid waste generated in 25 years $= 200{,}000 \times 365 \times 25 = 1.825 \times 10^9$ kg.

$$\text{Compaction Ratio } 3.8 = \frac{\text{Volume before compaction}}{\text{Volume after compaction}}$$

$$\text{Volume before compaction } = \frac{1.825 \times 10^9}{110} = 1.66 \times 10^7 \text{ m}^3$$

$$\text{Volume of waste after compaction } = \frac{1.66 \times 10^7}{3.8} = 4.37 \times 10^6 \text{ m}^3$$

Cover : $\frac{\text{Sw + Cover}}{\text{Sw}} = 1 + \frac{\text{Cover}}{\text{Sw}} = 1.5, \therefore \frac{\text{Cover}}{\text{Sw}} = 0.5$

Or, Cover $= 0.5 \times 4.37 \times 10^6 = 2.185 \times 10^6 \text{ m}^3$

Total volume of landfill required $= (4.37 + 2.185) \times 10^6 = 6.555 \times 10^6 \text{ m}^3$

MULTIPLE CHOICE QUESTIONS (MCQs)

1. Which of the following is not a fossil fuel?
   (a) Coal (b) Petrol (c) Natural gas (d) Bio-gas
2. Carcinogenic hazardous waste causes,
   (a) Asthama (b) Blindness (c) Cancer (d) Liver Failure
3. Optimum pH for micro-organism to flourish is?
   (a) 6.5 to 7.5 (b) 5.5 to 6.5 (c) Less than 6 (d) More than 8
4. In a composting process which of the following is not a product?
   (a) Water (b) Carbon dioxide
   (c) Methane (d) Compost
5. The average thickness of ozone (O_3) layer is?
   (a) 0.01 mm (b) 0.6mm (c) 3 mm (d) 1 cm
6. Which of the following gases damage Ozone layer?
   (a) CO (b) CO_2 (c) NO*x* (d) CFC
7. What is the threshold of pain due to sound intensity?
   (a) 100 dB (b) 120 dB (c) 200 dB (d) 80 dB
8. The Specific Absorption Rate (SAR) of cell phones is normally less than.
   (a) 1 (b) 5 (c) 10 (d) 100
9. Some of the non-toxic metals normally found in natural water are,
   (a) Arsenic, Lead and Mercury (b) Calcium, Sodium and Silver
   (c) Cadmium, Chromium and Copper (d) Iron, Manganese and Magnesium
10. MPN count is a measure of?
    (a) Coliform bacteria (b) BOD
    (c) DO (d) Hardness
11. Bio-chemical Oxygen Demand (BOD) of waste water is a measure of?
    (a) Total concentration of bio-chemicals
    (b) Total concentration of organic matter
    (c) Concentration of bio-degradable organic matter
    (d) Concentration of chemically oxidizable matter
12. Zero hardness of water is achieved by?
    (a) Lime soda Process (b) Excess lime treatment
    (c) Ion Exchange Treatment (d) Alum & Lime Treatment
13. The most important water quality parameter for domestic use of water is,
    (a) Carbonate hardness (b) Non-carbonate Hardness
    (c) Chlorides (d) Coliform group of organisms
14. In Natural water hardness is mainly caused by
    (a) Ca^{++} and Mn^{++} (b) Ca^{++} and Fe^{++}
    (c) Na^{++} and K^{++} (d) Ca^{++} and Mg^{++}

15. Which of the following chemical is used for coagulation?
(a) Aluminum chloride (b) Ammonium chloride
(c) Aluminum sulphate (d) Copper sulphate

16. In a domestic wastewater sample, COD and BOD were measured. Generally which of the following statement is true for their relative magnitude?
(a) COD = BOD (b) COD > BOD (c) BOD > COD (d) None of these

17. The removal of dissolved organic matter occurs in,
(a) Slow Sand filters (b) Trickling filters
(c) Rapid sand filters (d) Dual Media Filters

18. The hydraulic retention time of the wastewater in aeration tank is ?
(Given Volume of Tank = $200m^3$ and Quantity of waste water 1 MLD)
(a) 12 minutes (b) 4 Hours and 48 minutes
(c) 10 Hours (d) 24 Hours

19. Which of the following greenhouse gases is contributor due to incomplete combustion of coal?
(a) Oxides of Nitrogen (b) Methane
(c) Carbon monoxide (c) Carbon dioxide

20. Which part of a solar cooker is responsible for greenhouse effect?
(a) Black coating inside (b) Mirror
(c) Glass sheet (d) Outer cover

21. Which of the following is not useful in reducing the use of fossil fuels?
(a) Using Fluorescent Tubes (b) Using petrol in vehicles
(c) Using Solar Devices (d) Improving the fuel efficiency

22. Organic Solid waste when incinerated, the percent of residue is approximately.
(a) 5 to 10% (b) 11 to 20% (c) 21 to 30% (d) 31 to 50%

23. During 2010-2020 which form of energy has recorded maximum growth?
(a) Hydro (b) Wind (c) Bio (d) Solar

24. For excellent water quality the Water Quality Index (WQI) should be between,
(a) 0 – 25 (b) 50 – 70 (c) 70 – 90 (d) 90 – 100

25. One carbon credit corresponds to how much carbon dioxide (in metric tons).
(a) 1 (b) 100 (c) 10 (d) 10,000

26. In a Fission reaction which particle is released?
(a) Proton (b) Neutron (c) Electron (d) All these

27. The basic difference between bio-mass and fossil fuel is of?
(a) Pollution caused (b) Energy content
(c) Time scale (d) Simple use

28. What kind of project is Namami Gange?
(a) EIA (b) Development
(c) Restoration (d) Water Harvesting

29. In the waste Hierarchy which is the base of the pyramid?
(a) Disposal (b) Recovery (c) Reuse (d) Recycling

30. In a composting process carbon/nitrogen Ratio's (C/N Ratio) normal average range is,
(a) 10 : 1 (b) 20 : 1 (c) 40 : 1 (d) 30 : 1

31. According to International Energy Agency (IEA) what is the approximate share of Nuclear Energy?
(a) 5 % (b) 27 % (c) 32 % (d) 1.8 %

32. Which of the following is not a tangible loss?
(a) Loss of Crop (b) Loss of Building
(c) Loss of cattles (d) Environment degradation

33. Among No, SO_2, O_3 and Soot which of the following is a secondary pollutant?
(a) O_3 (b) NO (c) SO_2 (d) Soot

34. The most used Nuclear Fuel in the world is ?
(a) Thorium – 232 (b) Uranium – 238
(c) Uranium – 235 (d) Plutonium – 239

35. Water distribution systems are sized to meet?
(a) Maximum hourly demand (b) Maximum daily and fire demand
(c) Average hourly demand (d) Average daily and Fire demand

36. In anaerobic digestion process, the first step is?
(a) Methanogenesis (b) Hydrolysis
(c) Acidogenesis (d) Acetogenesis

37. In disinfection which of the following forms of chlorine is most effective in killing the pathogenic bacteria?
(a) Cl (b) OCl (c) NH_4Cl (d) HOCl

38. A soil formation through which only seepage is possible being partly permeable and capable of giving insignificant yield, is classified as,
(a) Aquifer (b) Aquitard (c) Aquifuge (d) Aquiclude

39. Which of the following sewage treatment options requires maximum land area for a given discharge?
(a) Trickling filter (b) Anaerobic pond
(c) Oxidation ditch (d) Oxidation pond

40. If 25000 vehicles are being used in a city with an average travel of 15000 km per vehicle. The NO_x emission rate is 2g/km. per vehicle. The total annual release of NO_x will be?
(a) 75×10^4 kg (b) 48×10^4 kg (c) 150×10^4 kg (d) 37.5×10^4 kg

41. The sound pressure (in μPa) of the faintiest sound that a normal healthy person can hear is?
(a) 0.2 (b) 2 (c) 20 (d) 55

42. The drop manholes are provided in sewerage system when there is,
(a) Change in alignment of sewer line
(b) Change in size of sewers
(c) Change in elevation of ground level
(d) Change from gravity to pressure system

43. The unit in which both sedimentation and digestion processes of sludge take place simultaneously is?
(a) Skimming Tank (b) Imhoff Tank
(c) Deritus Tank (d) Digestion Tank

44. Match List-I with List-II and select the correct answer. Use the codes given.

List-I		List-II	
A.	Grit Chamber	1.	Zone Settling
B.	Secondary Settling tank	2.	Stoke's Lan
C.	Activated Sludge process	3.	Aerobic
D.	Trickling filter	4.	Contact Stabilization

Codes:

	A	B	C	D
(a)	1	2	3	4
(b)	2	1	3	4
(c)	1	2	4	3
(d)	2	1	4	3

45. A coastal city produces MSW with high moisture content, high organic materials, low calorific value and low inorganic matter. Which is the most suitable disposal option?
(a) Composting (b) Dumping in sea
(c) Incineration (d) Landfill

46. MPN Index is a measure of which of the following.
(a) Coliform bacteria (b) D.O. Content
(c) BOD (d) Hardness

47. Which of the following is **not** present in Acid Rain ?
(a) HNO_3 (b) H_2SO_4 (c) H_2CO_3 (d) CH_3COOH

48. Following observations have been made for the elevation and temperature to ascertain the stability of the atmosphere :

Elevation in (m)	Temperature (in °C)
10	15.5
60	15.0
130	14.3

The atmosphere is classified as,
(a) Stable (b) Unstable (c) Neutral (d) Inverse

(**Note :** $ELR_{1-2} = \dfrac{15.5 - 15}{(60-10)\,10^{-3}} = 10^{\circ}C/km$; $ELR_{2-3} = \dfrac{15 - 14.3}{(130-60) \times 10^{-3}} = 10^{\circ}C/km$

Since ELR > ALR ($9.8^{\circ}C/km$), Atmosphere is **unstable**.

(ELR : Environmental Lapse Rate, ALR : Adiabatic Lapse Rate.)

49. Particulate matter (fly ash) carried out in effluent gases from the furnaces burning fuels are better removed by.

(*a*) Cotton bag house filter (*b*) Electrostatic precipitator (ESP)

(*c*) Cyclone (*d*) Wet scrubber

(**Note :** Electrostatic Precipitators are generally used in thermal power plants to remove fly ash.)

Name of series	Minimum particle size that can be removed in μm	Efficiency in %
Gravitational settling chamber	> 50	< 50
Cyclone collector	5.25	50 to 90
Wet scrubbers (Cyclonic scrubbers)	> 2.5 μm	< 80
ESP	> 1 μm	95 to 99
Fabric filters	< 1 μm	> 99

50. The most important type of species involved in the degradation of organic matter in case of Activated Sludge Process is

(*a*) autotrophs (*b*) heterotrophs (*c*) prototrophs (*d*) photo-trophs

51. 'Chipko Movement' was associated with :

(*a*) Soil conservation (*b*) Wild life conservation

(*c*) Forest conservation (*d*) Human society

52. Threshold concentration of SO_2 in any industrial activity is **not** permitted beyond :

(*a*) 2 ppm (*b*) 3 ppm (*c*) 4 ppm (*d*) 5 ppm

MCQs Answers key

1	*d*	10	*a*	19	*c*	28	*c*	37	*d*	46	*a*
2	*c*	11	*c*	20	*c*	29	*a*	38	*b*	47	*d*
3	*a*	12	*c*	21	*b*	30	*d*	39	*d*	48	*b*
4	*c*	13	*d*	22	*c*	31	*a*	40	*a*	49	*b*
5	*c*	14	*d*	23	*d*	32	*d*	41	*c*	50	*b*
6	*d*	15	*c*	24	*d*	33	*a*	42	*c*	51	*c*
7	*b*	16	*b*	25	*a*	34	*c*	43	*b*	52	*d*
8	*a*	17	*b*	26	*b*	35	*b*	44	*b*		
9	*d*	18	*b*	27	*c*	36	*b*	45	*a*		

Index